FLUORESCENCE SENSORS and BIOSENSORS

FLUORESCENCE SENSORS and BIOSENSORS

Edited by
Richard B. Thompson

CRC Press
Taylor & Francis Group
Boca Raton London New York

CRC Press is an imprint of the
Taylor & Francis Group, an **informa** business

CRC Press
Taylor & Francis Group
6000 Broken Sound Parkway NW, Suite 300
Boca Raton, FL 33487-2742

First issued in paperback 2019

ISNB-13: 978-0-8247-2737-6 (hbk)
ISNB-13: 978-0-367-39151-5 (hbk)

Library of Congress Cataloging-in-Publication Data

Fluorescence sensors and biosensors / edited by Richard B. Thompson.
     p. cm.
  Includes bibliographical references and index.
  ISBN 0-8247-2737-1 (alk. paper)
  1. Biosensors. 2. Fluorescent probes. 3. Fluorescence. I. Thompson, Richard B.

R857.B54F55 2005
610'.28--dc22
                                               2005048617

Visit the Taylor & Francis Web site at
http://www.taylorandfrancis.com

and the CRC Press Web site at
http://www.crcpress.com

Preface

This volume originated in the realization that an edited volume on developments in fluorescence sensing and biosensing was sorely needed. In our frame of reference, fluorescence-based sensors transduce the presence or level of some chemical analyte as a change in fluorescence; biosensors use biological (or biomimetic) molecules to effect the recognition of the analyte. This book is exclusively devoted to fluorescence (and phosphorescence)-based sensing, which I view as the most powerful approach for chemical sensing for three reasons. Fluorescence has high intrinsic sensitivity, with experiments studying the emission of single molecules becoming commonplace. Fluorescence transduction is very flexible, with many different phenomena being employed herein to transduce analyte as changes in fluorescence. Finally, fluorescence sensing is a peerless research tool for addressing biological questions because it permits analytes to be imaged remotely either in the microscope or through fiber optics. The enormous impact of fluorescence techniques may be judged by the growth of the company Molecular Probes, which in 25 years grew from a firm offering a handful of fluorescent dyes to one offering thousands of fluorescent products (and which sold for $300 million in 2003). These fluorophores have appeared in thousands of scholarly papers during this period, and fluorescence has emerged as a mainstream research and development technology in fields such as fluorescence-activated cell sorting, DNA sequencing, high-throughput screening, and clinical diagnostics. Thus, I decided to solicit chapters on many of the most important new developments in the field, with a view to highlighting technologies with the broadest potential impact. The chapters are roughly grouped under three headings: new recognition or transduction approaches, other new technology, and selected applications.

Tuan Vo-Dinh was kind enough to contribute an overview of the field from his unique perspective of leadership over decades. This was especially good of him in view of the release of his comprehensive *Biomedical Photonics Handbook* (CRC Press, 2003). Drs. Yang and Ellington have been the leaders for some time in the development of aptamers (oligoribonucleotides) for sensing, which is an especially powerful technique for obtaining recognition chemistry for analytes that have no other means of being "seen." Dr. Daunert and her colleagues provide a succinct overview of one of the most important biorecognition chemistry platforms, that based on periplasmic binding proteins. Dr. Yao and his colleagues comprehensively review the use of the molecular beacon approach for DNA recognition, which has become a dominant method for transducing the recognition of DNA sequences in array formats. Dr. Blagoi and his colleagues present the use of resonance energy transfer itself in sensing, an approach that is used by several of the other authors. Finally, the editor and his colleagues summarize the issues and future prospects for the use of the carbonic anhydrase recognition platform for metal ion determination and imaging.

There have been many new developments in fluorophores for fluorescent labeling and sensing as well as fundamentally new approaches for sensor and assay construction; three of these are highlighted in this section. Dr. Geddes and his colleagues make abundantly clear the current fascination and practical interest in metal-enhanced fluorescence, where, under certain conditions, fluorophores exhibit almost magically changed emission. Dr. Savitsky introduces a series of phosphorescent labels and presents persuasive evidence of their advantages in a series of clinically relevant assays. Dr. Anzenbacher describes a series of fluorescent probes selective for anions, a heretofore underappreciated (and underserved from the standpoint of analytical methods) group of analytes compared with metal ions. Finally, Drs. Chang and Yager introduce us to their unique technology for fluid handling on a nanoscale, which is a central issue in many analytical and clinical determinations.

The last section of the book deals with application of fluorescence sensing technology to some rather practical problems. Dr. Herron gives a very lucid summary of the work of the Utah group on the development of planar waveguide biosensors and their application to clinical diagnostics. Dr. Tolosa and her colleagues discuss the issues in adapting fluorescence-based sensing approaches to very demanding problems in biochemical production by fermentation. Finally, several of my distinguished colleagues and I discuss some of the myriad practical problems in measuring analytes such as free zinc ion at ultratrace levels in biological specimens.

We hope you enjoy the book and find it valuable as well.

Richard Thompson

About the Editor

Richard B. Thompson, Ph.D. is Associate Professor of Biochemistry and Molecular Biology at the University of Maryland School of Medicine in Baltimore. He was educated at Northwestern University (B.A.) and the University of Illinois (Ph.D., biochemistry), was a postdoctoral fellow at the University of Maryland, and was a National Research Council Associate at and later joined the staff of the Naval Research Laboratory in Washington, D.C., before joining the faculty in Maryland.

Dr. Thompson's primary professional interest is in the area of fluorescence-based biosensors, particularly those employing fiber optics. He has published numerous articles in refereed journals and symposium proceedings, as well as book chapters. He has served on review panels for agencies including the National Institutes of Health, National Science Foundation, Department of Energy, and Department of Defense, as well as three journal editorial boards. He has cochaired and coedited the proceedings of the conference series "Advances in Fluorescence Sensing Technology" as well as a CD-ROM compendium of 1000 papers on fluorescence science and technology and other proceedings. He received the U.S. Navy Special Act Award for his work in support of Operation Desert Storm. He is a member of the Biophysical Society, American Association for the Advancement of Science, American Society for Biochemistry and Molecular Biology, American Chemical Society, Society of Photo-Optical Instrumentation Engineers, and U.S. Naval Institute.

Acknowledgments

The Editor wishes to express his gratitude not only to the contributors for their respective chapters but also to his co-workers for their efforts, his sponsors for their support, and his colleagues for their interest. Special thanks are owed to his wife Karen and his family for their patience and to Krystyna Gryczynska for drawing many of the figures.

Contributors

Pavel Anzenbacher, Jr., Ph.D.
Department of Chemistry and Center
for Photochemical Sciences
Bowling Green State University
Bowling Green, Ohio

Kadir Aslan, Ph.D.
University of Maryland
Biotechnology Institute
Baltimore, Maryland

Mark E. Astill, M.S.
ARUP Institute for Clinical and
Experimental Pathology
ARUP Laboratories
Salt Lake City, Utah

Gabriela Blagoi, B.S.
Department of Chemistry
University of New Orleans
New Orleans, Louisiana

Rebecca A. Bozym, B.S.
Department of Biochemistry and
Molecular Biology
University of Maryland School
of Medicine
Baltimore, Maryland

Stacy Z. Brown, B.S.
Department of Pharmaceutics and
Pharmaceutical Chemistry
University of Utah
Salt Lake City, Utah

Hugh N. Chang, M.S.E.
Department of Bioengineering
University of Washington
Seattle, Washington

Douglas A. Christensen, Ph.D.
Department of Bioengineering
University of Utah
Salt Lake City, Utah

Michele L. Cramer, M.S.
Department of Biochemistry and
Molecular Biology
University of Maryland School
of Medicine
Baltimore, Maryland

Sylvia Daunert, Ph.D.
Department of Chemistry
University of Kentucky
Lexington, Kentucky

Sapna K. Deo, Ph.D.
Department of Chemistry
University of Kentucky
Lexington, Kentucky

Jacob D. Durtschi, B.S.
Department of Pharmaceutics and
Pharmaceutical Chemistry
University of Utah
Salt Lake City, Utah

Andrew Ellington, Ph.D.
Department of Chemistry and Institute
for Cellular and Molecular Biology
University of Texas at Austin
Austin, Texas

Carol A. Fierke, Ph.D.
Department of Chemistry
University of Michigan
Ann Arbor, Michigan

Christopher J. Frederickson, Ph.D.
NeuroBioTex, Inc.
Houston, Texas

Chris D. Geddes, Ph.D.
University of Maryland
Biotechnology Institute
Baltimore, Maryland

Ignacy Gryczynski, Ph.D.
Department of Biochemistry and
  Molecular Biology
University of Maryland School
  of Medicine
Baltimore, Maryland

Anson V. Hatch, Ph.D.
Department of Bioengineering
University of Washington
Seattle, Washington

Kenneth R. Hawkins, B.S.
Department of Bioengineering
University of Washington
Seattle, Washington

James N. Herron, Ph.D.
Department of Pharmaceutics and
  Pharmaceutical Chemistry
University of Utah
Salt Lake City, Utah

Michal Hershfinkel, Ph.D.
Zlotowski Center for Neuroscience
Ben-Gurion University of the Negev
Beer Sheva, Israel

M. Shelly John, Ph.D.
Department of Chemistry and UF Brain
  Institute
University of Florida
Gainesville, Florida

Karolina Jursíková, M.S.
Department of Chemistry and Center
  for Photochemical Sciences
Bowling Green State University
Bowling Green, Ohio

Yordan Kostov, Ph.D.
Department of Chemical and
  Biochemical Engineering
University of Maryland Baltimore
  County
Catonsville, Maryland

Joseph R. Lakowicz, Ph.D.
University of Maryland
Biotechnology Institute
Baltimore, Maryland

Frances S. Ligler, D.Phil., D.Sc.
Center for Bio/Molecular Science
  and Engineering
U. S. Naval Research Laboratory
Washington, D.C.

Govind Rao, Ph.D.
Department of Chemical and
  Biochemical Engineering
University of Maryland Baltimore
  County
Catonsville, Maryland

Nitsa Rosenzweig, Ph.D.
Department of Chemistry
University of New Orleans
New Orleans, Louisiana

Zeev Rosenzweig, Ph.D.
Department of Chemistry
University of New Orleans
New Orleans, Louisiana

Alexander P. Savitsky, Ph.D., D.Sc.
A. N. Bach Institute of Biochemistry
Russian Academy of Science
Moscow, Russia

Bethel V. Sharma, B.S.
Department of Chemistry
University of Kentucky
Lexington, Kentucky

Suresh S. Shrestha, Ph.D.
Department of Chemistry
University of Kentucky
Lexington, Kentucky

Eric M. Simon, M.E.
Dexterity Design, Inc.
Salt Lake City, Utah

Richard S. Smith, Ph.D.
Department of Bioengineering
University of Utah
Salt Lake City, Utah

Andrea V. Stoddard, B.S.
Department of Chemistry
University of Michigan
Ann Arbor, Michigan

Chris R. Taitt, Ph.D.
Center for Bio/Molecular Science
  and Engineering
U. S. Naval Research Laboratory
Washington, D.C.

Weihong Tan, Ph.D.
Department of Chemistry and UF Brain
  Institute
University of Florida
Gainesville, Florida

Lyndon Tan, B.S.
Lumenal Technologies, L.P.
Salt Lake City, Utah

Alan H. Terry, M.S.
Department of Pharmaceutics and
  Pharmaceutical Chemistry
University of Utah
Salt Lake City, Utah

Richard B. Thompson, Ph.D.
Department of Biochemistry and
  Molecular Biology
University of Maryland School
  of Medicine
Baltimore, Maryland

Samuel E. Tolley, Ph.D.
Department of Bioengineering
University of Utah
Salt Lake City, Utah

Leah Tolosa, Ph.D.
Department of Chemical and
  Biochemical Engineering
University of Maryland Baltimore
  County
Catonsville, Maryland

Tuan Vo-Dinh, Ph.D.
Center for Advanced Biomedical
  Photonics
Oak Ridge National Laboratory
Oak Ridge, Tennessee

Hsu-Kun Wang, M.S.
Department of Pharmaceutics and
  Pharmaceutical Chemistry
University of Utah
Salt Lake City, Utah

Nissa K. Westerberg, Ph.D.
Department of Biochemistry and
  Molecular Biology
University of Maryland School
  of Medicine
Baltimore, Maryland

Paul Yager, Ph.D.
Department of Bioengineering
University of Washington
Seattle, Washington

Litao Yang, M.S.
Department of Chemistry and Institute
 for Cellular and Molecular Biology
University of Texas at Austin
Austin, Texas

Gang Yao, Ph.D.
Department of Chemistry
 and UF Brain Institute
University of Florida
Gainesville, Florida

Contents

Dedication

To my parents,
Charles and Alice Thompson,
for their unwavering support

1 Introduction

T. Vo-Dinh, Ph.D.

Nature has been performing biosensing since the dawn of time, using the sensory nerve cells of the nose to detect scents and those of the tongue to taste dissolved substances. As time has progressed, so has our level of understanding about the function of living organisms in detecting trace amounts of biochemicals in complex systems. The abilities of biological organisms to recognize foreign substances are unparalleled and have to some extent been mimicked by researchers in the development of biosensors. Using bioreceptors from biological organisms or receptors that have been patterned after biological systems, scientists have developed a new means of chemical analysis that often has the high selectivity of biological recognition systems. These biorecognition elements in combination with various transduction methods have helped to create the rapidly expanding fields of bioanalysis and related technologies known as biosensors and biochemical sensors.

Two basic operating principles of biosensors involve "biological recognition" and "sensing." Therefore, a biosensor can be generally defined as a device that consists of two basic components connected in series: (1) a biological recognition system, often called a bioreceptor, and (2) a transducer. The basic principle of a biosensor is to detect this molecular recognition and transform it into another type of signal using a transducer. The main purpose of the recognition system is to provide the sensor with a high degree of selectivity for the analyte to be measured. The interaction of the analyte with the bioreceptor is designed to produce an effect measured by the transducer, which converts the information into a measurable effect such as an electrical signal.

Biosensors can be classified by either their bioreceptor or their transducer type. A bioreceptor is a biological molecular species (e.g., an antibody, an enzyme, a protein, or a nucleic acid) or a living biological system (e.g., cells, tissue, or whole organisms) that utilizes a biochemical mechanism for recognition. The sampling component of a biosensor contains a biosensitive layer. The layer can either contain bioreceptors or be made of bioreceptors covalently attached to the transducer. Bioreceptors are the key to specificity for biosensor technologies. They are responsible for binding the analyte of interest to the sensor for the measurement. These bioreceptors may take many forms, and the different bioreceptors that have been used are as numerous as the different analytes that have been monitored using biosensors. The most common forms of bioreceptors used in biosensing are based on (1) antibody/antigen interactions, (2) nucleic acid interactions, (3) enzymatic interactions, (4) cellular interactions (i.e., microorganisms, proteins), and (5) interactions using biomimetic materials (i.e., synthetic bioreceptors). For transducer classification, the previously mentioned techniques (optical, electrochemical, and mass-sensitive) are used.

Recent advances in nanotechnology and photonics have led to a new generation of nanobiosensor devices for probing the cell machinery and elucidating intimate life processes occurring at the molecular level, which were heretofore invisible to human inquiry. *In vivo* tracking of biochemical processes within intracellular environments is now possible with the use of fluorescent molecular probes. With powerful microscopic tools using near-field optics, it is now possible to explore the biochemical processes and submicroscopic structures of living cells at unprecedented resolution. Nanoparticle probes that have fluorescent indicator dye molecules embedded in 20–200-nm-diameter polymer or sol-gel spheres have been developed for cell imaging. The encapsulating spheres protect the indicator dye from cellular degradation and the cell from the toxic effects of the dyes while allowing cellular ions to penetrate and interact with the indicator dye. Following injection of nanoparticle probes, the cells are illuminated with an excitation light source and the resulting fluorescence signal from the indicator dye of the nanoprobes is detected. The indicator dye can exhibit the presence of a constituent of interest; for example, allowing researchers to track how and where metal ions are stored and released in the cell. Similarly, quantum dot probes are capable of monitoring individual chemical species inside a living cell.

Metal nanoparticle-based systems have been developed as a new kind of probe for gene diagnostics using surface-enhanced Raman scattering (SERS) and surface-enhanced luminescence (SEL) detection. The SERS effect can enhance the Raman and luminescence signals of molecules adsorbed on metallic nanostructures due to strong electromagnetic field enhancement caused by "plasmons," which are the quanta associated with longitudinal waves propagating in matter through the collective motion of large numbers of electrons. Incident light irradiating these so-called plasmonic-active nanostructures excites conduction electrons in the metal and induces excitation of surface plasmons, leading to enormous electromagnetic enhancement. This enhancement effect has spurred recent interest in the research field of plasmonics, which refers to the investigation, development, and applications of enhanced electromagnetic properties of metallic nanostructures in biochemical analyses. For example, a new DNA assay platform using SERS labels, referred to as the SERGen assay, involves an array of oligonucleotide capture probes immobilized directly on metal nanoparticles or nanoparticles covered with a thin metal layer. Individual plasmonics-based nanoparticles and nanoshells, which consist of polymeric nanospheres covered with a silver layer, have been developed for chemical and biological analysis. Optically active nanoshells and composites of thermally sensitive hydrogels have been developed in order to photothermally modulate drug delivery.

Many diseases are known to arise as a result of the gradual accumulation of genetic and molecular changes in single cells. The identification of molecular changes and related biomarkers in single cells remains a high priority in medical research. *In vivo* detection of the molecular alterations that distinguish any particular diseased cell from a normal cell could ultimately help to identify the nature of the disease and predict the pathologic behavior of that disease cell as well as its responsiveness to treatment. Understanding the profile of molecular changes in a living cell for any particular disease will lead to the possibility of correlating the resulting phenotype of that disease with molecular events.

A recent advance in the field of biosensors has been the development of optical nanobiosensors, which have dimensions on the nanometer scale. Nanobiosensors having fiber optics with tip diameters ranging between 20 and 40 nm have been developed for single-cell analysis. Because the diameter of the fiber tip is significantly smaller than the wavelength of light (300–500 nm) used for excitation of the analyte, the photons cannot escape normally from the tip of the fiber. Instead, in a fiber-optic nanosensor, after the photons have traveled as far down the fiber as possible, excitons or evanescent fields continue to travel through the remainder of the tip, providing excitation for the fluorescent species of interest present in the sensing layer within 100 nm of the sensing tip. As a result, only species that are extremely close to the fiber's tip can be excited, thereby precluding the excitation of interfering fluorescent species in other locations within a sample. This near-field excitation mode is extremely important in a cell, because it prevents other fluorescent species outside the near-field excitation volume from being excited. Therefore, the nanoscale size of this new class of nanosensors also allows for measurements in the smallest of environments. One such environment that has generated a great deal of interest is that of individual cells. Using these nanosensors, it is possible to probe individual chemical species in specific locations within a cell.

Combined with the molecular recognition of antibody probes, such optical nanosensors are capable of providing selectivity in location and analyte parameters. Nanosensors therefore offer significant improvements over such methods in many cases, as they allow the user to perform analyses in whatever location he or she desires without need for homogeneous dispersion of the indicator dye. Using these nanobiosensors, it has become possible to probe individual chemical species in specific locations throughout a living cell. Several studies have demonstrated the use of fiber-optic nanobiosensors for *in vivo* measurements of single cells. By performing similar measurements inside various subcellular compartments or organelles, it may be possible to obtain critical information about the location of biomarkers of disease processes or molecular signaling events within single living cells.

The marriage of genomics, electronics, biomaterials, and photonics is expected to revolutionize many areas of sensor and biosensor technology in the 21st century. This ultimate technology convergence will expand our horizons for the discovery of new biosensing tools for detection, diagnosis, and prevention studies and also new targets for therapeutic development. It is now possible to develop nanocarriers for targeted delivery of drugs that have their shells conjugated with antibodies for targeting antigens and fluorescent chromophores for *in vivo* tracking.

Since the first biosensors were reported in the early 1960s, there has been accelerated growth of research activities in this area. Biosensors are used in a wide variety of applications, primarily in three major areas: biological monitoring, biomedical diagnostics, and environmental sensing applications. With the increasing interest in practicality and cost-effectiveness in environmental and medical diagnostics, I believe this is a most appropriate time for a comprehensive treatment that deals with the various biosensor methods, devices, and applications. Dr. Richard Thompson has edited an exceptional book that contains the latest developments in sensor and biosensor technologies described by many experts in the field, and it should be of considerable use to anyone who wishes to develop, apply, and use various types of fluorescence sensor biosensor methods and instruments.

I am proud to have prepared an introduction for such an excellent book and an outstanding colleague.

2 Prospects for the *De Novo* Design of Nucleic Acid Biosensors

Litao Yang, M.S., and Andrew D. Ellington, Ph.D.

CONTENTS

2.1 INTRODUCTION

Nucleic acid biosensors are typically considered to function solely by hybridization. In this regard, the rules for base pairing are well known and can be used to fine tune melting temperatures and mismatch discrimination. However, nucleic acids may take on three-dimensional shapes that can bind to nonnucleic acid analytes. By manipulating the secondary structures of such complex nucleic acid receptors, it is possible to generate a variety of biosensors. By using the same thermodynamic rules for base-pairing analyses that gave rise to the ability to engineer nucleic acid hybridization, it should prove possible to engineer the sensing and signaling properties of other types of nucleic acid biosensors.

2.2 *IN VITRO* SELECTION OF FUNCTIONAL NUCLEIC ACIDS

Nucleic acid binding species (or aptamers, derived from the Latin word *aptus*, meaning "fitted") can be selected from random pools *in vitro* by a process known as systematic evolution of ligands by exponential enrichment (SELEX) (Figure 2.1).[1–5] In short, an

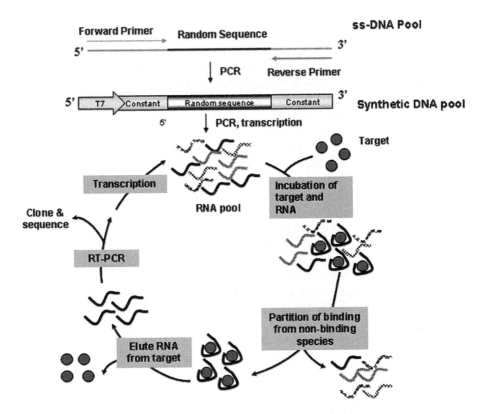

FIGURE 2.1 Schematic of the *in vitro* selection of RNA aptamers. A synthetic, single-stranded DNA (ssDNA) pool is PCR amplified and transcribed *in vitro* using T7 RNA polymerase to generate a RNA pool. Binding species are separated from nonbinding species following incubation with a target molecule. Binding species are amplified by reverse transcription, PCR, and *in vitro* transcription, and the delimited pool is used for the next round of selection. Binding conditions over iterative rounds of selection are made increasingly stringent in order to isolate tight and highly specific aptamers.

oligonucleotide template that contains a core of randomized residues flanked by constant regions is generated by chemical synthesis. The initial pool can be amplified by polymerase chain reaction (PCR) and transcribed into RNA if one of the constant regions contains a convenient promoter sequence, such as the promoter for T7 RNA polymerase. Binding species are sieved from nonbinding species by any of a variety of affinity techniques, including filter capture, gel mobility shift, or affinity chromatography. Following selection, the delimited pool is amplified by some combination of reverse transcription, PCR, and *in vitro* transcription, depending on whether the original pool was an RNA or DNA pool. After multiple cycles of selection and amplification, those species that have the best affinity for a given target molecule are typically isolated (although other considerations, such as replicability, may also skew the results of a selection experiment). Catalytic nucleic acids can be selected by similar schemes as reviewed by Joyce.[6] In these instances, a nucleic acid is typically released from or added to a solid support based on its ability to cleave or add an oligonucleotide substrate.

Aptamers can function as binding reagents and reporter molecules but are significantly different from antisense nucleic acids, which function merely through sequence-dependent hybridization. In contrast, aptamers fold into complex three-dimensional binding pockets that can specifically recognize a variety of ligand classes, including inorganic ions, small organics, low-molecular-weight biochemical species such as metabolites and peptides, proteins, and even the surfaces of supramolecular structures such as viruses and cells.

2.3 NUCLEIC ACID BIOSENSORS

There is a variety of means by which selected nucleic acids can be adapted to function as biosensors. These strategies fall into two basic categories: the simple capture of analytes or conformational transduction of analyte recognition.

2.3.1 FLUORESCENT SIGNALS RESULTING FROM ANALYTE CAPTURE

Given that aptamers are binding reagents, they can be utilized as simple antibody substitutes (i.e., affinity reagents). Anti-thrombin aptamers have been immobilized on the surface of glass slides to capture and detect thrombin in affinity matrix-assisted laser desorption ionization (MALDI). Binding to thrombin was found to be specific, and nonspecific proteins such as human serum albumin could be removed by rinsing the slides.[7]

Many of the classic roles that antibodies have occupied in biochemistry can be readily assumed by aptamers. For example, aptamers may be used in sandwich assays, just as antibodies are in enzyme-linked immunosorbent assays (ELISAs), creating an enzyme-linked oligonucleotide assay (ELONA).[8] Aptamers may presumably function either as capture reagents in sandwich assays, with subsequent detection by antibody-enzyme conjugates, or as secondary binding reagents, in which case the aptamer would be linked to a reporter enzyme, or as both a capture and detection reagent.

Small molecules may also be detected by aptamer conjugates using competition assays.[9] An aptamer competition assay for cholic acid led to the detection of the target in a linear range between 0.1 and 5 mM.[10] Aptamers have been used in place of antibodies in assays such as Western blots and in the affinity purification of protein targets or complexes.[11,12]

Aptamers have also been used in imaging technologies in a manner similar to antibodies. For example, fluoresceinated anti-human neutrophil elastase (HNE) aptamers have been used to detect HNE-coated beads in flow cytometry.[13] Anti-human CD4 aptamers conjugated with either fluorescein or phycoerythrin could stain human CD4-expressing cells in flow cytometry.[14] Aptamers can be used for diagnostic imaging as well as fluorescence-activated cell sorting (FACS). Radiolabeled anti-HNE aptamers were used to image inflammation *in vivo* in a rat model and were found to achieve a higher target:background ratio in less time than comparable IgG reagents, the typical choice for imaging inflammation in a clinical setting.[15]

Fluorescently labeled aptamers may also be generally used as reporters. Fluorescently labeled anti-IgE aptamers were used to detect IgE during capillary electrophoresis with laser-induced fluorescence (CE-LIF).[16] The detection limit for IgE was calculated to be 46 pM, and the target protein could be detected in serum. The anti-thrombin aptamer was used under similar conditions to detect thrombin with a detection limit of 40 nM.

Ligands, rather than aptamers, may be fluorescently labeled for detection or diagnostic assays. Biotinylated anti-L-adenosine RNA aptamers were immobilized on avidin-coated multimode optical fibers and used to capture fluorescein isothiocyanate (FITC)-labeled L-adenosine, which was then detected and quantitated by measuring total internal reflection fluorescence. The responsivity of this device was sufficient for the determination of kinetic and equilibrium binding constants in real time.[17]

2.3.2 FLUORESCENT SIGNALS RESULTING FROM CONFORMATIONAL TRANSDUCTION

By far the most interesting aspect of adapting aptamers to function as biosensors is the exploitation of conformational transduction mechanisms. Such mechanisms are also being applied to protein biosensors.[18] For example, the maltose-binding protein undergoes a hinge-bending motion upon binding to maltose; by introducing fluorescent or electrochemical labels that are sensitive to changes in the chemical environment, these ligand-induced conformational changes can be directly reported.[19] Unfortunately, although extant conformational changes in proteins can be exploited, they are hard to design from scratch. In contrast, because nucleic acid structure is largely based on the simple rules that govern Watson-Crick base pairing, conformational transduction can be readily engineered and, as this chapter suggests, perhaps even predicted.

2.3.2.1 Small Motions

Nucleic acids frequently undergo small but significant conformational changes upon ligand binding.[20–23] This property can be used to directly transduce ligand recognition into optical signals. The chemical microenvironments of dye molecules conjugated to aptamers will change following ligand binding and conformational change, resulting in changes in fluorescent signals, such as fluorescence intensity.

The Ellington lab has previously introduced fluorescein into conformationally labile positions in both RNA[24] and DNA anti-adenosine aptamers.[25] The tertiary structures of these aptamers had been determined by nuclear magnetic resonance (NMR) analysis, and it was known that they underwent adenosine triphosphate (ATP)-dependent conformational changes.[26–28] The fluorescently labeled anti-ATP aptamers not only could specifically bind to ATP relative to other related ligands, such as guanosine triphosphate (GTP), but their fluorescence intensities tracked ATP concentration (Figure 2.2).[29] However, the apparent K_d values for the signaling aptamer were 30 μM (DNA aptamer) and 300 μM (RNA aptamer), whereas those of the parental aptamers were 6 μM (DNA aptamer)[25] and 6–8 μM (RNA aptamer).[24] The increase in apparent K_d values may indicate that there is an inherent loss of binding affinity, either due to derivatization with the fluorophore or because energy is required for the conformational transduction.

Signaling aptamers have also been generated based on their abilities to form dye conjugates. Merino and Weeks[30] synthesized an anti-ATP aptamer with a single 2′-amine-substituted cytidine at a position that was predicted to be flexible in the absence of ATP but constrained following binding. In fact, the 2′-amine modified

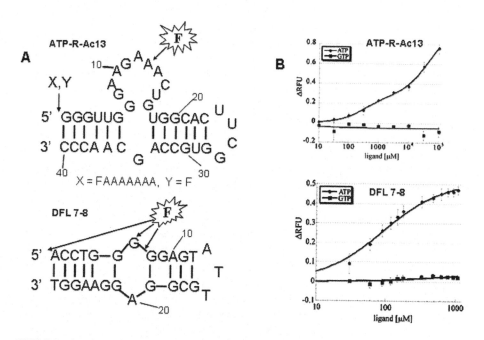

FIGURE 2.2 Responsivity of the designed signaling aptamers ATP-R-Ac13 (RNA) and DFL 7-8 (DNA). (a) Secondary structures of the two aptamers with dye incorporation sites indicated. (b) Response curves for the signaling aptamers in a variety of ATP (circle) and GTP (square) concentrations. Data are from Jhaveri et al.[29]

aptamer reacted better with fluorescamine (FCM) in the unliganded, flexible structure than in the liganded, constrained structure. Because the aptamer had already been labeled with Texas Red at its 3′ end, fluorescence resonance energy transfer (FRET) could be detected with the conjugated FCM. The fluorescence at 615 nm (due to FRET from FCM to Texas Red) was reduced by the addition of ATP. The $K_{1/2}$ values for ATP in simplified and urine solutions were 390 µM and 430 (±70) µM, respectively.

In the above examples, the fluorescence reporters were rationally placed in an aptamer's secondary or tertiary structure. However, structural information about an aptamer may not always be available, and this may limit the applicability of signaling aptamers, especially in the development of large-scale sensor arrays. To overcome these problems, we devised a method that better coupled selection with signal transduction.[31] Selections for binding to ATP were carried out starting with a 51-residue, random sequence pool that incorporated fluorescein-12-UTP (a fluorescein moiety connected to the 5 position of uridine by a 12-atom spacer) in place of uridine triphosphate (UTP). In order to reduce potential background fluorescence, the composition of the pool was skewed to contain very few uridine residues (31.3% A, C, and G; 6% fluorescein-U). Among the variants that emerged after 11 rounds of selection were several that could signal the presence of ATP by a ligand-dependent increase in fluorescence intensity. The best selected signaling aptamer could be used to quantitate ATP concentrations as low as 25 µM. This signaling aptamer contained only two uridine residues, and only one of these proved to be important for signaling. The single, functional fluorescein-12-UTP residue could be replaced with other UTP analogs, such as Cascade Blue-7-UTP and Rhodamine Green-5-UTP, and the resultant modified aptamers still showed ATP-dependent increases in fluorescence. Based on these results, it was thought that the single uridine in the selected aptamer might function as a general conformational "switch" that transduced molecular recognition into optical signals.

Fluorescent reporters can also noncovalently interact with an aptamer as well as be covalently conjugated. Recently, Ho and Leclerc mixed a water-soluble, cationic polythiophene derivative (polymer 1) with an anti-thrombin aptamer (Figure 2.3).[32] The complex could transduce thrombin recognition into an optical (colorimetric or fluorometric) signal and human thrombin could be detected in the femtomole range. The ease with which aptamers can be converted to biosensors using a noncovalently bound dye might increase the use of aptamers in high-throughput screening.

Because signaling aptamers transduce ligand recognition into fluorescent signals, they should more specifically detect target molecules than other fluorescently labeled aptamers, which signal irrespectively of where and how they bind. However, labeling may perturb aptamer structure and function, resulting in lower sensitivities. In addition, because the free energy of ligand binding is taken up in part in the free energy of the conformational transition, their sensitivities may be inherently limited. Finally, most of the signaling aptamers that we have examined require some knowledge of aptamer structure and a precise placement of fluorescent reporters, which make these reagents less useful for high-throughput or array methods. The development of selection methods partially obviates this problem, but the selected signaling aptamers still have to be screened individually for signaling function.

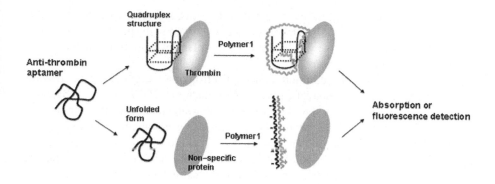

FIGURE 2.3 (See color insert following page 142.) Schematic description of detection of thrombin using an anti-thrombin aptamer and the chromic properties of polymer 1. Polymer 1 displays optical changes when it interacts with ssDNA. In solution it has an absorption maximum at 402 nm (yellow) and folds into a random-coil conformation. When polymer 1 complexes with unfolded DNA, its color shifts to red-violet due to the formation of a stoichiometric polyelectrolyte complex, which is insoluble and tends to aggregate. However, in the presence of thrombin, the bound form of the anti-thrombin aptamer is a quadruplex. The conformation assumed by polymer 1 with the protein–quadruplex complex shifts its color to orange. The fluorescence of polymer 1 with an aggregated and planar form is easily quenched. Therefore, the orange, partially planar form with less aggregation has much higher fluorescence intensity than the red-violet, highly conjugated form. The change of polymer 1 is detected by measuring absorbance or fluorescence intensity at different wavelengths.

2.3.2.2   Large Motions

Aptamers typically have a molecular weight of 10,000 to 30,000 Da, about the size of small proteins. Therefore, the formation of protein–aptamer complexes can yield large increases in aggregate molecular weight and size, which in turn can cause changes in rotational diffusion rate. Assuming an aptamer is labeled with a fluorophore, this phenomenon can be detected by measuring changes in fluorescence anisotropy.

As alluded to above, an aptamer that binds to the blood-clotting protein thrombin was one of the first aptamers selected from a single-stranded DNA library and remains one of the simplest to work with. The minimal size of this aptamer is only 15 nucleotides.[33] The aptamer folds into a compact chair-form quadruplex structure as determined by NMR and x-ray crystallography[34–36] and can inhibit the activity of thrombin *in vitro*.[37] The 5′ end of the aptamer was labeled with FITC and the 3′ end was appended with an alkyl amine group that allowed the aptamer to be covalently attached to a glass slide.[38] Evanescent wave-induced fluorescence anisotropy was then employed to measure binding between thrombin in solution and the immobilized aptamer. Thrombin binding did not affect the total fluorescence of the aptamer. As was the case with the solution-phase signaling aptamers, FITC labeling increased the apparent dissociation constant roughly 10-fold relative to an unlabeled aptamer,[39] although this diminution in sensitivity was reversed following immobilization. The labeled, immobilized aptamer could sensitively detect thrombin amounts as low as 0.7 amol in a 140-pL interrogated volume (a concentration of 5 nM). Both the immobilized and free FITC-labeled aptamer exhibited high specificity for thrombin over elastase, another serine protease with an isoelectric point and molecular weight similar to those of thrombin.

Tan's group has also used fluorescence anisotropy measurements to interrogate a fluorescein-labeled anti-platelet-derived growth factor (PDGF) aptamer.[40] Because the fluorescein label was distal to the protein-binding site, there was little diminution in the binding affinity of the aptamer conjugate. Using this method, binding could be monitored in real time and in homogeneous solutions. The anisotropy of the aptamer increased more than twofold upon binding PDGF, but there were no apparent changes in the presence of other extracellular proteins such as albumin, hemoglobin, myoblobin, or lysozyme. The assay can be performed in only a few seconds and can detect PDGF concentrations as low as 0.22 nM.

One of the advantages of measuring anisotropy is that reporters can be introduced at almost any position in an aptamer to allow monitoring of real-time changes in analyte concentration. However, the limitation of this method is that the difference in fluorescence anisotropy before and after binding may be hard to detect for small ligands.

2.3.2.3 Global Structural Rearrangements

An alternative approach to transducing conformational changes into optical signals is to rely on large-scale reorganizations of aptamer conformation rather than small conformational changes or tumbling in solution. An excellent precedent for such biosensors exists in the form of molecular beacons. Molecular beacons are short, synthetic oligonucleotides that form a stem-loop structure in which the loop region is complementary to a target sequence. The two ends of the stem-loop contain a conjugated fluor-quencher pair. In the absence of a target biomolecule, the stem brings the fluorophore and quencher near each other to turn fluorescence "off" (Figure 2.4a).[41,42] Upon binding of an oligonucleotide target to the loop region, the beacon undergoes a conformational change that opens the stem, frees the fluorophore from the adjacent quencher, and produces a fluorescent signal. Molecular beacons can also signal via changes in FRET.

Since Tyagi and Kramer originally developed "molecular beacons" in 1996 for nucleic acid detection,[43] molecular beacons have been adapted as extremely useful tools in several applications, such as real-time detection of nucleic acid hybridization in living cells,[44] allele discrimination,[45] single-base mismatch detection,[46] and as sensitive DNA biosensors.[47,48]

Besides reporting complementary sequences, molecular beacons have also been used for protein detection. In particular, proteins that bind to single-stranded DNA can lead to the denaturation of the hairpin structure and a concomitant increase in fluorescence. For example, both *Escherichia coli* single-stranded DNA binding protein (SSB) and rat liver lactate dehydrogenase (LDH)-5 isoenzyme can be detected with conventional molecular beacons.[49,50] A molecular beacon that contained a run of 15 thymidine residues in its loop and 5 base pairs (bp) in its stem could be activated 28-fold by saturating concentrations of RecA.[51] More specific protein-detecting molecular beacons can also be designed. The Adeno-associated virus Rep78/Rep68 protein interacts

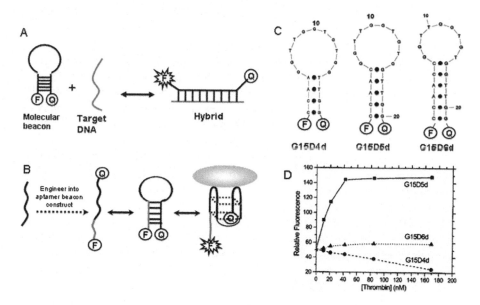

FIGURE 2.4 Mechanisms of molecular and aptamer beacons. (a) A molecular beacon forms a stem-loop structure that contains a fluorophore and quencher at each end. Hybridization of target DNA with the loop region of the molecular beacon leads to dequenching of the fluorophore. (b) Mechanism of a thrombin aptamer beacon. A thrombin aptamer was designed to form an aptamer beacon by putting additional residues at the 5′ end to form a stable stem-loop structure. Equilibrium exists between beaconlike and G-quartet conformations. Thrombin binding causes a conformational change of the aptamer domain and stabilizes the fluorescent, target-binding structure. (c) Predicted secondary structures of three aptamer beacons with varied stem lengths. (d) Aptamer beacon fluorescence emission plotted at different concentrations of thrombin. Data are from Hamaguchi et al.[53]

with a 23-bp Rep binding element in a sequence-dependent manner.[52] When a molecular beacon that contained the appropriate recognition sequence was constructed, Rep78 could activate the beacon. The utility of the various protein-triggered beacons was emphasized by the fact that they could be monitored in real time. Nonetheless, this type of protein-sensing molecular beacon is of necessity limited to proteins that can bind single-stranded DNA, nonspecifically or specifically.

A more general method for the generation of protein-sensing molecular beacons involves engineering aptamers to undergo a protein-dependent conformational change. In collaboration with Marty Stanton at Brandeis University, we hypothesized that by adding sequences to the anti-thrombin aptamer, it should be possible to form a stable, nonbinding stem-loop and consequently destabilize the native binding structure.[53] By appending a fluorescent reporter and quencher to the 5′ and 3′ ends of the extended, anti-thrombin aptamer, we created a structure that was initially quenched but that in the presence of thrombin formed the native quadruplex structure, resulting in a fluorescent signal (Figure 2.4b). Three aptamer beacons were designed with varied stem lengths to have different predicted free energies of stem-loop formation (Figure 2.4c). The best aptamer beacon, G15D5d, could detect the changes in thrombin concentration in real time and again was not activated by other serine proteases in the blood-clotting cascade, such as Factor IX and Factor Xa, which have 37% identity to thrombin in their catalytic domains. By plotting fluorescence vs. protein concentration, the apparent K_d of the beacon complex was 10 nM (Figure 2.4d). One of the most exciting aspects of this work is that it should be generalizable; that is, the strategy can in fact be extended to other aptamers to study other analytes, such as proteins and small molecules, including RNA aptamers. We will further examine the ability to design nucleic acid biosensors below.

Another type of aptamer beacon was designed that relied on folding, conformational change, and quenching or FRET, rather than dequenching.[54] The researchers noted that even in the absence of the protein target, equilibrium existed between the random coil state and the quadruplex state of the thrombin aptamer. Again, target binding shifts the equilibrium in favor of the quadruplex state, in this instance leading to a significant decrease in fluorescence (fluorescein-DABCYL aptamer) or an increase in fluorescence (fluorescein-coumarin). The apparent K_d values and limits of detection for these two types of aptamer beacons were 5.20 ± 0.49 nM and 373 ± 30 pM, vs. 4.87 ± 0.55 nM and 429 ± 63 pM, respectively.

In a more recent report, the same group generated another folding aptamer beacon by appending a fluorophore and a quencher at the termini of the anti-PDGF aptamer (Figure 2.5).[55] Under physiological conditions and in the absence of PDGF, the aptamer is largely denatured and the two termini are far apart. In the presence of PDGF, the structure of the aptamer is stabilized and the two termini are brought together, resulting in fluorescence quenching. The PDGF aptamer beacon successfully detected this tumor-related protein in human tumor cell culture media.

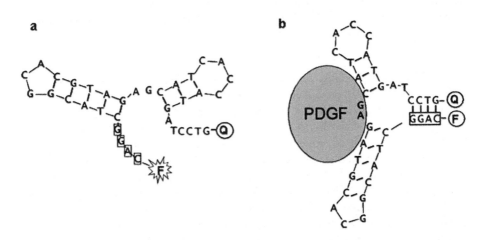

FIGURE 2.5 A folding aptamer beacon. The secondary structure of the anti-PDGF aptamer is shown in the absence (a) and presence (b) of PDGF. In the absence of PDGF, the two termini of the aptamer are far apart. PDGF binding causes a conformational change of the aptamer, and a helix composed of eight residues at the two termini is stabilized, resulting in fluorescence quenching.

Stojanovic et al. have also generated folding aptamer beacons for small organic ligands.[56] These researchers truncated one of the stems of a three-way junction of an anti-cocaine aptamer that forms the cocaine binding pocket, resulting in a desta-bilized, open conformation. The two separated ends were labeled with fluorophore and quencher. However, upon cocaine binding, the aptamer favored the native three-way junction conformation and the two stems (fluorophore and quencher) were brought together. Therefore, the aptamer went from a fluorescent, unliganded form to a quenched, ligand-bound form. Up to 61% of the initial fluorescence was quenched in the presence of cocaine and the signaling aptamer was again selective for cocaine over its metabolites. Moreover, the sensor was able to detect cocaine in serum with a cocaine concentration of 10 to 4000 μM.

Although the design of aptamer beacons is more straightforward than the design of signaling aptamers, techniques for the selection and optimization of aptamer beacons would nonetheless still be valuable. To this end, we have developed a novel method that directly selects for aptamer beacon signaling (Figure 2.6).[57] A random sequence

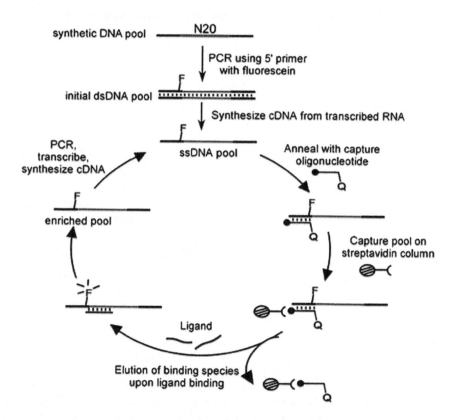

FIGURE 2.6 *In vitro* selection of molecular beacons. An N20 ssDNA pool was immobilized on a streptavidin column via biotin-labeled capture oligonucleotides. Interactions with introduced oligonucleotide targets induced a conformational change in some species and their subsequent release from the column. Eluted DNA was amplified by PCR, transcribed into RNA, and reverse transcribed to cDNA for subsequent rounds of selection.

DNA pool labeled with a fluorophore in its 5′ constant region was immobilized to an oligonucleotide affinity capture column. Upon addition of an oligonucleotide target, some members of the immobilized pool bound the target, folded into a new conformation, and were thereby released from the column. The eluted sequences were amplified and reapplied to the column. After nine rounds of selection and amplification, the population was significantly enriched in species that could be specifically eluted from the column by the target oligonucleotide. The oligonucleotide that was originally used to capture the selected sequences was then derivatized with a quencher. Upon addition of the target oligonucleotide, the selected binding sequences refolded, but in this instance, rather than being released from the column, they were released from the oligonucleotide–quencher conjugate, resulting in a 17-fold increase in fluorescence (Figure 2.7). The oligonucleotide-sensing sequences

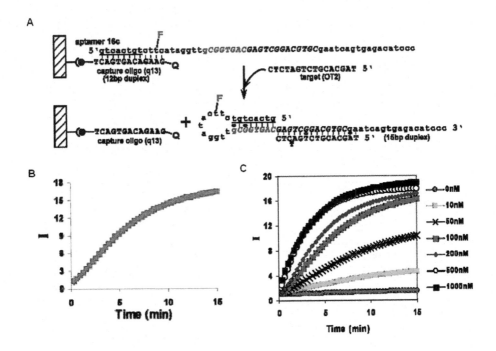

FIGURE 2.7 (See color insert following page 142.) Mechanism of elution and fluorescence responsivities of beacon 16c, one of the selected molecular beacons. (a) Sequences of 16c and its hypothesized mechanism of elution. The internal sequence (labeled with grey) is a weak competitor of the capture oligonucleotide for interaction with the 5′ end of the oligonucleotide. When the oligonucleotide target OT2 hybridizes with 16c, the newly formed helix stabilizes the internal hairpin structure and results in the release of the 16c from the immobilized capture oligonucleotide. (b) Time-dependent fluorescence response curves in the presence of a two-fold molar excess of OT2. Beacon 16c exhibited a 17-fold fluorescence increase. The y-axis "I" represents the signal:background ratio, which is the ratio of the fluorescence (background fluorescence subtracted) in the presence of OT2 and fluorescence (background fluorescence subtracted) in the absence of OT2. (c) Time-dependent fluorescence response curves of beacon 16c at varying concentrations of OT2. Data are from Rajendran and Ellington.[57]

discovered by this method had a significantly different mechanism for signaling than conventional molecular beacons.

2.3.2.4 Quaternary Structural Rearrangements

Beyond conformational changes that occur within a single aptamer, methods may be devised in which analytes bring together or break apart nucleic acid complexes, resulting in optical signaling. One method is very similar to the quenching aptamer beacons described above. Stojanovic et al. rationally designed fluorescently labeled heterodimeric signaling aptamers for cocaine and rATP detection.[58] Complementary strands in those aptamers were split into two separate pieces with fluorophore and quencher attached at the ends of the pieces. In the absence of the target ligand, the two oligomers mainly exist as monomers in solution. Ligand-dependent self-heterodimerization brings the 5′ fluorophore and 3′ quencher together and results in a quenched optical signal. The anti-cocaine aptamer was selective for cocaine (relative to its metabolites, such as benzoyl ecgonine and ecgonine methyl ester), whereas the anti-rATP aptamer was selective for rATP (as opposed to other ribonucleoside triphosphates such as rUTP, rGTP, and rCTP). The detection range for cocaine was 10 to 1250 μM, and that for rATP was 8 to 2000 μM. Both analytes could be simultaneously reported by labeling the anti-cocaine and rATP aptamer beacons with fluorescein and rhodamine, respectively.

An alternative strategy for ligand-mediated assembly involved the concurrent activation of a molecular beacon. Yamamoto et al. reported the selection of an aptamer that can bind the Tat protein of HIV-1.[59] The aptamer was then divided into two halves. One half was further designed to be a hairpin molecular beacon, with a fluorophore and quencher attached at the 5′ and 3′ ends, respectively. In the presence of Tat, the two halves of the aptamer are brought together and the molecular beacon is opened up, resulting in a fluorescent signal. The fluorescence is Tat dependent and works with both a Tat-derived peptide and the full-length Tat protein.

In similar strategies, antisense oligonucleotides were generated that were complementary to the given aptamers (Figure 2.8).[60] The oligonucleotide binds to the aptamer to form a duplex structure. The aptamer can be labeled with a fluorophore and the antisense oligonucleotide with a quenching moiety, leading to a quenched fluorescent signal in the absence of target. In the presence of ligand, the aptamer–target structure is stabilized and the equilibrium is concomitantly shifted away from the DNA–DNA duplex and toward the DNA–target complex. The structure-switching aptamer beacons can detect ATP ($K_d = 600$ μM) and thrombin ($K_d = 400$ nM). This method is particularly useful for the general design of aptamer biosensors, as it requires no prior knowledge of aptamer secondary or tertiary structure.

2.3.2.5 Aptazymes

Ligand-dependent conformational changes can be transduced directly into optical signals as well as into catalysis. Aptamers can be conjoined with nucleic acid enzymes (ribozymes) to generate so-called aptazymes. Aptazymes can potentially carry out the same range of reactions as have previously been seen for

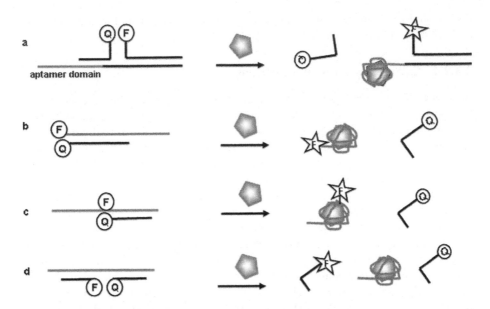

FIGURE 2.8 Mechanism of antisense-mediated, structure-switching aptamers. Tripartite (a,d) and bipartite (b,c) quenched systems are designed based on aptamer and antisense sequences. Upon ligand binding, the equilibrium between the aptamer and the complementary antisense strand is shifted and the complex is disrupted, resulting in dequenching of either a terminal or internal incorporated fluorophore.

ribozymes, including cleavage, ligation, phosphorylation, alkylation, and many others. In addition, aptazymes can turn over multiple substrates, just as ribozymes do, thus leading to an amplification of the initial ligand interaction.

To date, aptazymes have been constructed by taking advantage of the fact that many ribozymes contain stem-loop structures that are structurally important but whose sequence may be varied. By introducing aptamers in place of these stem-loops, the ligand-dependent conformational changes that aptamers normally undergo may potentially be transmitted to the catalytic cores of ribozymes, influencing their catalytic activities.

The first aptazyme was constructed in precisely this manner by Tang and Breaker.[61] These researchers joined an anti-adenosine aptamer that formed a hairpin secondary structure to the hammerhead ribozyme (Figure 2.9a). In the absence of ATP, the engineered hammerhead aptazyme could still efficiently catalyze self-cleavage. However, ATP binding caused a conformational change in the ligand-binding domain (aptamer) that in turn interfered with the catalytic domain (hammerhead).[62] At saturation, the catalytic rate of the ATP-modulated aptazyme was inhibited 180-fold. However, the aptazyme was not inhibited by dATP or other ribonucleoside triphosphates, as was the case with the original aptamer.

Because the ligand-binding site is different from the catalytic site, the activity of an aptazyme can also be modulated via changes in the connecting region between the two sites.[63] Although there are several possible mechanisms for activation, the most

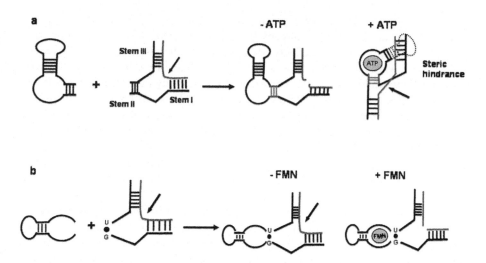

FIGURE 2.9 Rational design of regulated hammerhead ribozymes. (a) An anti-ATP aptamer was appended to the nonessential stem II of the hammerhead. Several constructs with different lengths and sequences at the joining region between the aptamer domain and the catalytic core were designed and tested for function in the presence or absence of ATP. ATP binding was found to inhibit the ribozyme's activity, probably due to a conformational change that led to steric hindrance of the catalytic core. (b) An FMN-activated ribozyme was designed by destabilizing stem II. In the absence of FMN, the enzyme loses activity. Upon FMN binding, the structure of the enzyme is stabilized and the enzyme regains activity. Arrows indicate the hammerhead cleavage site.

common is to rely upon ligand-induced stabilization of the joining region between the aptamer and the ribozyme. For example, the hammerhead ribozyme has a three-stem junction, and destabilization of stem II results in reduced catalytic ability. A weakened stem II was created by introducing G:U base pairs, and an anti-FMN aptamer was fused to the weakened stem (Figure 2.9b).[64] Ligand-binding then stabilized the catalytic domain and increased the catalytic rate of the ribozyme by up to 100-fold. When the FMN-binding domain was substituted with an anti-theophylline aptamer, the ribozyme was no longer responsive to FMN, but the resultant theophylline-dependent hammerhead aptazyme was activated by 40-fold. Similarly, Araki et al. appended an FMN-specific RNA aptamer to stem II of the hammerhead ribozyme via a designed linker sequence of one to six complementary base pairs.[65] The best construct had a stem length of 3 bp and showed about a 10-fold increase (K_{cat}) in the presence of saturating amounts of FMN; however, other constructs with longer linkers had little or no FMN dependence. Chemical modification experiments showed that binding of FMN to the aptamer domain induced conformational alterations in stem II that presumably affected catalytic activity.

Just as aptamer beacon strategies may rely on tertiary and quaternary as well as secondary structural rearrangements, similar strategies have been applied to the design of aptazymes. The *Tetrahymena* group I self-splicing intron has been altered to simultaneously interact with two arginine-rich motifs from the bacteriophage λN and HIV-1 Rev proteins.[66] The design was based on the structure of the starting intron and proteins, which had already been characterized by x-ray crystallography and NMR spectroscopy, respectively. Two peptide-binding RNA motifs, λBoxB and HIV-1 RRE, were introduced into the ribozyme P5b and P6 domains, and chimeric proteins (PepA and PepG) were constructed by conjoining the corresponding RNA-binding motifs, $λN_{1-19}$ and HIV Rev_{34-50}. PepA and PepG differed in terms of the linker sequences between the arginine-rich motifs, either four alanine or four glycine residues. Unfortunately, the designed proteins only very weakly affected the K_{cat} of the ribozyme. The four amino acids were then randomized and *in vivo* selection was carried out for mature reporter gene production. The selected, optimized protein (PepS) showed more efficient binding to the ribozyme. The K_d values for PepA, G, and S were 602, 141, and 63 nM, respectively.

Although rational design methods have proven to be very successful in generating aptazymes, *in vitro* selection can also be used to optimize catalytic activity. Soukup and Breaker appended the anti-FMN aptamer to the hammerhead ribozyme via a random N4-N4 linker (Figure 2.10).[67] Self-cleaving clones in the absence of FMN

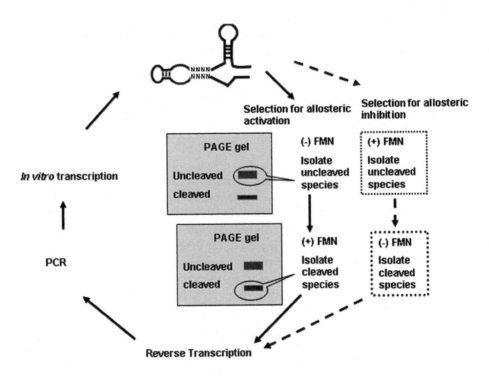

FIGURE 2.10 *In vitro* selection of *cis*-cleaving allosteric hammerhead ribozymes via randomization of a communication module. Selection for both activation and inhibition were carried out with the same pool. In both instances, the preselection (negative selection) eliminated self-cleaving species in either the absence or presence of FMN. Then a positive selection was performed to identify self-cleaving species in the presence or absence of FMN, respectively. Selected species were reverse transcribed, PCR amplified, and *in vitro* transcribed for the next round of selection.

were eliminated by so-called negative selection or preselection. Those variants that cleaved in the absence of FMN were separated from uncleaved variants by gel electrophoresis, and the uncleaved RNA was purified. The subsequent positive selection was then carried out in the presence of FMN, and cleaved variants were isolated and amplified by reverse transcription, PCR, and *in vitro* transcription. Selection for inhibition as well as activation was carried out by simply modifying the original selection protocol: FMN was introduced during the preselection but was excluded in the positive selection for variants with cleavage activity. After six rounds of selection and amplification, the activation and inhibition of the best classes of aptazymes were up to 270- and 600-fold, respectively, in the presence of FMN. Although various selected aptazymes may work via a number of mechanisms, one in particular was notable. Upon ligand binding, the base pairs in the joining region "slip" and new base pairs are formed that either assist or prevent the formation of the catalytic core of the hammerhead ribozyme. To examine whether the slip-structure mechanism is independent of the sequence and ligand specificity of the aptamer domain, new aptazymes were created by replacing the anti-FMN aptamer with either an anti-theophylline or an anti-ATP aptamer. The new aptazymes were specifically activated by 110- and 40-fold, respectively, in the presence of their cognate ligands. An additional selection was carried out in which the anti-theophylline aptamer was joined to the hammerhead via an N5–N5 linker, and aptazymes were selected that were from 1300- to 4700-fold activated by theophylline.[68] Once again, when the anti-theophylline aptamer was changed to an anti-FMN aptamer, theophylline-dependent activation was changed to FMN-dependent activation.

Effector specificity can also be changed by selection rather than design. The 25-residue theophylline-binding domain of one of the theophylline-dependent aptazymes was mutagenized, and allosteric ribozyme variants were selected that relied on 3-methylxanthine rather than theophylline. The specificity of the resultant aptazyme again indicates the remarkable abilities of RNA to form precise structures, because theophylline (1,3-dimethylxanthine) and 3-methylxanthine differ by only a single methyl group. The selected aptazyme had three substitutions, one of which was sufficient for changing effector specificity.

In order to move beyond mechanisms dependent only on the joining region, Koizumi et al.[69] generated a pool in which 25 random residues were appended to the hammerhead ribozyme via the previously selected communication module.[67] Variants were selected whose self-cleavage was promoted by 3′,5′-cyclic nucleotide monophosphates (cNMPs).[69] The cyclic nucleotides were mixed, and the selection was initially carried out with all four present; individual nucleotides were excluded from the mixture as particular cNMP-dependent aptazymes dominated the population. Cyclic guanosine monophosphate (cGMP)-, cyclic cytidine monophosphate (cCMP)-, and cyclic adenosine monophosphate (cAMP)-specific aptazymes were obtained after 18, 20, and 23 rounds, respectively, whereas no cyclic uridine monophosphate (cUMP)-dependent ribozymes were selected. The selected aptazymes exhibited up to 5000-fold activation. The effector-binding domains from the selected cGMP, cCMP, and cAMP aptazymes were further mutagenized and selection was used to generate artificial phylogeny data.[70] Furthermore, selection under low concentrations of cNMP (one tenth of that used in the previous selection) yielded a variant that showed an apparent K_d for cGMP that was about one tenth of that of the parental aptazyme. Once residues in the ligand-binding domain were defined, the ligand-binding domains were separated from the aptazymes and were still found to retain their ligand-binding properties.

Methods for the generation of aptazymes may be generalized beyond the hammerhead ribozyme. The Ellington lab has carried out experiments with a ribozyme ligase rather than a cleavase.[71–74] An allosteric ribozyme ligase (L1) was selected from a random sequence pool, and its catalytic activity was found to be activated 10,000-fold by a complementary DNA (cDNA) primer that was present during the selection.[71] This result is a classic example of the adage that "you get what you select for." Secondary structural analysis of the selected ligase revealed that the 3′ terminus base-paired with the substrate-binding domain (much as the pseudosubstrates of some protein kinases bind to their active sites) and that binding of the oligonucleotide effector resulted in the formation of an alternative structure that relieved this inhibition. The oligonucleotide-binding site of the selected aptazyme ligase could be rationally altered, and the aptazyme could transmogrify oligonucleotide recognition into signal amplification via PCR. Just as Breaker and his coworkers conjoined aptamers to catalysts, anti-adenosine, theophylline, and flavin aptamers were joined to the L1 ligase to create aptazymes.[72] In the presence of 1 mM ATP or theophylline, the designed aptazymes were activated by 800- and 1600-fold, respectively (Figure 2.11a and 2.11b). However, there was little apparent activation by flavin. To remedy this, a "communication module" was selected and an aptazyme (Figure 2.11b) was selected that was activated up to 260-fold by FMN.

In vitro selection methods were extended to the development of protein-activated aptazymes (Figure 2.12).[73] Variants responsive to either a tyrosyl transfer RNA (tRNA) synthetase (Cyt18) (Figure 2.11a) or hen egg white lysozyme were selected from an N50 pool appended to the L1 ligase. The resultant protein-dependent aptazymes were activated several thousand-fold by their cognate protein effectors. It has also proven possible to evolve peptide-dependent aptazymes. An HIV-1 Rev arginine-rich motif (ARM)-dependent aptazyme was selected from the L1-N50 pool.[74] The aptazyme ligase was activated more than 18,000-fold by the Rev ARM peptide (Figure 2.11a) and could also be specifically activated by the full-length HIV-1 Rev protein but not by most other arginine-rich motif peptides and proteins. However, protein dependent activation was not substantially improved following selection for protein activation from a doped sequence library based on the peptide-activated aptazyme.

Extant aptazymes are summarized in Table 2.1. Although most of the aptazymes to date have been ribozymes, DNA enzymes that are activated by ATP have also been generated by both design and selection.[75]

2.3.2.6 Aptazyme Assays

Aptazymes are particularly interesting as biosensors for two reasons. First, they can potentially be readily adapted to reagentless assay formats; second, they can potentially be adapted to arrays. There are nascent examples of both applications.

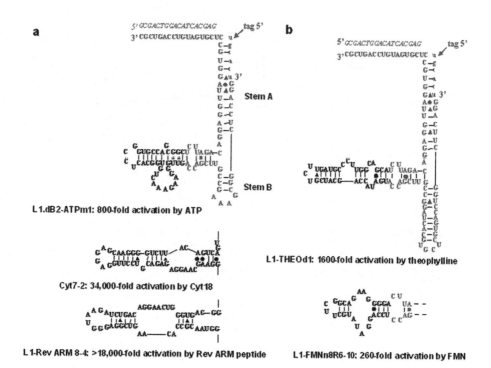

FIGURE 2.11 Secondary structures of various L1 aptazyme ligases. Aptazymes in (a) and (b) have a different stem B. These aptazymes were obtained by rational design and *in vitro* selection, as described in the text. Aptamer domains are drawn in black. Maximal activations by cognate ligands are shown below the structures.

In general, RNA and DNA catalysts have proven to be attractive tools for therapy and drug discovery, as reviewed by Steele et al.[76] Aptazymes may also be used in biomedical applications, for example, in assays for known drugs, cofactors, and their derivatives. Hammerhead aptazymes have been selected against an antibiotic, pefloxacin.[77] Ribozyme variants were inhibited in the presence of submicromolar levels of the drug. The aptazymes were differentially responsive to different moieties on the drug, implying that it might be possible to quickly derive a family of ribozymes to assay a variety of modifications introduced as a result of a medicinal chemistry program. Similarly, Ferguson et al. reported the selection of aptazymes that could sense the biomedically relevant compounds caffeine and aspartame at concentrations of 0.5 to 5 mM.[78] Srinivasan et al. identified ADP-specific aptazymes that could potentially be used as universal sensors of kinase activity for high-throughput screens.[79] Upon activation by ADP, the hammerhead aptazymes cleave an oligonucleotide substrate, resulting in a fluorescent signal.

Beyond simple assay procedures, it may even be possible to use aptazymes in drug discovery protocols. The Rev-binding element (RBE) of HIV-1 and an anti-Rev aptamer were fused to stem II or the 5′ end of the hammerhead ribozyme, respectively, in order to generate protein-responsive aptazymes.[80] Whereas the addition

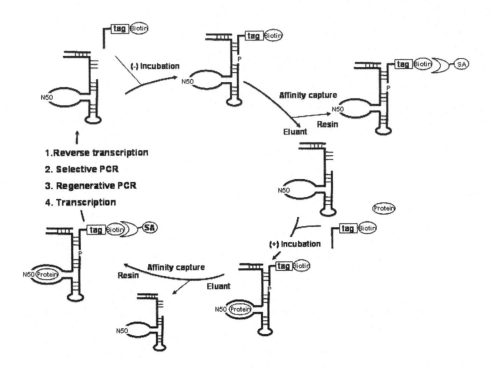

FIGURE 2.12 Selection of protein-dependent aptazyme ligases. Protein-dependent aptazyme ligases were selected using the L1-N50 pool and a biotinylated substrate. Prior to selection, a negative selection without target protein was performed in which non-protein-dependent reactive species were captured on a streptavidin-agarose resin. The eluents containing unligated ribozymes were positively selected by addition of target protein and captured on streptavidin-agarose resin. Ligated ribozymes were eluted and amplified for subsequent rounds of selection.

of the RBE to stem II did not affect the enzyme's catalytic ability, Rev peptide or Rev protein binding led to a conformational change that resulted in enzyme inhibition. Conversely, the 5′ appended anti-Rev aptamer interdicted the substrate-binding domain, and addition of Rev peptide or protein led to restoration of enzyme activity. These two formats were successfully adapted to screening small molecule libraries for Rev-binding compounds. The addition of a compound that led to restoration of activity could be easily read by cleavage of a fluorescent substrate. Three compounds identified from a library of 96 antibiotics could reverse either Rev-dependent inhibition or activation. Interactions between one of the compounds, coumermycin, and HIV-1 Rev were further confirmed by inhibition of HIV-1 replication *in vivo*. It is likely that these strategies could be readily extended to other proteins, as well. For example, an HIV-1 reverse transcriptase (HIV-1 RT) aptamer that was known to form a pseudoknot structure with RT was inserted into stem II of the hammerhead ribozyme, yielding an aptazyme that could be inhibited upon the addition of RT.[81] RT concentrations of 10.4 nM led to half-maximum inhibition of the enzyme. The allosteric ribozyme was specifically regulated by HIV-1 RT but not by other, noncognate

TABLE 2.1
Allosteric Ribozymes (Deoxyribozymes) Generated
by Different Strategies

Analyte	Enzyme	Regulation	Strategy	Reference
ATP	Hammerhead ribozyme	Inhibit	*In vitro* selection	116
ATP, theophylline	Hammerhead ribozyme	Inhibit	Rational design	61
FMN	Hammerhead ribozyme	Activate	Rational design	65
Oligonucleotides	Ribozyme ligase	Activate	*In vitro* selection	71
FMN, theophylline	Hammerhead ribozyme	Activate	Rational design	64
FMN, theophylline ATP	Hammerhead ribozyme	Activate, inhibit	Combination of rational design and *in vitro* selection	67
cNMP	Hammerhead ribozyme	Activate	*In vitro* selection	69
FMN	Hammerhead ribozyme	Activate	Rational design	65
ATP, theophylline, FMN	L1 ligase	Activate	Combination of rational design and *in vitro* selection	72
Theophylline, 3-methylxanthine	Hammerhead ribozyme	Activate	Selection for activation and altering the aptamer's ligand specificity	68
Doxycycline	Hammerhead ribozyme	Inhibit	*In vitro* selection	117
Pefloxacin	Hammerhead ribozyme	Inhibit	*In vitro* selection	77
Tyrosyl tRNA synthetase (Cyt18), lysozyme	L1 ligase	Activate	*In vitro* selection	73
FMN and theophylline	Hammerhead ribozyme	Cooperatively activate	Rational design	104
ATP	Deoxyribozyme ligase	Activate	Rational design and *in vitro* selection	75
Theophylline (a versatile communication module)	Hepatitis D virus (HDV) ribozyme	Activate	*In vitro* selection	118
Rev ARM	L1 ligase	Activate	*In vitro* selection	74
Caffeine, aspartame	Hammerhead ribozyme	Activate	*In vitro* selection	78
ADP	Hammerhead ribozyme	Activate	*In vitro* selection	79

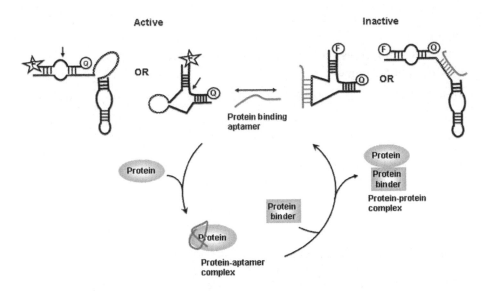

FIGURE 2.13 Detecting protein–protein interactions with an engineered ribozyme. The modified loop region of either the hammerhead or hairpin ribozyme was designed to be complementary with the protein-binding aptamer (grey) so that aptamer binding inactivates the ribozymes. In the presence of the aptamer's target protein (thrombin), the interaction between aptamer and protein causes the aptamer to release from the ribozymes; therefore, the substrate is cleaved to generate fluorescence. The addition of protein inhibitors disrupts the protein–aptamer complex and the enzyme is again inhibited by the released aptamers.

proteins. Interestingly, a DNA primer–template complex could bind RT and reverse the inhibition of cleavage activity.

These approaches were further extended to study protein–protein interactions (Figure 2.13).[80] In this instance, a sequence complementary to an anti-thrombin aptamer was introduced into the hammerhead or hairpin ribozymes. In the presence of the aptamer, a helix was formed that inactivated the ribozymes (this strategy is similar to several of the oligonucleotide-dependent ribozymes that will be described in the next section). In the presence of thrombin, the aptamer was "decoyed" away from the ribozyme, resulting in ribozyme-dependent cleavage of a substrate containing a fluor and quencher. Interactions between thrombin and a specific protein inhibitor, hirudin, disrupted the thrombin–aptamer complex, releasing the aptamer and again leading to inhibition of cleavage activity. This variation on the assay may eventually prove useful for screening protein libraries for interaction partners.

Assuming that aptazymes can be adapted to assays for drug development, target validation, or interaction partners, it will become necessary to further adapt them to high-throughput array formats. The development of aptazyme arrays may also enable broader applications in proteome or metabolome analysis. The use of aptazymes to monitor and quantify metabolites is of particular interest because there are few antibody assays that can directly report the presence of small organic molecules, especially in a multiplex format. Seetharaman et al. used seven different hammerhead aptazymes in an array format to detect various metal ions, enzyme cofactors, metabolites, and drug analytes.[82] Each aptazyme was immobilized in the array by first introducing a 5′ thiotriphosphate moiety that was subsequently reacted with a gold surface. The array could not only detect, discriminate between, and quantitate analytes in solution but could also detect naturally produced cAMP in bacterial culture media.

Aptazyme ligases have also been adapted to an array format. Our group has assayed combinations of aptazymes that detect small organics, peptides, and proteins in parallel (Figure 2.14).[83] The aptazyme substrates were labeled with biotin and the allosteric ribozymes were radioactive. Ligated products were retained in streptavidin-coated 96-well plates, and unligated materials were washed away. The aptazyme ligases in the array could specifically and simultaneously detect their cognate analytes, and the limits of detection of the best aptazymes were in the nanomolar range.

Cleavases and ligases each have advantages and disadvantages in array formats. Cleavases require only binary complex formation for activation, but the background rate of cleavage can strip an array of its corresponding sensors. Ligases require ternary complex formation (substrate and target) but have the advantage that the formation of a covalent bond can allow more stringent washing of the array to enhance signal:noise ratios.

2.4 PROSPECTS FOR THE DESIGN OF NUCLEIC ACID BIOSENSORS

One advantage that nucleic acid biosensors have over protein biosensors is that their structures are largely determined by the simple rules of Watson-Crick base pairing. Because of this, the stabilities, and in some cases the functionalities, of nucleic acid biosensors can be rationally altered. We have seen several examples of this presented above. For example, aptamer beacons were generated by engineering secondary structural changes, and aptazymes were engineered by destabilizing stems connecting ligand-binding and catalytic domains.

Design principles may prove to be most useful in the generation, optimization, and alteration of nucleic acid biosensors that can detect other nucleic acids. The specificities and responsivities of molecular beacons can, of course, be changed almost at will. In addition, a number of sequence-responsive aptazymes have been developed that rely on a number of different mechanisms. Ultimately, by refining the design process and linking together individual components into larger networks, schemes for nucleic acid computation may become not only feasible but also useful in the design of sequence-responsive medicines.

2.4.1 MECHANISMS FOR OLIGONUCLEOTIDE DEPENDENCE

2.4.1.1 Relief from Inhibition

First, ribozymes have been engineered so that the addition of an oligonucleotide effector results in a release from inhibition (similar to the findings with the L1 ligase that were mentioned above). Porta and Lizardi originally constructed an allosteric

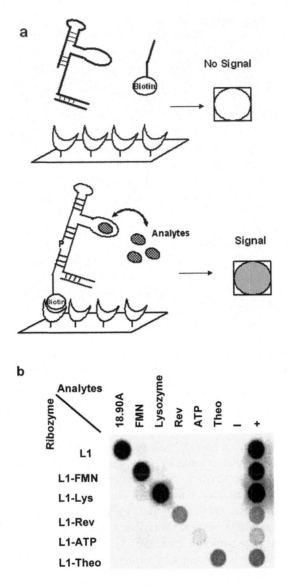

FIGURE 2.14 L1 ligase aptazyme array. (a) Working principle of the aptazyme ligase array. Biotin-labeled substrate oligonucleotide and radiolabeled L1 aptazyme ligases were used to detect an oligonucleotide effector (18.90A), small molecule effectors (FMN, ATP, and theophylline), and peptide (Rev ARM) and protein (lysozyme) effectors. In the presence of their cognate analytes, aptazyme ligation products were immobilized on streptavidin-coated 96-well plates, producing a radioactive signal, whereas unligated aptazymes were washed away. (b) Radioactive signals observed in the L1 aptazyme ligase array. "−" no effectors; "+" a mixture of all effectors. Data are from Hesselberth et al.[83]

hammerhead ribozyme that could be activated by an effector oligonucleotide.[84] The designed ribozyme contained an extension that hybridized with a portion of the catalytic core of the hammerhead, forming an alternative hairpin stem structure that denatured and inactivated the ribozyme. With the addition of a relatively long 35-nucleotide oligonucleotide that could hybridize to the extension, a conformational change occurred (much like in a molecular beacon) that restored structure and activity to the ribozyme. Although the activation achieved was only 10-fold, this early work presaged both the development of aptazymes, described above, and much of the work on oligonucleotide-dependent enzymes described in this section.

Burke and his coworkers have developed an almost identical strategy that they refer to as target ribozyme-attenuated probe (TRAP).[85,86] In this version of the "relief-from-inhibition" strategy, the oligonucleotide effector does not directly bind to the residues involved in the alternate, interfering secondary structure. Rather, the TRAP structure works more like a molecular beacon, in that hybridization to a loop causes a conformational change that destabilizes the interfering secondary structure (Figure 2.15a). This strategy is more flexible than the one originally espoused by Lizardi, because the loop sequence can be changed at will. Moreover, a 730-fold activation was observed in the presence of the effector oligonucleotide. An examination of the kinetics of the

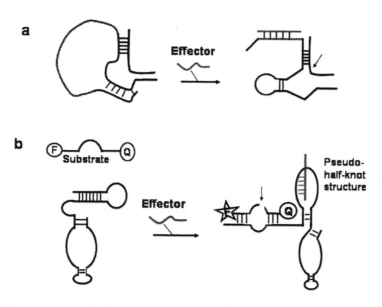

FIGURE 2.15 Relief-from-inhibition strategies for oligonucleotide-regulated ribozymes. (a) The hammerhead TRAP strategy. A molecular beacon structure was introduced into the ribozyme to inhibit the activity of the catalytic core. Upon addition of an oligonucleotide effector that can hybridize to the loop sequence, the active conformation was restored. (b) Pseudo-half-knot structures. A hairpin ribozyme was designed so that internal complementarity resulted in inactivation. In the presence of an oligonucleotide effector, a pseudo-half-knot structure was formed, freeing up the substrate-binding site.

TRAP activity showed that substrate-binding affinity is not significantly changed, irrespective of whether the oligonucleotide effector is present or absent. Most importantly, in terms of design principles, the better the binding of the oligonucleotide effector, the faster the cleavage rate, and the more stable the interfering secondary structure, the slower the cleavage rate. The switch between the active form and attenuated (inactive) form of the enzyme is the result of competition between the sense–antisense duplex and the attenuator–core duplex.

The relief-from-inhibition strategy has also been applied to the hairpin ribozyme.[87,88] In this instance, the added sequence interfered with the ability of the substrate to bind. Upon hybridization of a short RNA, such as a microRNA (miRNA), a pseudo-half-knot structure was formed (Figure 2.15b) and the ribozyme was activated. Nine different ribozyme constructs were tested against their corresponding miRNAs from *drosophila*, and no cross-reactivity to noncognate sequences was observed. In addition, lower standard deviations and signal:noise ratios were seen, as compared with detection by molecular beacons. However, the activation level (10-fold seen by iHP-let7) was much lower than for most other designed or selected allosteric ribozymes.

This strategy has also been applied by Najafi-Shoushtari et al. to the detection of *trp* leader messenger RNA (mRNA).[88] The observed activation was 1500-fold in the presence of *trp*-mRNA. Hairpin ribozymes that were inhibited as well as activated were constructed. However, in the presence of *trp*-RNA-binding attenuation protein, the *trp* leader mRNA was sequestered, resulting in disinhibition of the ribozyme.

Finally, inspired by molecular beacons, Stojanovic et al.[89] extended an RNA-cleaving deoxyribozyme (E6)[90] so that an inhibitory stem-loop was formed. In the presence of an effector oligonucleotide that hybridized to the loop, inhibition was relieved, similar to the TRAP strategy employed by Burke et al. Cleavage at the single ribonucleotide on a substrate that contained a 5′ fluorescein and 3′ quencher led to an analyte-dependence increase in fluorescence signal. The allosteric catalyst could discriminate against single base mutations in the effector sequence.

2.4.1.2 Structural Stabilization

Second, ribozymes have been engineered so that the addition of an oligonucleotide effector results in the formation of an otherwise destabilized structure. Kuwabara et al. have used oligonucleotide effectors to stabilize otherwise inactive ribozymes rather than to relieve inhibition. So-called maxizymes (<u>m</u>inimized, <u>a</u>ctive, <u>X-shaped</u> and intelligent ribo<u>zymes</u>) are composed of four pieces, one of which is the effector oligoribonucleotide.[91–93] A minizyme (shortened hammerhead ribozyme[94] that lacked the stem-loop II region but had equivalent activity to the wild type) was used as the basis for engineering the maxizyme. It was found that a homodimeric minizyme, which can cleave two identical substrates at the same time, was more active than each of the individual monomers. The maxizyme was similarly designed, but it is a heterodimer rather than a homodimer, with one catalytic core deleted in most cases. By changing one of the substrates to actually be a noncleavable pseudosubstrate, this pseudosubstrate can instead act as an effector oligonucleotide. Ultimately, two different maxizyme monomers are brought together by hybridization to the effector oligonucleotide sequence, capturing Mg^{2+} to form an active catalyst that can cleave target mRNAs both *in vitro* and *in vivo*.

A slightly different strategy has been adopted by Komatsu et al.[95-97] These researchers have engineered the monomeric hammerhead ribozyme to be dependent upon an oligonucleotide effector by first destabilizing stem II of the hammerhead, replacing it with a large, unstructured loop sequence. Upon hybridization of an oligonucleotide to this loop, stem II is re-created in *trans* (as a pseudo-half-knot structure) (Figure 2.16a), and the hammerhead core is activated. Because the precise sequence of stem II (now loop II) is relatively unimportant for the activity of the hammerhead ribozyme, a variety of different effector-activated ribozymes can be designed.[95] Oligonucleotides with modified nucleotides were used as effectors, and activations of up to 750-fold were observed.

The basic strategy of oligonucleotide stabilization of ribozyme structure has been extended from design to selection.[96,97] In this instance, the hairpin rather than

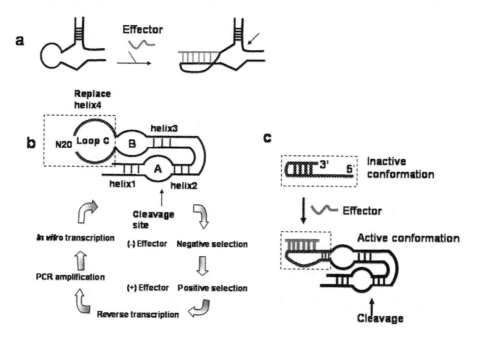

FIGURE 2.16 Structure stabilization strategies for oligonucleotide-regulated ribozymes. (a) Engineered structural stabilization. The hammerhead ribozyme was destabilized by engineering a nonstructured loop within the stem II region. An effector oligonucleotide complementary to part of the new loop allowed the formation of an active pseudo-half-knot structure. (b) Selection for structural stabilization. Because the sequence of helix 4 of the hairpin ribozyme is variable, this domain was replaced with a 20-residue random sequence loop. Selection was carried out to find the variants that could be activated by an effector oligonucleotide. (c) Mechanism of activation for selected ribozymes. In the absence of an effector oligonucleotide, the residues in the selected variant form a helix that disrupts the structure of internal loop B and renders the enzyme inactive. In the presence of the effector, the selected ribozyme forms an active pseudo-half-knot that reestablishes the structures of helix 4 and internal loop B.

the hammerhead ribozyme was engineered. The hairpin ribozyme contains two domains: each has two helices; the sequence of helix 4 can be varied. A 20-nucleotide random sequence was introduced into loop C (Figure 2.16b), and ribozymes were selected based on their ability to carry out a cleavage reaction in the presence of an effector oligonucleotide. After 12 rounds of both negative and positive selections, 34 active clones were identified; 18 of them had identical sequences, and all of them were predicted to bind to the effector oligonucleotide. As with the designed, effector-activated hammerhead ribozymes, the secondary structures of the selected effector-activated hairpin ribozymes are predicted to be pseudo-half-knots (Figure 2.16c). In contrast to the designed mechanism, however, in this instance hybridization to the effector oligonucleotide not only stabilizes the native secondary structure but also assists in disrupting an inhibitory secondary structure. This mechanism thus is in some ways a chimera of "relief from inhibition" and "structural stabilization." The observed activation is more than 300-fold in the best case. Interestingly, the selected module functioned similarly when introduced into stem II of the hammerhead ribozyme.

2.4.1.3 Structure Completion

Instead of directly inhibiting or activating a nucleic acid enzyme, it is possible to use oligonucleotide effectors to complete enzymes that otherwise lack critical sequence or structural components. In this regard, Wang et al. have built upon the 10-23 deoxyribozyme originally selected by Santoro and Joyce,[98] an enzyme in which the length and degree of Watson-Crick pairing between the substrate and substrate recognition arms affect activity and turnover. A so-called expansively regulated RNA-cleaving deoxyribozyme was built by designing an effector oligo-nucleotide that could bind to both the substrate and the deoxyribozyme, forming a stable three-way junction (Figure 2.17).[99-101] The substrate recognition arms of the

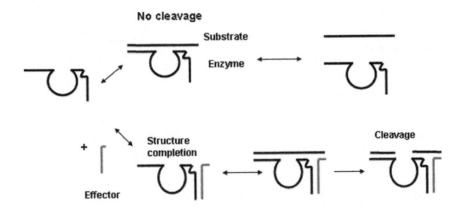

FIGURE 2.17 Structure completion strategies for oligonucleotide-dependent enzymes. The 10-23 RNA cleaving deoxyribozyme was inactivated by shortening its substrate-binding region. An effector oligonucleotide that can bind to both enzyme and substrate completes the substrate-binding region and allows substrate binding and subsequent cleavage.

deoxyribozyme form $6+3$ base pairs (the number of base pairs in stem A and stem B, respectively) with the substrate in the absence of the effector, but they form optimal $6+7$ base pairs in the presence of the effector.[99] Two unpaired purine nucleotides at the three-way RNA–DNA junction provided sufficient flexibility to ensure function in the adjacent stems. The observed rate constant (K_{obs}) was increased 55-fold with a DNA regulator and 250-fold with an RNA regulator of identical sequence. The K_{cat} value was also increased 260-fold in the presence of the RNA effector.

The "structure completion" strategy has also been applied[100] generally to a different deoxyribozyme, the 8-17 RNA cleavase[98]; to a so-called bipartite deoxyribozyme cleavase[102]; and to the hammerhead ribozyme. The catalytic activations for the 8-17 deoxyribozyme, bipartite deoxyribozyme, and hammerhead ribozyme variant were 31-, 20-, and 22-fold, respectively.

The structure completion strategy proved workable not only for oligonucleotide effectors but for other effectors as well. The perfectly Watson-Crick base-paired stem of the 10-23 deoxyribozyme was replaced with a partially base-paired stem that formed an anti-ATP aptamer.[101] In the presence of adenosine, the two half-aptamers bound to the deoxyribozyme, again forming a stable three-way junction. This Ado-DNAzyme I was optimized by varying the number of base pairs between the effector oligonucleotide (half-aptamer) and the substrate. Overall, the catalysis of Ado-DNAzyme I was found to be activated 147-fold in the presence of both DNA regulator and adenosine, whereas enhanced catalysis of only 4.6-fold was observed in the presence of the DNA regulator alone.

In perhaps the most compelling example of structure completion, researchers from Sirna Therapeutics divided the Bartel Class I ligase into two pieces and engineered the catalytic core by selection so that it could recognize a conserved portion of the hepatitis C virus (HCV) as one of the two pieces.[103] The evolved, completed ribozyme still was blazingly fast compared to most ribozymes, with a turnover rate of 69/min. Using a gel-based ligation assay with radioactive enzyme, as few as 6700 target molecules of an oligonucleotide that contained the conserved HCV sequence could be detected.

2.4.2 APPLICATIONS OF DESIGNED BIOSENSORS

2.4.2.1 Logic Gates

Ultimately, the use of design principles may allow the generation of nucleic acids that can make logical decisions. As we pointed out above, allosteric ribozymes essentially act as molecular transistors in which catalysis is "gated" by an effector. For example, the inhibition of the hammerhead ribozyme by ATP[61] may be seen as an inverter (NOT) function. Furthermore, a combination of ligands may allow ribozymes to integrate more complex signals. Anti-flavin and anti-theophylline aptamers were mounted on the hammerhead ribozyme in series.[104] The catalytic rate of the resultant aptazyme was enhanced in the presence of both ligands, and the two effector-binding domains appeared to exhibit cooperative interactions. The dual effector aptazyme is therefore, in essence, a molecular AND gate. Similarly, selected protein-dependent aptazymes derived from the L1 ligase also remain dependent upon

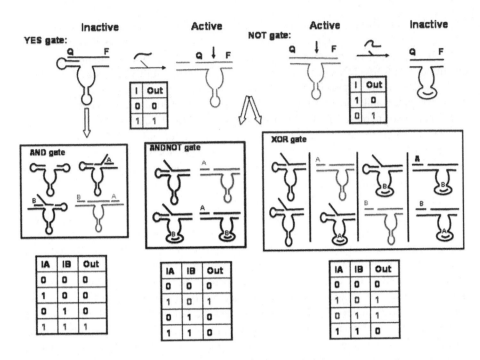

FIGURE 2.18 Generation of logic gates using oligonucleotide-dependent ribozymes. Molecular beaconlike stem-loop structures were added to the substrate-binding arm of the E6 deoxyribozyme cleavase to generate a YES gate. Binding of an oligonucleotide effector (input) to the loop region exposes substrate-binding sequences, and the enzyme's activity is monitored by separation of a fluor from a quencher in an oligonucleotide substrate. Hybridization of an effector oligonucleotide to an internal loop sequence distorts the stem structure required for the catalytic activity and results in inhibition of enzyme function. Unlike the "YES" gate, the resultant "NOT" gate inverts the input data. The YES and NOT gates were combined in a single molecule to form an AND NOT gate. Only in the presence of the input oligonucleotide (IA) complementary to the 5′ loop is there output in fluorescence signal (1); for all other input combinations, there is no increase in fluorescence. An XOR gate was also generated with two different AND NOT gates in parallel in response to two input oligonucleotides (IA and IB). Active substrate–ribozyme complexes are drawn in grey.

the oligonucleotide effector that the ligase originally required, and this is therefore also an example of a dual effector AND gate.[73]

Stojanovic et al. have shown that the ability to design nucleic acids that are dependent upon oligonucleotide effectors can greatly simplify the construction of a variety of nucleic acid logic gates.[105] Many of these gate structures, as shown in Figure 2.18, were derived from the E6 deoxyribozyme in a manner similar to their original molecular beaconlike construct,[89] as described above. Molecular beaconlike stem-loop structures could be added to both of the substrate-binding arms in parallel, leading to the generation of an AND gate (albeit on the 8-17 rather than the E6 deoxyribozyme core). The readout of the gate was the cleavage of an oligonucleotide substrate and the concomitant separation of a fluor from a quencher.

The loop adjacent to the catalytic core of E6 served as a docking site for an oligonucleotide effector, denaturing the active form of the deoxyribozyme and producing a NOT function. By combining the mechanisms of the AND and NOT gates, it proved possible to immediately construct an AND NOT gate. Indeed, given these two functions, it should prove possible to generate the entire set of Boolean logic, if not within a single molecule. Two different AND NOT gates in parallel served as an XOR gate.

2.4.2.2 Computation

In 1994, Adleman used oligonucleotides to solve a Hamiltonian path problem, a seminal implementation of DNA as a computational molecule.[106] Since then, there has been a variety of other attempts at DNA computation, most of them computationally disappointing.[107] However, the logic gates developed by Stojanovic and Stefanovic show promise in the development of so-called silicomimetic devices, in which biological inputs might be acted upon by simple, preset algorithms, yielding relevant biological outputs.[108] For example, based on the XOR and AND gates already developed, these authors generated a half-adder that can add two single binary digits (bits) and is in turn the key building block for a full adder. More amazingly, a deoxyribozyme-based molecular automaton was recently engineered to flawlessly play a game of tic-tac-toe against a human opponent[109] (commented upon in Tabor and Ellington[110]). The automaton was set up in nine sample wells (corresponding to squares on a tic-tac-toe board; Figure 2.19), each containing a subset of 23 different deoxyribozyme gates. Those gates were differentially regulated by combinations of eight oligonucleotide effector molecules (representing plays by the human opponent), but all enzymes recognized and cleaved the same fluorescent substrate (indicating the automaton's move). The system was predicated on DNA base pairing; effectors base-pair with deoxyribozymes, allosterically activating and inactivating their target enzymes. The automaton was designed such that individual effector oligonucleotides could potentially activate a subset of enzymes while inhibiting others. Up to three allosteric domains were incorporated into each enzyme, allowing for complex responsivity to accumulating effector molecules during the game. The judicious layout of deoxyribozymes within the nine wells engendered a logic that ensured that the automaton never lost to its human opponent (more than 100 empirical victories were achieved by the automaton).

2.4.2.3 Smart Drugs

Although raw computation may never be the forte of nucleic acids, designed nucleic acid biosensors that can make logical decisions may ultimately prove useful as "smart drugs" that sense and respond to particular disease states or as engineered circuits in cells. These prospects are particularly appealing because of the recent surge in knowledge regarding the role of simple base-pairing rules in controlling cellular metabolism. It may be very possible for a ribozyme automaton to, for example, sense a pathogenic mRNA or combination of mRNAs and then generate an appropriate antisense, small interfering RNA (siRNA), or miRNA treatment.

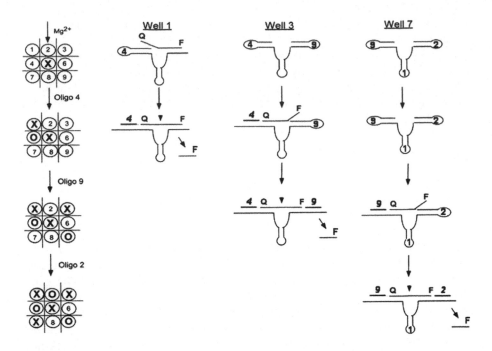

FIGURE 2.19 Deoxyribozyme tic-tac-toe. Upon the addition of Mg^{2+}, a nonallosteric deoxyribozyme in square (well) 5 cleaves its substrate, yielding a fluorescent signal and signifying the automaton's first move. The human player responds by adding a DNA oligonucleotide effector to all nine wells, in this case, an oligonucleotide indicating the choice of playing square 4. The effector signifying play in square 4 binds to and activates a preprogrammed, allosteric deoxyribozyme in square 1, which in turn cleaves (red) the substrate and indicates the automaton's response: an "X" in square 1. In all other wells, no activation occurs and no move is indicated. As the game continues, the addition of subsequent effectors activates specific deoxyribozymes in specific wells according to the underlying logic of the game. Deoxyribozymes with one, two, and three allosteric domains are shown in squares 1, 3, and 7, respectively. The unaffected domain of the deoxyribozyme in square 7 is an inhibitory domain that would participate in a different game from the one shown here. Should the player choose square 1, this deoxyribozyme would have been catalytically inactivated by oligonucleotide effector 1.

In this regard, Benenson et al. have already designed a finite automaton composed of DNA, T4 DNA ligase, ATP, and the restriction enzyme Fok I.[111] Fok I recognizes the sequence 5′ GGATG but cleaves at a distance from this site. A double-stranded DNA template serves as input and encodes a variety of states. Cleavage with Fok I exposes a unique sticky end for ligation to a particular double-stranded DNA. The ligated DNA encodes a Fok I recognition site and serves as a "transition rule" that can guide additional cleavage and ligation events. In each transition, the original input template is progressively digested until the end of the instruction set (the series of encoded states) is reached. The output is then read via the size of the remaining product.

This system has since been modified to respond to mRNA inputs and to ultimately release a nucleic acid drug.[112] The sensing of the mRNA inputs is simple but ingenious: essentially an mRNA will either hybridize to a transition molecule and inactivate it or hybridize to a precursor and thereby release a DNA strand necessary for the formation of a transition molecule. The set of computations are stochastic, rather than deterministic; that is, the latter only has one transition (e.g., input a: S0→S1) and one output, whereas the former has two competing transitions (e.g., input a: S0→S1 [probability 0.2] and S0→S0 [probability 0.8]), and the probability of the output is a product of the probabilities of all the single choices.[113] Two input molecules are operated on in parallel, ultimately releasing variable concentrations of either an antisense DNA or the complement of the antisense DNA. Thus, depending on the levels of various input mRNAs, differing amounts of an active antisense therapeutic will be generated. Systematic errors in the operation of the automaton (primarily caused by pipetting) were calculated to be 5%. Although this scheme has only been demonstrated *in vitro* and shows few prospects of ever having utility *in vivo*, it nonetheless points the way toward how autonomous ribozyme automata might similarly act.

Another potential way to implement therapeutic nucleic acid automata would be as designed genetic circuits, the difference being that designed circuits would interface more closely with the extant genetic machinery of the cell. There is now a large body of literature on synthetic biology and designed genetic circuits, such as "repressilators,"[114] but Isaacs et al. have begun to design these circuits using RNA biosensors.[115] A *cis*-repressing mRNA (crRNA) was constructed by placing a short sequence complementary to the ribosome-binding site (RBS) downstream of the promoter and upstream of the RBS. The loop region of the stem-loop formed by pairing between the anti-RBS and RBS is in turn available for hybridization (in the same molecular beaconlike fashion we commented on above). A noncoding, *trans*-activating RNA (taRNA), transcribed from a second promoter, can open the stem-loop and reexpose the RBS. The suppression and restoration of translation was measured using green fluorescent protein (GFP) expression. This promising system also worked by placing crRNA and taRNA under the control of other promoters. In this paradigm, the implanted circuitry senses the inputs to promoter sequences rather than the mRNAs themselves. However, it is clear that future implementations of this or other genetic circuits could easily rely on a combination of promoter, mRNA, and protein inputs, based on the variety of nucleic acid biosensors that have been described throughout this review.

ACKNOWLEDGMENTS

The research reported in this chapter was supported in part by the Countermeasures to Biological and Chemical Threats Program at the Institute for Advanced Technology, University of Texas at Austin, under contract number DAAD13-02-C-0079 from the Soldier Biological Chemical Command and/or contract number DAAD17-01-D-0001 from the U.S. Army Research Laboratory.

REFERENCES

1. Jayasena, S.D., Aptamers: an emerging class of molecules that rival antibodies in diagnostics. *Clin Chem,* 45, 1628–1650, 1999.
2. Wilson, D.S. and Szostak, J.W., *In vitro* selection of functional nucleic acids. *Annu Rev Biochem,* 68, 611–647, 1999.
3. Brody, E.N. and Gold, L., Aptamers as therapeutic and diagnostic agents. *J Biotechnol,* 74, 5–13, 2000.
4. Famulok, M., Mayer, G., and Blind, M., Nucleic acid aptamers—from selection *in vitro* to applications *in vivo. Acc Chem Res,* 33, 591–599, 2000.
5. Hesselberth, J., Robertson, M.P., Jhaveri, S., and Ellington, A.D., *In vitro* selection of nucleic acids for diagnostic applications. *J Biotechnol,* 74, 15–25, 2000.
6. Joyce, G.F., Directed evolution of nucleic acid enzymes. *Annu Rev Biochem,* 73, 791–836, 2004.
7. Dick, L.J.L. and McGown, L.B., Aptamer-enhanced laser desorption/ionization for affinity mass spectrometry. *Anal Chem,* 76, 3037–3041, 2004.
8. Drolet, D.W., Moon-McDermott, L., and Romig, T.S., An enzyme-linked oligonucleotide assay. *Nat Biotechnol,* 14, 1021–1025, 1996.
9. Ito, Y., Fujita, S., Kawazoe, N., and Imanishi, Y., Competitive binding assay for thyroxine using *in vitro* selected oligonucleotides. *Anal Chem,* 70, 3510–3512, 1998.
10. Kato, T., Yano, K., Ikebukuro, K., and Karube, I., Bioassay of bile acids using an enzyme-linked DNA aptamer. *Analyst,* 125, 1371–1373, 2000.
11. Murphy, M.B., Fuller, S.T., Richardson, P.M., and Doyle, S.A., An improved method for the *in vitro* evolution of aptamers and applications in protein detection and purification. *Nucleic Acids Res,* 31, e110, 2003.
12. Hartmuth, K., Vornlocher, H.P., and Luhrmann, R., Tobramycin affinity tag purification of spliceosomes. *Methods Mol Biol,* 257, 47–64, 2004.
13. Davis, K.A., Abrams, B., Lin, Y., and Jayasena, S.D., Use of a high affinity DNA ligand in flow cytometry. *Nucleic Acids Res,* 24, 702–706, 1996.
14. Davis, K.A., Lin, Y., Abrams, B., and Jayasena, S.D., Staining of cell surface human CD4 with 2-F-pyrimidine-containing RNA aptamers for flow cytometry. *Nucleic Acids Res,* 26, 3915–3924, 1998.
15. Charlton, J., Sennello, J., and Smith, D., *In vivo* imaging of inflammation using an aptamer inhibitor of human neutrophil elastase. *Chem Biol,* 4, 809–816, 1997.
16. German, I., Buchanan, D.D., and Kennedy, R.T., Aptamers as ligands in affinity probe capillary electrophoresis. *Anal Chem,* 70, 4540–4545, 1998.
17. Kleinjung, F., Klussmann, S., Erdmann, V.A., Scheller, F.W., Fuerste, J.P., and Bier, F.F., Binders in biosensors: high-affinity RNA for small analytes. *Anal Chem,* 70, 328–331, 1998.
18. Hellinga, H.W. and Marvin, J.S., Protein engineering and the development of generic biosensors. *Trends Biotechnol,* 16, 183–189, 1998.
19. Benson, D.E., Conrad, D.W., de Lorimier, R.M., Trammell, S.A., and Hellinga, H.W., Design of bioelectronic interfaces by exploiting hinge-bending motions in proteins. *Science,* 293, 1641–1644, 2001.
20. Burgstaller, P., Kochoyan, M., and Famulok, M., Structural probing and damage selection of citrulline- and arginine-specific RNA aptamers identify base positions required for binding. *Nucleic Acids Res,* 23, 4769–4776, 1995.
21. Ye, X., Gorin, A., Frederick, R., Hu, W., Majumdar, A., Xu, W., McLendon, G., Ellington, A., and Patel, D.J., RNA architecture dictates the conformations of a bound peptide. *Chem Biol,* 6, 657–669, 1999.

22. Hermann, T. and Patel, D.J., Adaptive recognition by nucleic acid aptamers. *Science,* 287, 820–825, 2000.
23. Patel, D.J. and Suri, A.K., Structure, recognition and discrimination in RNA aptamer complexes with cofactors, amino acids, drugs and aminoglycoside antibiotics. *Rev Mol Biotechnol,* 74, 39–60, 2000.
24. Sassanfar, M. and Szostak, J.W., An RNA motif that binds ATP. *Nature,* 364, 550–553, 1993.
25. Huizenga, D.E. and Szostak, J.W., A DNA aptamer that binds adenosine and ATP. *Biochemistry,* 34, 656–665, 1995.
26. Dieckmann, T., Suzuki, E., Nakamura, G.K., and Feigon, J., Solution structure of an ATP-binding RNA aptamer reveals a novel fold. *RNA,* 2, 628–640, 1996.
27. Jiang, F., Kumar. R.A., Jones, R.A., and Patel, D.J., Structural basis of RNA folding and recognition in an AMP–RNA aptamer complex. *Nature,* 382, 183–186, 1996.
28. Lin, C.H. and Patel, D.J., Structural basis of DNA folding and recognition in an AMP–DNA aptamer complex: distinct architectures but common recognition motifs for DNA and RNA aptamers complexed to AMP. *Chem Biol,* 4, 817–832, 1997.
29. Jhaveri, S., Kirby, R., Conrad, R., Maglott, E.J., Bowser, M., Kennedy, R.T., Glick, G., and Ellington, A.D., Designed signaling aptamers that transduce molecular recognition to changes in fluorescence intensity. *J Am Chem Soc,* 122, 2469–2473, 2000.
30. Merino, E.J. and Weeks, K.M., Fluorogenic resolution of ligand binding by a nucleic acid aptamer. *J Am Chem Soc,* 125, 12370–12371, 2003.
31. Jhaveri, S., Rajendran, M., and Ellington, A.D., *In vitro* selection of signaling aptamers. *Nat Biotechnol,* 18, 1293–1297, 2000.
32. Ho, H.A. and Leclerc, M., Optical sensors based on hybrid aptamer/conjugated polymer complexes. *J Am Chem Soc,* 126, 1384–1387, 2004.
33. Wu, Q., Tsiang, M., and Sadler, J.E., Localization of the single-stranded DNA binding site in the thrombin anion-binding exosite. *J Biol Chem,* 267, 24408–24412, 1992.
34. Macaya, R.F., Schultze, P., Smith, F.W., Roe, J.A., and Feigon, J., Thrombin-binding DNA aptamer forms a unimolecular quadruplex structure in solution. *Proc Natl Acad Sci USA,* 90, 3745–3749, 1993.
35. Padmanabhan, K., Padmanabhan, K.P., Ferrara, J.D., Sadler, J.E., and Tulinsky, A., The structure of alpha-thrombin inhibited by a 15-mer single-stranded DNA aptamer. *J Biol Chem,* 268, 17651–17654, 1993.
36. Schultze, P., Macaya, R.F., and Feigon, J., Three-dimensional solution structure of the thrombin-binding DNA aptamer d(GGTTGGTGTGGTTGG). *J Mol Biol,* 235, 1532–1547, 1994.
37. Bock, L.C., Griffin, L.C., Latham, J.A., Vermaas, E.H., and Toole, J.J., Selection of single-stranded DNA molecules that bind and inhibit human thrombin. *Nature,* 355, 564–566, 1992.
38. Potyrailo, R.A., Conrad, R.C., Ellington, A.D., and Hieftje, G.M., Adapting selected nucleic acid ligands (aptamers) to biosensors. *Anal Chem,* 70, 3419–3425, 1998.
39. Macaya, R.F., Waldron, J.A., Beutel, B.A., Gao, H., Joesten, M.E., Yang, M., Patel, R., Bertelsen, A.H., and Cook, A.F., Structural and functional characterization of potent antithrombotic oligonucleotides possessing both quadruplex and duplex motifs. *Biochemistry,* 34, 4478–4492, 1995.
40. Fang, X., Cao, Z., Beck, T., and Tan, W., Molecular aptamer for real-time oncoprotein platelet-derived growth factor monitoring by fluorescence anisotropy. *Anal Chem,* 73, 5752–5757, 2001.
41. Morrison, L.E., Homogeneous detection of specific DNA sequences by fluorescence quenching and energy transfer. *J Fluorescence,* 9, 187–196, 1999.

42. Fang, X., Li, J.J., Perlette, J., Tan, W., and Wang, K., Molecular beacons: novel fluorescent probes. *Anal Chem*, 72, 747A–753A, 2000.

43. Tyagi, S. and Kramer, F.R., Molecular beacons: probes that fluoresce upon hybridization. *Nat Biotechnol*, 14, 303–308, 1996.

44. Sokol, D.L., Zhang, X., Lu, P., and Gewirtz, A.M., Real time detection of DNA. RNA hybridization in living cells. *Proc Natl Acad Sci USA*, 95, 11538–11543, 1998.

45. Tyagi, S., Bratu, D.P., and Kramer, F.R., Multicolor molecular beacons for allele discrimination. *Nat Biotechnol*, 16, 49–53, 1998.

46. Dubertret, B., Calame, M., and Libchaber, A.J., Single-mismatch detection using gold-quenched fluorescent oligonucleotides. *Nat Biotechnol*, 19, 365–370, 2001.

47. Liu, X. and Tan, W., A fiber-optic evanescent wave DNA biosensor based on novel molecular beacons. *Anal Chem*, 71, 5054–5059, 1999.

48. Liu, X., Farmerie, W., Schuster, S., and Tan, W., Molecular beacons for DNA biosensors with micrometer to submicrometer dimensions. *Anal Biochem*, 283, 56–63, 2000.

49. Li, J.J., Fang, X., Schuster, S.M., and Tan, W., Molecular beacons: a novel approach to detect protein–DNA interactions. *Angew Chem Int Ed Engl*, 39, 1049–1052, 2000.

50. Fang, X., Li, J.J., and Tan, W., Using molecular beacons to probe molecular interactions between lactate dehydrogenase and single-stranded DNA. *Anal Chem*, 72, 3280–3285, 2000.

51. Bar-Ziv, R. and Libchaber, A., Effects of DNA sequence and structure on binding of RecA to single-stranded DNA. *Proc Natl Acad Sci USA*, 98, 9068–9073, 2001.

52. Lou, H.J., Brister, J.R., Li, J.J., Chen, W., Muzyczka, N., and Tan, W., Adeno-associated virus Rep78/Rep68 promotes localized melting of the rep binding element in the absence of adenosine triphosphate. *Chembiochem*, 5, 324–332, 2004.

53. Hamaguchi, N., Ellington, A., and Stanton, M., Aptamer beacons for the direct detection of proteins. *Anal Biochem*, 294, 126–131, 2001.

54. Li, J.J., Fang, X., and Tan, W., Molecular aptamer beacons for real-time protein recognition. *Biochem Biophys Res Commun*, 292, 31–40, 2002.

55. Fang, X., Sen, A., Vicens, M., and Tan, W., Synthetic DNA aptamers to detect protein molecular variants in a high-throughput fluorescence quenching assay. *Chembiochem*, 4, 829–834, 2003.

56. Stojanovic, M.N., de Prada, P., and Landry, D.W., Aptamer-based folding fluorescent sensor for cocaine. *J Am Chem Soc*, 123, 4928–4931, 2001.

57. Rajendran, M. and Ellington, A.D., *In vitro* selection of molecular beacons. *Nucleic Acids Res*, 31, 5700–5713, 2003.

58. Stojanovic, M.N., de Prada, P., and Landry, D.W., Fluorescent sensors based on aptamer self-assembly. *J Am Chem Soc* 2000;122:11547–11548.

59. Yamamoto, R., Baba, T., and Kumar, P.K., Molecular beacon aptamer fluoresces in the presence of Tat protein of HIV-1. *Genes Cells*, 5, 389–396, 2000.

60. Nutiu, R. and Li, Y., Structure-switching signaling aptamers. *J Am Chem Soc*, 125, 4771–4778, 2003.

61. Tang, J. and Breaker, R.R., Rational design of allosteric ribozymes. *Chem Biol*, 4, 453–459, 1997.

62. Tang, J. and Breaker, R.R., Mechanism for allosteric inhibition of an ATP-sensitive ribozyme. *Nucleic Acids Res*, 26, 4214–4221, 1998.

63. Breaker, R.R., Engineered allosteric ribozymes as biosensor components. *Curr Opin Biotechnol*, 13, 31–39, 2002.

64. Soukup, G.A. and Breaker, R.R., Design of allosteric hammerhead ribozymes activated by ligand-induced structure stabilization. *Structure Fold Des*, 7, 783–791, 1999.

65. Araki, M., Okuno, Y., Hara, Y., and Sugiura, Y., Allosteric regulation of a ribozyme activity through ligand-induced conformational change. *Nucleic Acids Res,* 26, 3379–3384, 1998.
66. Atsumi, S., Ikawa, Y., Shiraishi, H., and Inoue, T., Design and development of a catalytic ribonucleoprotein. *EMBO J,* 20, 5453–5460, 2001.
67. Soukup, G.A. and Breaker, R.R., Engineering precision RNA molecular switches. *Proc Natl Acad Sci USA,* 96, 3584–3589, 1999.
68. Soukup, G.A., Emilsson, G.A., and Breaker, R.R., Altering molecular recognition of RNA aptamers by allosteric selection. *J Mol Biol,* 298, 623–632, 2000.
69. Koizumi, M., Soukup, G.A., Kerr, J.N., and Breaker, R.R., Allosteric selection of ribozymes that respond to the second messengers cGMP and cAMP. *Nat Struct Biol,* 6, 1062–1071, 1999.
70. Soukup, G.A., DeRose, E.C., Koizumi, M., and Breaker, R.R., Generating new ligand-binding RNAs by affinity maturation and disintegration of allosteric ribozymes. *RNA,* 7, 524–536, 2001.
71. Robertson, M.P. and Ellington, A.D., *In vitro* selection of an allosteric ribozyme that transduces analytes to amplicons. *Nat Biotechnol,* 17, 62–66, 1999.
72. Robertson, M.P. and Ellington, A.D., Design and optimization of effector-activated ribozyme ligases. *Nucleic Acids Res,* 28, 1751–1759, 2000.
73. Robertson, M.P. and Ellington, A.D., *In vitro* selection of nucleoprotein enzymes. *Nat Biotechnol,* 19, 650–655, 2001.
74. Robertson, M.P., Knudsen, S.M., and Ellington, A.D., *In vitro* selection of ribozymes dependent on peptides for activity. *RNA,* 10, 114–127, 2004.
75. Levy, M. and Ellington, A.D., ATP-dependent allosteric DNA enzymes. *Chem Biol,* 9, 417–426, 2002.
76. Steele, D., Kertsburg, A., and Soukup, G.A., Engineered catalytic RNA and DNA: new biochemical tools for drug discovery and design. *Am J Pharmacogenomics,* 3, 131–144, 2003.
77. Piganeau, N., Thuillier, V., and Famulok, M., *In vitro* selection of allosteric ribozymes: theory and experimental validation. *J Mol Biol,* 312, 1177–1190, 2001.
78. Ferguson, A., Boomer, R.M., Kurz, M., Keene, S.C., Diener, J.L., Keefe, A.D., Wilson, C., and Cload, S.T., A novel strategy for selection of allosteric ribozymes yields RiboReporter sensors for caffeine and aspartame. *Nucleic Acids Res,* 32, 1756–1766, 2004.
79. Srinivasan, J., Cload, S.T., Hamaguchi, N., Kurz, J., Keene, S., Kurz, M., Boomer, R.M., Blanchard, J., Epstein, D., Wilson, C., and Diener, J.L., ADP-specific sensors enable universal assay of protein kinase activity. *Chem Biol,* 11, 499–508, 2004.
80. Hartig, J.S., Najafi-Shoushtari, S.H., Grune, I., Yan, A., Ellington, A.D., and Famulok, M., Protein-dependent ribozymes report molecular interactions in real time. *Nat Biotechnol,* 20, 717–722, 2002.
81. Hartig, J.S. and Famulok, M., Reporter ribozymes for real-time analysis of domain-specific interactions in biomolecules: HIV-1 reverse transcriptase and the primer-template complex. *Angew Chem Int Ed Engl,* 41, 4263–4266, 2002.
82. Seetharaman, S., Zivarts, M., Sudarsan, N., and Breaker, R.R., Immobilized RNA switches for the analysis of complex chemical and biological mixtures. *Nat Biotechnol,* 19, 336–341, 2001.
83. Hesselberth, J.R., Robertson, M.P., Knudsen, S.M., and Ellington, A.D., Simultaneous detection of diverse analytes with an aptazyme ligase array. *Anal Biochem,* 312, 106–112, 2003.
84. Porta, H. and Lizardi, P.M., An allosteric hammerhead ribozyme. *Biotechnology (NY),* 13, 161–164, 1995.

85. Burke, D.H., Ozerova, N.D., and Nilsen-Hamilton, M., Allosteric hammerhead ribozyme TRAPs. *Biochemistry,* 41, 6588–6594, 2002.
86. Saksmerprome, V. and Burke, D.H., Structural flexibility and the thermodynamics of helix exchange constrain attenuation and allosteric activation of hammerhead ribozyme TRAPs. *Biochemistry,* 42, 13879–13886, 2003.
87. Hartig, J.S., Grune, I., Najafi-Shoushtari, S.H., and Famulok, M., Sequence-specific detection of microRNAs by signal-amplifying ribozymes. *J Am Chem Soc,* 126, 722–723, 2004.
88. Najafi-Shoushtari, S.H., Mayer, G., and Famulok, M., Sensing complex regulatory networks by conformationally controlled hairpin ribozymes. *Nucleic Acids Res,* 32, 3212–3219, 2004.
89. Stojanovic, M.N., de Prada, P., and Landry, D.W., Catalytic molecular beacons. *Chembiochem,* 2, 411–415, 2001.
90. Breaker, R.R. and Joyce, G.F., A DNA enzyme with Mg(2+)-dependent RNA phosphoesterase activity. *Chem Biol,* 2, 655–660, 1995.
91. Kuwabara, T., Warashina, M., Tanabe, T., Tani, K., Asano, S., and Taira, K., A novel allosterically trans-activated ribozyme, the maxizyme, with exceptional specificity *in vitro* and *in vivo. Mol Cell,* 2, 617–627, 1998.
92. Tanabe, T., Kuwabara, T., Warashina, M., Tani, K., Taira, K., and Asano, S., Oncogene inactivation in a mouse model. *Nature,* 406, 473–474, 2000.
93. Soda, Y., Tani, K., Bai, Y., Saiki, M., Chen, M., Izawa, K., Kobayashi, S., Takahashi, S., Uchimaru, K., Kuwabara, T., Warashina, M., Tanabe, T., Miyoshi, H., Sugita, K., Nakazawa, S., Tojo, A., Taira, K., and Asano, S., A novel maxizyme vector targeting a bcr-abl fusion gene induced specific cell death in Philadelphia chromosome-positive acute lymphoblastic leukemia. *Blood,* 104, 356–363, 2004.
94. Amontov, S., Nishikawa, S., and Taira, K., Dependence on Mg^{2+} ions of the activities of dimeric hammerhead minizymes. *FEBS Lett,* 386, 99–102, 1996.
95. Komatsu, Y., Yamashita, S., Kazama, N., Nobuoka, K., and Ohtsuka, E., Construction of new ribozymes requiring short regulator oligonucleotides as a cofactor. *J Mol Biol,* 299, 1231–1243, 2000.
96. Komatsu, Y., Nobuoka, K., Karino-Abe, N., Matsuda, A., and Ohtsuka, E., *In vitro* selection of hairpin ribozymes activated with short oligonucleotides. *Biochemistry,* 41, 9090–9098, 2002.
97. Komatsu, Y., Regulation of ribozyme activity with short oligonucleotides. *Biol Pharm Bull,* 27, 457–462, 2004.
98. Santoro SW, Joyce GF., A general purpose RNA-cleaving DNA enzyme. *Proc Natl Acad Sci USA,* 94, 4262–4266, 1997.
99. Wang, D.Y. and Sen, D., A novel mode of regulation of an RNA-cleaving DNAzyme by effectors that bind to both enzyme and substrate. *J Mol Biol,* 310, 723–734, 2001.
100. Wang, D.Y., Lai, B.H., Feldman, A.R., and Sen, D., A general approach for the use of oligonucleotide effectors to regulate the catalysis of RNA-cleaving ribozymes and DNAzymes. *Nucleic Acids Res,* 30, 1735–1742, 2002.
101. Wang, D.Y., Lai, B.H., and Sen, D., A general strategy for effector-mediated control of RNA-cleaving ribozymes and DNA enzymes. *J Mol Biol,* 318, 33–43, 2002.
102. Feldman, A.R. and Sen, D., A new and efficient DNA enzyme for the sequence-specific cleavage of RNA. *J Mol Biol,* 313, 283–294, 2001.
103. Vaish, N.K., Jadhav, V.R., Kossen, K., Pasko, C., Andrews, L.E., McSwiggen, J.A., Polisky, B., and Seiwert, S.D., Zeptomole detection of a viral nucleic acid using a target-activated ribozyme. *RNA,* 9, 1058–1072, 2003.

104. Jose, A.M., Soukup, G.A., and Breaker, R.R., Cooperative binding of effectors by an allosteric ribozyme. *Nucleic Acids Res,* 29, 1631–1637, 2001.
105. Stojanovic, M.N., Mitchell, T.E., and Stefanovic, D., Deoxyribozyme-based logic gates. *J Am Chem Soc,* 124, 3555–3561, 2002.
106. Adleman, L.M., Molecular computation of solutions to combinatorial problems. *Science,* 266, 1021–1024, 1994.
107. Cox, J.C. and Ellington, A.D., DNA computation function. *Curr Biol,* 11, R336, 2001.
108. Stojanovic, M.N. and Stefanovic, D., Deoxyribozyme-based half-adder. *J Am Chem Soc,* 125, 6673–6676, 2003.
109. Stojanovic, M.N. and Stefanovic, D., A deoxyribozyme-based molecular automaton. *Nat Biotechnol,* 21, 1069–1074, 2003.
110. Tabor, J.J. and Ellington, A.D., Playing to win at DNA computation. *Nat Biotechnol,* 21, 1013–1015, 2003.
111. Benenson, Y., Paz-Elizur, T., Adar, R., Keinan, E., Livneh, Z., and Shapiro, E., Programmable and autonomous computing machine made of biomolecules. *Nature,* 414, 430–434, 2001.
112. Benenson, Y., Gil, B., Ben-Dor, U., Adar, R., and Shapiro, E., An autonomous molecular computer for logical control of gene expression. *Nature,* 429, 423–429, 2004.
113. Adar, R., Benenson, Y., Linshiz, G., Rosner, A., Tishby, N., and Shapiro, E., Stochastic computing with biomolecular automata. *Proc Natl Acad Sci USA,* 101, 9960–9965, 2004.
114. Hasty, J., McMillen, D., and Collins, J.J., Engineered gene circuits. *Nature,* 420, 224–230, 2002.
115. Isaacs, F.J., Dwyer, D.J., Ding, C., Pervouchine, D.D., Cantor, C.R., and Collins, J.J., Engineered riboregulators enable post-transcriptional control of gene expression. *Nat Biotechnol,* 22, 841–847, 2004.
116. Tang, J. and Breaker, R.R., Examination of the catalytic fitness of the hammerhead ribozyme by *in vitro* selection. *RNA,* 3, 914–925, 1997.
117. Piganeau, N., Jenne, A., Thuillier, V., and Famulok, M., An allosteric ribozyme regulated by doxycyline. *Angew Chem Int Ed Engl,* 40, 3503, 2001.
118. Kertsburg, A. and Soukup, G.A., A versatile communication module for controlling RNA folding and catalysis. *Nucleic Acids Res,* 30, 4599–4606, 2002.

3 Biosensors Based on Periplasmic Binding Proteins

Bethel V. Sharma, B.S., Suresh S. Shrestha, Ph.D., Sapna K. Deo, Ph.D., and Sylvia Daunert, Ph.D.

CONTENTS

3.1 INTRODUCTION

The increasing demand for sensing devices in areas such as monitoring pollution levels in the environment, rapid diagnostics in the health care arena, and space exploration has inspired the development of new biosensors. Biosensors are sensing devices consisting of a selective biological recognition element coupled with a signal-transducing component. The interaction of the biorecognition element with the analyte provides the basis for the sensor development.[1,2] The selectivity of the biosensor is determined by the recognition element used. Enzymes, receptors, antibodies, and nucleic acids are commonly used as biorecognition elements.[3] Binding proteins can also serve as good recognition elements for sensor development. This review focuses on sensing systems that use periplasmic binding proteins as their recognition elements.

3.1.1 Periplasmic Binding Proteins as Biosensing Elements

The members of the superfamily of *Escherichia coli* periplasmic binding proteins have a high affinity for a wide variety of small molecule ligands.[4] This property offers a range of specificities for which to design sensing systems utilizing the proteins' natural affinities for many analytes of interest.[5] These proteins are isolated from the periplasmic space of Gram-negative bacteria, where they serve as the key receptor for the uptake of nutrients to the cell.[6] They are involved in the active transport of a ligand and/or bacterial chemotaxis toward or away from a stimulus.[6–8] In addition, selected periplasmic binding proteins have been reported to act as chaperones that assist in the folding of denatured or unfolded proteins in the cell.[9]

These proteins, along with membrane-bound complexes, provide a means for transporting essential nutrients into the cytoplasm from the periplasm of bacteria.[7] The first periplasmic binding protein to be identified and characterized was the sulfate binding protein from *Salmonella typhimurium* in 1965.[10] Since then, many more periplasmic binding proteins have been discovered and studied. There are more than two dozen periplasmic binding proteins that have been identified, and several of them have been well characterized.[5,11] A partial list of these proteins is presented in Table 3.1. Periplasmic binding proteins exist that are specific for amino acids, oxoanions, metals, sugars, vitamins, and even oligopeptides.[12] The sequence homology between the different proteins is quite low, but the overall three-dimensional structure of the members of the family of proteins is strikingly similar. All of these proteins demonstrate two distinct globular domains in their tertiary structures that are connected by either two or three short polypeptide segments. The ligand binding site is located in the cleft between the two lobes[13] (see Figure 3.1a). These proteins have been found to have high affinity and specificity for their respective ligands, with binding constants in the micromolar to submicromolar range.[6,13,14] In the presence of the ligand, the two lobes come together and trap the ligand inside the binding pocket in a manner similar to the way a Venus flytrap engulfs its prey. Hence, the binding mechanism of these proteins is commonly referred to as a Venus flytrap or hinge-motion mechanism.[14]

3.1.2 Strategies for the Design of Periplasmic Binding Protein-Based Biosensors

As mentioned above, the interaction between the binding protein and the ligand results in a change in the conformation of the protein. This constitutes the basis for the development of sensing systems using binding proteins. There are two main strategies that may be used to develop a biosensor from a binding protein: rational design and construction of "hybrid proteins." Rational design involves making a prediction based on a judicious study of the protein crystal structure or that of a structurally similar protein.[1,2,15–17]

The first example of such an approach involved the rational design of a calcium biosensing system based on the protein calmodulin.[1] That particular system demonstrated that a nanomolar limit of detection for calcium could be attained using a rationally designed, site-specific, fluorescently labeled protein construct as the

TABLE 3.1
Binding Affinity of Selected Periplasmic Binding Proteins

Binding Protein Ligand	K_d (μM)
Amino acids	
Arginine/ornithine	0.03–0.1
Lysine/arginine/ornithine	0.15 5
Cysteine	0.01
Glutamine	0.1–0.3
Glutamate/aspartate	0.8–6
Histidine	0.15–1.5
Leucine	0.7
Leucine/isoleucine/valine	0.2–2
Peptides	
Oligopeptide	—
Dipeptide	—
Carbohydrates	
Galactose/glucose	0.2
Maltose	1
Ribose	0.13
Xylose	0.6
Anions	
Citrate	1–2.6
Phosphate	0.8
Sulfate	0.02
Vitamins	
Cobalamin	0.005
Thiamine	0.03–0.1
Metals	
Nickel*	10

Adapted from Sohanpal et al.[5] and
*Heddle et al.[36]

biosensing entity. This structure–function approach allows one to determine where on the protein molecule to attach a label or what area of the structure to modify to obtain more efficient binding (see Figure 3.2). When a binding protein is labeled with a reporter molecule (e.g., a fluorescent label or electroactive probe), this ligand-induced conformational change in the protein generates a change in the environment of the reporter group attached to the protein, and hence a change in the signal produced by the reporter is observed. In addition, the interaction of a ligand with the binding protein can affect the reporter moiety in the protein directly, resulting in a

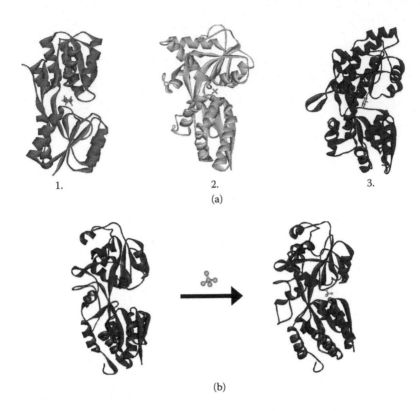

FIGURE 3.1 (See color insert following page 142.) (a) Ribbon diagrams of selected peri-plasmic binding proteins with their respective ligands. Brookhaven Protein Data Bank[62] files for crystal structure coordinates are indicated in parentheses below. Structures rendered using Weblab ViewerLite software. (1) Ribose binding protein (2DRI); (2) sulfate binding protein (1SBP); (3) maltose binding protein (1ANF). (b) Ribbon diagrams denote the conformational change that occurs upon binding phosphate for PBP (1OIB and 1IXH, respectively).

change in the signal generated. Fluorescent molecules are commonly used as the reporter groups in the binding protein–based sensing systems. The fluorescence signal can also be detected by monitoring changes in the intensity, spectra, lifetime, or anisotropy of the intrinsic tryptophanyl fluorescence or the fluorescence of an extrinsic label.

The creation of hybrid proteins, where one component of the molecule is respon-sible for ligand binding and the other for signal generation, allows for the use of selected features of two different proteins in combination for a synergistic effect. For example, the fluorescent properties of green fluorescent protein (GFP) have been widely exploited as a signaling molecule for both *in vitro* and *in vivo* biochemical assays.[18] By creating chimeras of a binding protein and a fluorescent protein, a molecule with both a desired binding specificity and intrinsic signaling moiety can be constructed (an example can be seen in Figure 3.3). Genetically fusing the sequences of the two proteins in a plasmid construct also allows for the manufacture

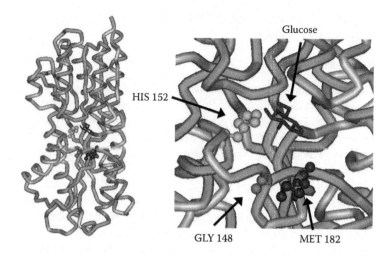

FIGURE 3.2 (See color insert following page 142.) Model of the GBP structure (left) in its closed conformation while bound to glucose and a close-up view (right) of the selected sites indicating rationally designed labeling sites for fluorescent tags. The sites were chosen so as to be near the glucose binding site without interfering with the glucose binding activity.

of these hybrid proteins in living cells. Several examples of such hybrid proteins may be found in recent literature, with a few specifically using periplasmic binding proteins.[19–24] In the following sections we will discuss both types of sensor constructs as the different sensing systems are presented.

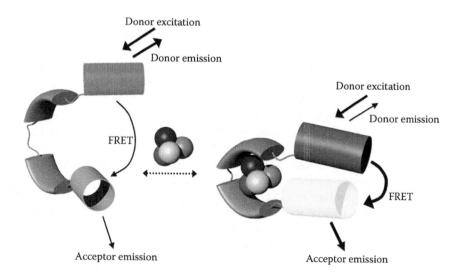

FIGURE 3.3 (See color insert following page 142.) Schematic depicting a typical "hybrid protein" FRET system. As the binding protein closes around the ligand, the FRET pair donor and acceptor come closer, which increases the amount of nonradiative energy transfer.

3.2   SENSING SYSTEMS BASED ON PERIPLASMIC BINDING PROTEINS

3.2.1   PHOSPHATE

Effluents from industries and automobiles account for the major source of pollution in the environment. The availability of binding proteins that specifically recognize a wide range of environmental pollutants provides an excellent opportunity to develop sensing systems for environmental contaminants.

Phosphate is a noteworthy environmental pollutant and its determination is necessary to ensure safe drinking water and protect aquatic life. Inorganic phosphate is principally monitored in natural waters.[25] Phosphate is also important in various biological processes, as many of them involve adenosine triphosphate (ATP) hydrolysis. A sensing system for inorganic phosphate was developed by Salins et al.[26] using the phosphate binding protein (PBP) isolated from *E. coli*.

Phosphate binding protein is one of the periplasmic binding proteins found in the periplasmic space of Gram-negative bacteria. The x-ray crystal structure of PBP in its phosphate-bound and free forms has been reported (shown in Figure 3.1b).[27] PBP consists of 321 amino acid residues, and its crystal structure shows that it has two globular domains connected by three short polypeptide segments that act as a flexible hinge. Each PBP molecule binds a single phosphate molecule with a K_d of approximately 0.8 µM.[28] The phosphate sensing system developed by Salins et al. was based on the conformational change that PBP undergoes upon binding phosphate.[29] An environment-sensitive fluorescent probe was attached site-specifically to a unique cysteine introduced genetically on PBP at position 197 employing site-directed mutagenesis. A maleimide derivative of a coumarin fluorescent probe, N-[2-(1-maleimidyl)ethyl]-7(diethylamino)coumarin-3-carboxamide (MDCC), was conjugated to the unique cysteine. In the presence of phosphate, the fluorescent probe-labeled PBP showed a 93% increase in the fluorescence signal. A calibration plot generated for phosphate indicated a detection limit of 4.2×10^{-7} M using this sensing system (Figure 3.4). A selectivity study was also performed that showed no significant response to other structurally similar anions.

The mechanism of sensing was investigated by Lundgren et al.[30] using steady-state and time-resolved fluorescence spectroscopy and demonstrated that a change in the local environment of the probe in PBP occurred upon phosphate binding. In addition, the anisotropy measurements indicated that local structural changes occurred upon binding of phosphate to PBP. Dynamic quenching measurements were performed using a PBP labeled with acrylodan through the mutant cysteine residue inserted at position 197. As can be seen in Figure 3.5, for both the free and bound labeled protein, iodide was used as a quencher to indicate changes in the intensity of the acrylodan over varying concentrations of quencher. This Stern–Volmer plot was used to determine that the label at position 197 was more accessible to the quencher when the PBP was in the bound conformation. The authors postulated that this study increases our understanding of the optimization of the placement of the signal transduction function in the rational design of biosensors using periplasmic binding proteins.

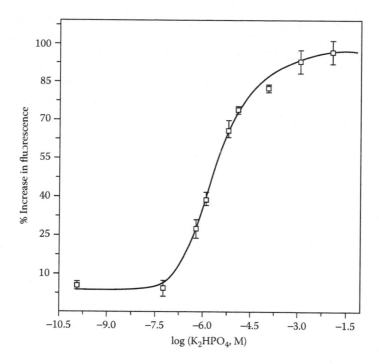

FIGURE 3.4 Calibration plot for phosphate. (Adapted from Salins et al.[29])

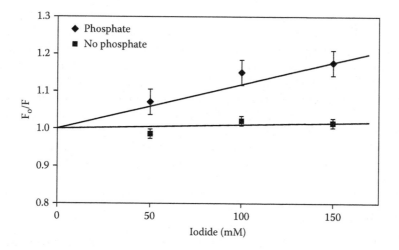

FIGURE 3.5 Stern–Volmer plot for free (■) and phosphate-bound (♦) PBP. (Adapted from Lundgren et al.[30])

To demonstrate the applicability of the PBP-based sensing system for phosphate in a variety of analytical platforms, this system was also adapted to a fiber-optic platform. In this case, MDCC-labeled PBP was entrapped behind a dialysis membrane attached to a fiber-optic bundle. A calibration curve for phosphate was constructed using the fiber-optic setup and a detection limit of 1.8×10^{-6} M was achieved for phosphate.[31] This type of system should find applications in portable, miniaturized monitoring systems that may be employed on site and in field studies.

3.2.2 NICKEL

Quantifying amounts of metal ions is very important in environmental monitoring, wastewater treatment, and industrial process monitoring. The uses of nickel compounds include nickel plating, coloring ceramics, certain types of batteries, and as a catalyst for chemical reactions. Nickel and its various compounds have no distinctive odor or taste. However, exposure to nickel can cause several respiratory illnesses and dermatitis in humans.[32,33] The U.S. Environmental Protection Agency (EPA) recommends that children's drinking water contain no more than 0.04 mg/l (6.81×10^{-7} M) of nickel for 1 to 10 days of exposure.[32]

A sensing system for nickel was developed by Salins et al.[34,35] using the periplasmic nickel binding protein (NBP) from *E. coli*. This sensing system is based on the conformational change observed in the NBP upon binding to nickel. Native NBP contains a unique cysteine at position 15. The NBP is expressed from *E. coli* strain B834 containing the plasmid p8602 that carries the *nikA* gene, which encodes the NBP. The fluorescent probe MDCC is conjugated to the protein through the sulfhydryl of the cysteine at position 15. In the presence of nickel, a quenching in the fluorescence signal of MDCC is observed.

A calibration plot was constructed for nickel using labeled NBP and a detection limit of 8×10^{-8} M was achieved. The response of the system to structurally similar metals demonstrated that only cobalt was capable of quenching the fluorescence of the probe with a magnitude similar to nickel. Nevertheless, the detection limit for cobalt, 1.6×10^{-6} M, is more than one order of magnitude higher than that for nickel. This assay for nickel was adapted to a 96-well fluorescence microtiter plate reader and also to a fiber-optic setup using protein that was immobilized behind a dialysis membrane at the tip of a fiber optic. A detection limit of 1×10^{-6} M was achieved for nickel using the fiber-optic setup. The crystal structures of both the ligand free and bound forms of this protein have recently been solved and will no doubt enhance the ability to rationally design advanced biosensing moieties incorporating NBP in the future.[36]

3.2.3 GLUCOSE

The market for clinical diagnostics is increasing every year and today sales are estimated to be well over $20 billion.[37] Products on the market include tests for blood sugar, prostate cancer, and osteoporosis, among others. Traditionally, testing is conducted at hospitals, clinics, or a physician's office. However, advancements in sensing technologies have made testing at a patient's home more feasible and increasingly popular. There are numerous test kits available on the market for various medical conditions, including diabetes, malaria, pregnancy, cholesterol, and blood in the stool (an indicator of possible colorectal cancer), so people can test in the privacy of their own home. Binding proteins, with their intrinsic selectivity and sensitivity, can be used as biorecognition elements in sensing system development for clinical diagnostics as well.

Glucose is one of the essential molecules that has primary relevance in any discussion regarding clinical diagnostics. Blood glucose levels need to be frequently monitored by diabetic patients. There are many types of glucose sensors that are commercially available. Glucose-monitoring devices alone account for a large percentage of the testing kits sold on the market. Because of the overwhelming number of diabetic patients around the world, efforts to develop improved, easy-to-use glucose-measurement devices continue.[37] To address this need, a sensing system for glucose was developed by Salins et al.[38] based on the galactose/glucose binding protein (GBP) isolated from the periplasm of *E. coli*. GBP is a product of the *mglB* gene and is involved in the uptake of glucose and galactose in *E. coli*.[39] GBP has the typical periplasmic binding protein structure, with two globular domains separated by a flexible hinge formed from three polypeptide segments. GBP shows similar affinity toward glucose and galactose, with binding constants of 0.2 μM and 0.4 μM, respectively.[40]

Salins et al. employed two different strategies for the development of the sensing system for glucose. In the first strategy, unique cysteines were introduced at three different sites (Gly148, His152, and Met182) on wild-type GBP, which otherwise lacks any native cysteine residues (see Figure 3.2). Four different environmentally sensitive fluorescent probes, namely, 6-acryloyl-2-dimethylaminonaphthalene (acrylodan), N-((2-(iodoacetoxy)ethyl)-N-methyl)amino-7-nitrobenz-2-oxa-1,3-diazole (IANBD), MDCC, and 5-((((2-iodoacetyl)amino)ethyl)amino)naphthalene-1-sulfonic acid (IAEDANS) were conjugated to the mutant GBP through the sulfhydryl of the unique cysteine. In the presence of increasing amounts of glucose and galactose, a quenching in the fluorescence signal was observed from the labeled mutant proteins. A 30% quenching in fluorescence was observed from MDCC-labeled mutant GBP152 in the presence of glucose and a detection limit of 1×10^{-6} M was achieved. A separate study was performed using the G148C mutant of GBP. As shown in Figure 3.6a, a glucose calibration curve can be generated using the MDCC-labeled mutant GBP148 after storage at 37°C, which corresponds to body temperature. In Figure 3.6b, the reversible, dynamic response of this same protein may be seen. This stability was observed for more than 3 months, making this protein amenable for *in vivo* biosensing.[41]

The wild-type GBP contains a binding site for calcium in addition to the binding site for glucose/galactose. In a second strategy undertaken by the same group of investigators, the calcium ion in the native protein was replaced by a lanthanide ion through extensive dialysis.[42] Lanthanides possess intrinsic fluorescent properties and, hence, can be used as reporter molecules. Two lanthanide ions, terbium and europium, were investigated by monitoring fluorescence emitted by these ions in the presence of glucose. The terbium-substituted protein showed an 85% and 92% enhancement in the fluorescence from terbium in the presence of glucose and

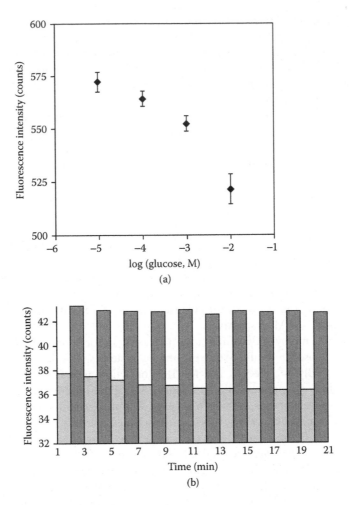

FIGURE 3.6 (a) Calibration study for glucose performed using 148GBP that had been stored at 37°C to demonstrate the temperature stability of the system. (b) Response of the 148GBP biosensing system for glucose in a fiber-optic configuration to alternating solutions of buffer (■) and a saturating concentration of glucose (■) to demonstrate signal reproducibility (unpublished results).

galactose, respectively. However, when europium was used as a reporter, the increase in fluorescence was significantly smaller. Calibration plots were generated for glucose employing both the strategies mentioned above, and a detection limit in the submicromolar range was obtained for glucose. As GBP also binds to galactose, this sensing system may be affected by interference from galactose in glucose sensing. However, this should not pose a problem in the clinical analysis of blood glucose, as normal free galactose levels are negligible by comparison.

Similarly, GBP has also been utilized by Tolosa et al.[43] to develop a fluorescence lifetime-based sensing system for glucose. Fluorescence lifetime-based sensing systems have the advantage of being protein concentration independent, and they provide a wide dynamic range. A unique cysteine was introduced into GBP at position 26 by site-directed mutagenesis. The environment-sensitive fluorescent probe 2-(4'-iodoacetamidoanilino)naphthalene-6-sulfonic acid (I-ANS) was attached to the protein through the sulfhydryl group of the cysteine. In the presence of increasing amounts of glucose, the fluorescence intensity of I-ANS-labeled GBP was found to decrease by almost twofold. To develop a lifetime based sensing system, I ANS labeled GBP was coupled with the long-lifetime fluorescent probe $[Ru(bpy)_3]^{2+}$. The ruthenium probe was dissolved in heated polyvinyl alcohol and painted on the outside of the cuvette containing I-ANS-labeled GBP. The modulation of the combined emission of I-ANS-labeled GBP and $[Ru(bpy)_3]^{2+}$ at low frequencies depended upon the fractional fluorescence intensity of I-ANS-labeled GBP. The modulation measured at 2.1 MHz of the I-ANS-labeled GBP was found to decrease with increasing glucose. As the modulation may be measured accurately to 0.005 MHz, glucose can be measured with an accuracy of ± 0.2 μM. The calibration curve generated for glucose against the change in modulation showed that micromole concentrations of glucose could be monitored with this system. With the choice of an appropriate fluorescent probe, a greater change in intensity may be obtained from GBP in the presence of glucose. A greater change in intensity results in a greater change in modulation, which provides a more accurate determination of glucose concentration.

Another development using GBP as the biosensing element was introduced by Marvin and Hellinga.[44] Their system also employs an optical detection methodology in which a fluorescence allosteric signal transducer function is engineered into the protein. The design is based on similar work using the structure of the maltose binding protein (MBP), which, unlike GBP, has had crystal structures solved for both the bound and unbound states.[17] Single cysteine mutants were made using overlap polymerase chain reaction (PCR) methods, and purified protein was labeled with the thiol-reactive fluorophores acrylodan and IANBD. Six positions were chosen for cysteine mutations and labeling with each of the fluorophores: N15, H152 (nonallosteric sites) and L255, D257, P294, and V296 (allosteric sites). Glucose or galactose was titrated into a 50-nM solution of labeled protein in a 0.1-M NaCl, 50-mM phosphate buffer, at pH 7.0. The fluorescence emission was measured at 540-nm and 520 nm for the IANBD- and acrylodan-labeled samples, respectively. The four proteins with mutations in the hinge region showed a change in fluorescence with ligand binding. This matched with the prediction that the selected sites in the hinge region were allosterically linked to the glucose binding cleft. Acrylodan at position 255 showed the largest change, which was a twofold decrease in fluorescence. In addition, the sugar binding constants for both galactose and glucose were not altered by more than a factor of four. This showed that the ligand-binding event could be transduced to a fluorescence-signaling event with minimal effects on the binding properties and without the use of analyte-consuming methods, which are most often used in home glucose testing kits.

More recently, a different way of introducing the binding properties of GBP to a fluorescent signaling moiety was presented. GBP was genetically fused to a cyan fluorescent protein (CFP) at the N-terminus and a yellow fluorescent protein (YFP) at the C-terminus to construct a fusion protein with both binding and signal transduction capabilities.[19] The idea of designing fusion proteins that incorporate one protein with a signal transduction function and another with a binding capacity to create literal "fluorescent protein biosensors" has been described.[18] This system was designed for *in vivo* measurements of glucose within living cells. The hinge-binding motion in solution of the GBP moved the attached CFP and YFP far enough apart to decrease the nonradiative fluorescence resonance energy transfer (FRET) between them upon glucose binding (a similar model is presented in Figure 3.3). The glucose binding constant for this fusion protein of GBP and two flanking GFP variants was found to be 170 nM. This system also exhibited a maximum ratio of emissions intensities of the donor and acceptor FRET pair of 0.23. An affinity mutant of the fusion protein was created that had a binding constant of 0.59 mM and a larger maximum ratio change of 0.29. Substrate selectivity decreased at higher concentrations of various sugars and alcohols tested, with the exception of trehalose, which showed no change. The affinity mutant was tested *in vivo* in a COS-7 mammalian cell line.

Fluorescence microscopic imaging and digital photography of the images allowed ratios to be calculated for the two emission wavelengths over entire cells. Some differences in distribution throughout the cells were observed. If these differences were due to changes in cell thickness, this could be corrected by the use of confocal imaging. It took only 10 seconds to see a response in glucose-deprived cells once 10 mM glucose was added. A control sensor was also developed by constructing a mutant that did not show a change in the ratio up to 100 mM glucose or after 10 minutes. Measurement of cytosolic levels of glucose upon external exposure showed that a ratiometric change could be measured with addition of increasing amounts of glucose up to 10 mM and for nearly an hour.

A glucose indicator protein (GIP), created by Ye and Schultz,[24] is a fusion of GFP to either end of the GBP structure. This design is another way of using GFP with a FRET-based signal transduction mechanism grafted onto the GBP structure itself. The donor protein was a GFP and the FRET acceptor was a YFP. The GFP used has a maximum excitation wavelength at 395 nm and emission at 510 nm, whereas the YFP has its excitation maximum at 513 nm and emission at 527 nm. Thus, significant overlap occurs to achieve FRET if the two fluorophores can come close enough to each other while attached to opposite termini of the folded GBP. The entire fusion protein gene was ligated together beginning at the 5′ end with the YFP sequence, followed by the GBP, and last the GFP sequence at the 3′ end.

To test the ability of the GIP to respond to changes in glucose concentration, glucose was titrated into a 40-μg sample of GIP in a 400-μl volume. Samples were incubated for 5 minutes and the fluorescence signal was measured using an excitation wavelength of 395 nm and an emission wavelength of 527 nm. In addition, a dialysis hollow fiber was created using regenerated cellulose with a molecular weight cutoff (MWCO) of 10 kDa. This type of setup was used to model a microsensor that could be used for continuous glucose measurement. The final dimensions of the dialysis tube encapsulating the microsensor were 1.5 cm long with an inside diameter of 190 μm. The microsensor was placed within a 7-μl quartz microcuvette perfusion cell. All solutions flowed through the cell by gravity and the fluorescence signal was measured as before, with and without glucose. The fluorescence spectrum of the YFP was not altered by the novel configuration, but some energy transfer within the GIP did occur in the absence of glucose. Based on the ability of the GBP to adopt two different, stable conformations, the change of conformation could alter the distance between the donor and acceptor and change the amount of energy transfer that could occur.

A linear relationship between a reduction in FRET and an increase in glucose concentration could be observed up to 20 μM. The apparent K_m for the GIP was approximately 5 μM. For the hollow-fiber microsensor, a decrease in the FRET could also be observed when a 10-μM glucose solution was passed through the flow cell, and the baseline signal could be regained after passing through a sugar-free phosphate buffered saline (PBS) solution for 100 seconds. Thus, this system exhibited both reproducibility and reversibility while the protein was encapsulated, which could perhaps delay the response time beyond what could be achieved in solution. This system may also be amenable to use within whole living cells but not necessarily for use *in vivo* using these UV wavelengths. This problem may be overcome by selecting a more appropriate FRET pair.

3.2.4 MALTOSE

Binding proteins have been utilized to develop sensing systems for biomolecules such as sugars and amino acids. The periplasmic MBP is involved in the transport of maltose, maltotriose, and maltodextrins, with typical binding constants in the micromolar range.[45] Similar to other binding proteins, MBP also undergoes a conformational change with ligand binding as a result of a hinge motion in the protein that can be utilized for its binding characteristics.[46] The binding pocket of MBP is rich in aromatic amino acids, and the binding of a ligand induces major changes in the fluorescence and UV absorption spectra.[47–49] In a preliminary study by Zhou and Cass,[50] MBP was used as a sensing element for the development of a maltose sensor based on nonradiative energy transfer. For that, MBP was labeled with IAEDANS. A tryptophan residue of MBP and IAEDANS were used as the donor–acceptor pair for the energy transfer study. An increase in transfer efficiency from 29% to 42% was obtained upon addition of maltose.

A reagentless sensing system for maltose using MBP was developed later by the same research group.[51] In that work, a unique cysteine was introduced at position 337 in MBP using site-directed mutagenesis. Two different environmentally sensitive fluorescent probes, IANBD and acrylodan, were conjugated to MBP through the sulfhydryl of a unique cysteine. In the presence of increasing amounts of maltose, an increase in the fluorescence intensity was observed from the fluorescent probe labeled MBP. There was a 160% increase in the fluorescence from the IANBD-labeled MBP in the presence of 200 μM maltose. Similarly, an 80% increase in the signal from acrylodan-labeled MBP was obtained in the presence of 5 μM maltose. These results demonstrate that by choosing an appropriate label, one may create MBPs sensitive to different ranges of maltose concentration that are useful in the reagentless optical sensing of maltose.

Another system designed for maltose sensing was to allow for FRET between two coupled GFP variants that were fused to the ends of the MBP sequence. Fehr et al.[20] called this a fluorescent indicator protein or FLIP system. The system was designed by inserting the *E. coli malE* gene sequence for MBP between the sequences of two GFP variants: cyan fluorescent protein (CFP) at the N-terminus and YFP at the C-terminus. The protein was expressed in *E. coli* BL21-Gold (DE3) cells for subsequent purification and *in vitro* studies, and the expression vector was inserted into *Saccharomyces cerevisiae* SuSy7/ura3 cells for *in vivo* imaging studies. The YFP–CFP FRET pair was excited at 440 nm to excite the CFP FRET donor, and after energy transfer the YFP acceptor emission was measured at 530 nm. The fusion of an intact MBP within this system showed no ratiometric change between the emission intensities; however, the deletion of the first five amino acids allowed for a FRET ratio of 0.2. This was likely due to an increase in reversibility of maltose binding. Site-directed mutations in the MBP sequence to create alanine residues at positions W62, W230, and W340 decrease the binding affinity of the protein for maltooligosaccharides, which confounds the maltose binding results in a mixed system and also creates a broader range of affinities.

Confocal imaging of maltose uptake in yeast cells was tested because maltose is not immediately phosphorylated once inside these cells. A cell line deficient in maltose uptake was used. Images showed the fluorescent protein successfully expressed in the cytosol vs. the vacuole. Specificity was shown when addition of sucrose to the cells had no affect. Future modifications of this system will be beneficial to increase the ratio change and increase the dynamic range of the sensor.

Another homogeneous maltose biosensing assay was developed recently by Medintz et al.[52] at the U.S. Naval Research Laboratory. Their system used a FRET interaction between a fluorescent substituent attached to the MBP and a free acceptor molecule. The MBP was mutated to contain a single cysteine residue at position 95 for labeling purposes. The FRET pair in this case was a Cy3.5-labeled MBP (the donor) and a Cy5-labeled β-cyclodextrin (β-CD) molecule (the acceptor). The labeled β-CD binds to the MBP, bringing the two fluorescent probes close enough that FRET can occur. Upon binding maltose, however, the β-CD is displaced and the donor signal increases. Another system was used with Cy3 and QSY9 (a dark quencher) as the donor and acceptor, respectively, and the displacement by maltose instead resulted in a decrease in donor fluorescence. The detection limits of the FRET complex were from 100 nM to 50 μM for maltose.

3.2.5 RIBOSE

The development of a biosensing method for ribose has also been developed based on the *E. coli* ribose binding protein (RBP). Ribose is important in biological systems because it is a component of nucleic acids and certain amino acids and can act as an energy source.[23] This system arose from the need to analyze the uptake and distribution of the sugar in living cells. This time, Fehr et al.[19] created a FRET-based ribose nanosensor by creating a fusion protein of RBP flanked by the two GFP variants, CFP and YFP, simply by inserting the RBP gene into the construct used for GBP. Affinity mutants were also created for this system in order to increase the range of affinities available for ribose binding. Selectivity experiments showed that in addition to ribose, allose, erythrose, and ribulose were recognized by both affinity mutants, but only at high concentrations. The sensors were also expressed in the cytosol of COS-7 cells, and emission intensity ratios were measured and recorded by a charge-coupled device (CCD) camera. The *in vivo* binding studies showed reversibility of the ribose binding, but the process was slow, taking more than 15 minutes to return to baseline. However, the general use of this type of system could be extended to other proteins and analytes.

3.2.6 AMINO ACIDS

Amino acids are another class of biomolecules that can be detected using binding proteins. A variety of pathological conditions such as histidinemia, cystinuria, and phenylketonuria has been linked to defects in amino acid metabolism and may result in an increased amount of amino acids in the urinary output.[53] A binding assay for cystine was developed by Oshima et al.[54] using a periplasmic cystine-binding protein from *E. coli*. Cystine is believed to have an antioxidant effect in the body. For the purpose of the assay, radiolabeled cystine was allowed to bind to the cystine binding protein. The binding of cystine was determined by measuring radioactivity in the cystine binding protein. The assay can measure down to 10 pmol of cystine. Similarly, an L-glutamine binding protein-based assay was developed by Willis and Seegmiller[55] using radiolabeled L-glutamine that was able to measure close to 20 pmol of L-glutamine.

A reagentless optical sensing system was developed for glutamine by Dattelbaum and Lakowicz[56] based on the *E. coli* glutamine binding protein (GlnBP). Glutamine is used as an energy and nitrogen source when culturing eukaryotic cells. Catabolism of glutamine leads to the buildup of ammonia gas, which is toxic to the growing culture; hence, monitoring glutamine is essential throughout the process of cell growth cycles.[57] In addition, glutamine is known to enhance muscle growth and increase body mass. GlnBP offers high specificity and sensitivity for binding glutamine ($K_d = 0.2$ μM) and therefore may be used as the binding element for developing sensing systems for glutamine. Unique cysteines were introduced into the protein at positions 138, 179, and 209 by site-directed mutagenesis. Environment-sensitive fluorescent probes were attached through the unique sulfhydryl groups.

The labeled mutant, GlnBP179, showed the greatest change in fluorescence emission. Mutant GlnBP179 labeled with acrylodan and 2-(4′(iodoactamido)anilino)naph-thalene-6-sulfonic acid (IAANS) resulted in a 65% and 35% decrease in the signal, respectively, in the presence of 6.4 μM glutamine (see Figure 3.7). The detection limit for glutamine was in the micromolar range. The selectivity study showed no significant change in the fluorescence intensity from structurally similar amino acids, such as glutamate, asparagine, and arginine. Time-resolved studies showed a 2.4-fold decrease in mean lifetime of acrylodan-labeled mutant GlnBP179 upon binding of glutamine, indicating the feasibility of developing lifetime-based assays for glutamine.

Another assay for the detection of glutamine was developed using the same S179C GlnBP mutant that was used in the study mentioned above.[58] This time, the

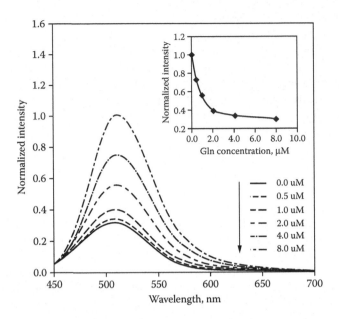

FIGURE 3.7 Emission spectra of acrylodan-labeled GlnBP excited at 360 nm with increasing concentrations of glutamine. Inset: Changes in emission intensity at 515 nm upon addition of glutamine. (Reprinted with permission from Ge et al.[57])

authors labeled the cysteine 179 with acrylodan (Acr) and labeled the N-terminus of the protein with ruthenium bis-(2,2′-bipyridyl)-1,10-phenanthroline-9-isothiocy-anate (Ru). The dual-labeled Ru-GlnBP-Acr protein contained both a reporter of glutamine binding and a long-lived reference that did not change upon glutamine binding. The acrylodan fluorescence was quenched by glutamine binding activity. When exciting the acrylodan at 360 nm, the emissions intensity at 515 nm steadily decreases as the concentration of glutamine in the sample increases. Meanwhile, the intensity of the Ru label remains constant at 610 nm.

The ratiometric approach is unique in this case and allows for correction of possible differences in source intensity, path length, and sample positioning. Frequency domain data also showed a decrease in the lifetime of acrylodan, along with a consistent lifetime of Ru as the glutamine concentration increased. A possible drawback of the assay is the greater-than-100% labeling efficiency of the ruthenium, which is likely due to the nonspecific labeling of other proteins present in the samples, which could contribute to batch-to-batch differences in performance of the sensing system.

3.3 SUMMARY AND PROSPECTS

Binding proteins, in general, are a class of proteins that have a relatively high binding affinity for their respective ligands. These proteins bind their ligands in a very selective manner, with affinity constants in the micromolar range. Most of these binding proteins are known to undergo a ligand-induced conformational change,[7,59] making them ideal for sensing system development. These proteins may be produced recombinantly in large quantities and may be manipulated genetically to promote attachment of reporter molecules. They may also be genetically or chemically modified for signal transduction or immobilization purposes.[5,60,61] Most of these proteins are stable at room temperature. These characteristics make binding proteins ideal as recognition elements in analytical sensing systems. Binding protein-based biosensing systems have applications in environmental monitoring, clinical diagnostics, and high-throughput screening. Furthermore, these binding proteins may be adapted to various sensing platforms, such as fiber-optic chemical sensors, high-density microwell plates used for high-throughput screening, and microfluidic systems.

Molecular biology techniques offer unique opportunities to manipulate genes and produce recombinant forms of these proteins with desired properties. In order to form a complete biosensing unit, it is necessary to add a signal transduction function to the binding properties of a selected periplasmic binding protein.[5] Enhanced binding affinity may also be achieved through genetic engineering, increasing the performance of the sensing system. Other enhancements include altering selectivities, increasing stability, or mutating the protein to produce a binding protein with an intrinsic signal transduction function.

Finally, the biosensing systems described above, when applied to state-of-the-art technologies, such as lab-on-a chip devices, provide sensing that is smaller, faster, and less expensive. Thus, binding proteins that undergo a change in conformation upon binding to specific ligands serve as a good starting point for the development of numerous types of detection methodologies as the recognition elements for biosensors. A variety of such binding proteins is widely present in living systems and may be used for developing assays for numerous analytes. Several of these proteins are found in the periplasm of bacteria. X-ray crystal structures of many of these binding proteins are available, which aids in sensing system development for target analytes.

In addition to the binding proteins discussed in this review, there are many other binding proteins whose structures have been well characterized. For example, the molybdate binding protein, the arabinose binding protein, the histidine binding protein, the ribose binding protein, and the dipeptide binding protein are well known and characterized and could be used to develop sensing systems for their respective ligands. Moreover, knowledge of the interactions between these binding proteins and their respective ligands provides a wealth of information that could aid in drug design and discovery. Many drugs work by interacting with receptors on the cell surface. Studies have found that there are considerable homologies between binding proteins and receptors in terms of binding mechanisms to their ligands.[8] Hence, these binding proteins may have additional applications in pharmaceutical discovery and optimization assays as well as use in diagnostics applications. Furthermore, the studies conducted to understand ligand–protein interactions may aid in designing recognition sites for other clinically and environmentally important molecules.

On the horizon are new techniques that allow for the systematic reorganization of native binding sites to confer unique specificities to the same molecule.[16,60]

These structure-based computational methods may potentially allow for tailoring of a particular binding property into a receptor of interest. The future holds great promise for the ultimate binding protein sensor application, by utilizing designer proteins that incorporate a desired combination of binding affinity and signal transduction function into a single sensing element.

ACKNOWLEDGMENTS

The authors would like to acknowledge support for this research from the National Aeronautics and Space Administration, the National Institutes of Health, the National Science Foundation, and the National Science Foundation Integrative Graduate Education and Research Traineeship (IGERT) program.

REFERENCES

1. Schauer-Vukasinovic, V., Cullen, L., and Daunert, S., Rational design of a calcium sensing system based on induced conformational changes of calmodulin. *J Am Chem Soc,* 119, 11102–11103, 1997.
2. Hellinga, H.W. and Marvin, J.S., Protein engineering and the development of generic biosensors. *Trends Biotechnol,* 16, 183–189, 1998.
3. Iqbal, S.S., Mayo, M.W., Bruno, J.G., Bronk, B.V., Batt, C.A., and Chambers, J.P., A review of molecular recognition technologies for detection of biological threat agents. *Biosens Bioelectron,* 15, 549–578, 2000.
4. de Lorimier, R.M., Smith, J.J., Dwyer, M.A., Looger, L.L., Sali, K.M., Paavola, C.D., Rizk, S.S., Sadigov, S., Conrad, D.W., Loew, L., and Hellinga, H.W., Construction of a fluorescent biosensor family. *Protein Sci,* 11, 2655–2675, 2002.
5. Sohanpal, K., Watsuji, T., Zhou, L.Q., and Cass, A.E.G., Reagentless fluorescence sensors based upon specific binding-proteins. *Sensor Actuat B,* 11, 547–552, 1993.
6. De Wolf, F.A. and Brett, G.M., Ligand-binding proteins: their potential for application in systems for controlled delivery and uptake of ligands. *Pharmacol Rev,* 52, 207–236, 2000.
7. Ames, G.F., Bacterial periplasmic transport systems: structure, mechanism, and evolution. *Annu Rev Biochem,* 55, 397–425, 1986.
8. Felder, C.B., Graul, R.C., Lee, A.Y., Merkle, H.P., and Sadee, W., The Venus flytrap of periplasmic binding proteins: an ancient protein module present in multiple drug receptors. *AAPS PharmSci,* 1, E2, 1999.
9. Richarme, G. and Caldas, T.D., Chaperone properties of the bacterial periplasmic substrate-binding proteins. *J Biol Chem,* 272, 15607–15612, 1997.
10. Dreyfuss, J. and Pardee, A.B., Evidence for a sulfate-binding site external to cell membrane of *Salmonella typhimurium. Biochim Biophys Acta,* 104, 308–310, 1965.
11. Hiles, I.D., Gallagher, M.P., Jamieson, D.J., and Higgins, C.F., Molecular characterization of the oligopeptide permease of *Salmonella typhimurium. J Mol Biol,* 195, 125–142, 1987.
12. Quiocho, F.A. and Ledvina, P.S., Atomic structure and specificity of bacterial periplasmic receptors for active transport and chemotaxis: variation of common themes. *Mol Microbiol,* 20, 17–25, 1996.

13. Pflugrath, J.W. and Quiocho, F.A., The 2 A resolution structure of the sulfate-binding protein involved in active transport in *Salmonella typhimurium. J Mol Biol,* 200, 163–180, 1988.

14. Sack, J.S., Saper, M.A., and Quiocho, F.A., Periplasmic binding protein structure and function. Refined X-ray structures of the leucine/isoleucine/valine-binding protein and its complex with leucine. *J Mol Biol,* 206, 171–191, 1989.

15. Hellinga, H.W., Rational protein design: combining theory and experiment. *Proc Natl Acad Sci USA,* 94, 10015–10017, 1997.

16. Marvin, J.S. and Hellinga, H.W., Conversion of a maltose receptor into a zinc biosensor by computational design. *Proc Natl Acad Sci USA,* 98, 4955–4960, 2001.

17. Marvin, J.S., Corcoran, E.E., Hattangadi, N.A., Zhang, J.V., Gere, S.A., and Hellinga, H.W., The rational design of allosteric interactions in a monomeric protein and its applications to the construction of biosensors. *Proc Natl Acad Sci USA,* 94, 4366–4371, 1997.

18. Tsien, R.Y., The green fluorescent protein. *Annu Rev Biochem,* 67, 509–544, 1998.

19. Fehr, M., Lalonde, S., Lager, I., Wolff, M.W., and Frommer, W.B., *In vivo* imaging of the dynamics of glucose uptake in the cytosol of COS-7 cells by fluorescent nanosensors. *J Biol Chem,* 278, 19127–19133, 2003.

20. Fehr, M., Frommer, W.B., and Lalonde, S., Visualization of maltose uptake in living yeast cells by fluorescent nanosensors. *Proc Natl Acad Sci USA,* 99, 9846–9851, 2002.

21. Dikici, E., Deo, S.K., and Daunert, S., Drug detection based on the conformational changes of calmodulin and the fluorescence of its enhanced green fluorescent protein fusion partner. *Anal Chim Acta,* 500, 237–245, 2003.

22. Puckett, L.G., Lewis, J.C., Bachas, L.G., and Daunert, S., Development of an assay for beta-lactam hydrolysis using the pH-dependence of enhanced green fluorescent protein. *Anal Biochem,* 309, 224–231, 2002.

23. Lager, I., Fehr, M., Frommer, W.B., and Lalonde, S., Development of a fluorescent nanosensor for ribose. *FEBS Lett,* 553, 85–89, 2003.

24. Ye, K.M. and Schultz, J.S., Genetic engineering of an allosterically based glucose indicator protein for continuous glucose monitoring by fluorescence resonance energy transfer. *Anal Chem,* 75, 3451–3459, 2003.

25. Mayewski, P.A., Spencer, M.J., Lyons, W.B., and Twickler, M.S., Seasonal and spatial trends in south Greenland snow chemistry. *Atmos Environ,* 21, 863–869, 1987.

26. Salins, L.L.E., Schauer-Vukasinovic, V., and Daunert, S., Optical sensing systems based on biomoleculer recognition of recombinant proteins. *Proc SPIE Int Soc Opt Eng,* 3270, 16–24, 1998.

27. Ledvina, P.S., Yao, N., Choudhary, A., and Quiocho, F.A., Negative electrostatic surface potential of protein sites specific for anionic ligands. *Proc Natl Acad Sci USA,* 93, 6786–6791, 1996.

28. Brune, M., Hunter, J.L., Corrie, J.E.T., and Webb, M.R., Direct, real-time measurement of rapid inorganic phosphate release using a novel fluorescent probe and its application to actomyosin subfragment 1 ATPase. *Biochemistry,* 33, 8262–8271, 1994.

29. Salins, L.L.E., Deo, S., and Daunert, S., Phosphate binding protein as the biorecognition element in a biosensor for phosphate. *Sensor Actuat B,* 97, 81–89, 2004.

30. Lundgren, J.S., Salins, L.L., Kaneva, I., and Daunert, S., A dynamical investigation of acrylodan-labeled mutant phosphate binding protein. *Anal Chem,* 71, 589–595, 1999.

31. Salins, L.L.E., Wenner, B.R., and Daunert, S., Fiber optic biosensor for phosphate based on the analyte-induced conformational change of genetically engineered phosphate binding protein. *Abstr Pap Am Chem S,* 217, U792, 1999.

32. Agency for Toxic Substances and Disease Registry, *Toxicological Profile for Nickel*. Atlanta: Agency for Toxic Substances and Disease Registry, 2003.

33. Denkhaus, E. and Salnikow, K., Nickel essentiality, toxicity, and carcinogenicity. *Crit Rev Oncol Hematol,* 42, 35–56, 2002.

34. Salins, L.L.E., Shrestha, S., and Daunert, S., Fluorescent biosensing systems based on analyte-induced conformational changes of generically engineered periplasmic bonding proteins. In: *Chemical and Biological Sensors for Environmental Monitoring,* Mulchandani, A. and Sadik, O.A., eds. ACS Symposium Series 762. New York: Oxford University Press, 2000:87–101.

35. Salins, L.L.E., Goldsmith, E.S., Ensor, C.M., and Daunert, S., A fluorescence-based sensing system for the environmental monitoring of nickel using the nickel binding protein from *Escherichia coli. Anal Bioanal Chem,* 372, 174–180, 2002.

36. Heddle, J., Scott, D.J., Unzai, S., Park, S.Y., and Tame, J.R., Crystal structures of the liganded and unliganded nickel-binding protein NikA from *Escherichia coli. J Biol Chem,* 278, 50322–50329, 2003.

37. Markin, R.S. and Whalen, S.A., Laboratory automation: trajectory, technology, and tactics. *Clin Chem,* 46, 764–771, 2000.

38. Salins, L.L.E., Ware, R.A., Ensor, C.M., and Daunert, S., A novel reagentless sensing system for measuring glucose based on the galactose/glucose-binding protein. *Anal Biochem,* 294, 19–26, 2001.

39. Scholle, A., Vreemann, J., Blank, V., Nold, A., Boos, W., and Manson, M.D., Sequence of the MglB gene from *Escherichia coli* K12—comparison of wild-type and mutant galactose chemoreceptors. *Mol Gen Genet,* 208, 247–253, 1987.

40. Mahoney, W.C., Hogg, R.W., and Hermodson, M.A., The amino acid sequence of the D-galactose-binding protein from *Escherichia coli* B/r. *J Biol Chem,* 256, 4350–4356, 1981.

41. Deo, S.K. and Daunert, S., unpublished data.

42. Wenner, B.R., Shreshtha, S., Sharma, B.V., Lai, S., and Daunert, S., Genetically designed biosensing systems for high-throughput screening of pharmaceuticals, clinical diagnostics, and environmental monitoring. In: *Advances in Fluorescence Sensing Technology.* San Jose, CA: SPIE–The International Society for Optical Engineering, 2001.

43. Tolosa, L., Gryczynski, I., Eichhorn, L.R., Dattelbaum, J.D., Castellano, F.N., Rao, G., and Lakowicz, J.R., Glucose sensor for low-cost lifetime-based sensing using a genetically engineered protein. *Anal Biochem,* 267, 114–120, 1999.

44. Marvin, J.S. and Hellinga, H.W., Engineering biosensors by introducing fluorescent allosteric signal transducers: construction of a novel glucose sensor. *J Am Chem Soc,* 120, 7–11, 1998.

45. Shilton, B.H., Flocco, M.M., Nilsson, M., and Mowbray, S.L., Conformational changes of three periplasmic receptors for bacterial chemotaxis and transport: the maltose-, glucose/galactose- and ribose-binding proteins. *J Mol Biol,* 264, 350–363, 1996.

46. Bedouelle, H., Renard, M., Belkadi, L., and England, P., Harnessing malE for the study of antigen/antibody recognitions. *Res Microbiol,* 153, 395–398, 2002.

47. Szmelcman, S., Schwartz, M., Silhavy, T.J., and Boos, W., Maltose transport in *Escherichia coli* K12—comparison of transport kinetics in wild-type and lambda-resistant mutants with dissociation constants of maltose-binding protein as measured by fluorescence quenching. *Eur J Biochem,* 65, 13–19, 1976.

48. Spurlino, J.C., Lu, G.Y., and Quiocho, F.A., The 2.3-A resolution structure of the maltose-binding or maltodextrin-binding protein, a primary receptor of bacterial active transport and chemotaxis. *J Biol Chem,* 266, 5202–5219, 1991.
49. Gehring, K., Bao, K., and Nikaido, H., UV difference spectroscopy of ligand-binding to maltose-binding protein. *FEBS Lett,* 300, 33–38, 1992.
50. Zhou, L.Q. and Cass, A.E.G., Periplasmic binding-protein based biosensors. 1. Preliminary study of maltose binding protein as sensing element for maltose biosensor. *Biosens Bioelectron,* 6, 445–450, 1991.
51. Gilardi, G., Zhou, L.Q., Hibbert, L., and Cass, A.E.G., Engineering the maltose-binding protein for reagentless fluorescence sensing. *Anal Chem,* 66, 3840–3847, 1994.
52. Medintz, I.L., Goldman, E.R., Lassman, M.E., and Mauro, J.M., A fluorescence resonance energy transfer sensor based on maltose binding protein. *Bioconjugate Chem,* 14, 909–918, 2003.
53. Wilcken, B., Smith, A., and Brown, D.A., Urine screening for aminoacidopathies: is it beneficial? Results of a long-term follow-up of cases detected by screening one million babies. *J Pediatr,* 97, 492–497, 1980.
54. Oshima, R.G., Willis, R.C., Furlong, C.E., and Schneide, J.A., Binding assays for amino acids. The utilization of a cystine binding protein from *Escherichia coli* for the determination of acid-soluble cystine in small physiological samples. *J Biol Chem,* 249, 6033–6039, 1974.
55. Willis, R.C. and Seegmiller, J.E., Filtration assay specific for determination of small quantities of L-glutamine. *Anal Biochem,* 72, 66–77, 1976.
56. Dattelbaum, J.D. and Lakowicz, J.R., Optical determination of glutamine using a genetically engineered protein. *Anal Biochem,* 291, 89–95, 2001.
57. Ge, X.D., Tolosa, L., Simpson, J., and Rao, G., Genetically engineered binding proteins as biosensors for fermentation and cell culture. *Biotechnol Bioeng,* 84, 723–731, 2003.
58. Tolosa, L., Ge, X.D., and Rao, G., Reagentless optical sensing of glutamine using a dual-emitting glutamine-binding protein. *Anal Biochem,* 314, 199–205, 2003.
59. Zukin, R.S., Hartig, P.R., and Koshland, D.E. Jr., Use of a distant reporter group as evidence for a conformational change in a sensory receptor. *Proc Natl Acad Sci USA,* 74, 1932–1936, 1977.
60. Looger, L.L., Dwyer, M.A., Smith, J.J., and Hellinga, H.W., Computational design of receptor and sensor proteins with novel functions. *Nature,* 423, 185–190, 2003.
61. Marvin, J.S. and Hellinga, H.W., Manipulation of ligand binding affinity by exploitation of conformational coupling. *Nat Struct Biol,* 8, 795–798, 2001.
62. Berman, H.M., Bhat, T.N., Bourne, P.E., Feng, Z., Gilliland, G., Weissig, H., and Westbrook, J., The Protein Data Bank and the challenge of structural genomics. *Nat Struct Biol,* 7(suppl), 957–959, 2000.

4 Molecular Beacon DNA Probes Based on Fluorescence Biosensing

Gang Yao, Ph.D., M. Shelly John, Ph.D., and Weihong Tan, Ph.D.

CONTENTS

4.1 INTRODUCTION

Over the past decade, the biosensor research field has experienced profound growth in many respects. The introduction of various biorecognition schemes and technological advancements has greatly improved the detection and identification capability of many of these biosensors. All these efforts are designed to enhance the sensitivity and specificity in detecting chemical and biological species. Many of these biosensors have found applications in medical diagnostics, intracellular measurements, and environmental and food technologies.[1,2] In addition, biosensors may be designed for the rapid detection of new strains of microorganisms, which helps minimize the spread of infectious diseases. Moreover, biosensors provide discrete or continuous signals in either qualitative or quantitative monitoring. They can detect, record, and transmit information regarding changes in or the presence of various biochemicals or biological materials in the environment. The transduction mechanisms in most biosensors are simple and straightforward. A biosensor consists of a substrate to support the biorecognition element and an external stimulus to interrogate the presence of analyte molecules or species that interact specifically with the recognition element. The transformed information, in the form of optical or electrical signals, is detected with a detection unit. Based on the signal being measured, biosensors are usually classified into four different groups: electrochemical,[3,4] optical,[5-7] mass-sensitive,[8,9] and thermal.[10]

Among these techniques, optical biosensors offer a number of advantages, such as high sensitivity and selectivity. Optical sensors have the potential for rapid detection as well as remote-sensing capability. Moreover, the transformed signals are isolated from electromagnetic interferences. Optical sensors generally utilize an alteration in fundamental optical properties such as phase, amplitude, or frequency of the light that stimulates the transducing component. Among these, the most important optical manifestations are based on the frequency change associated with biosensing that utilizes fluorescence. The objective of this chapter is to introduce molecular beacons (MBs) as a new biorecognizing element to detect various biological molecules. Subsequent sections provide an overview of the applications, synthesis, and characterization of MBs. Finally, we introduce various design mechanisms utilized for the development of biosensors.

4.2   FLUORESCENCE BIOSENSORS

Fluorescence biosensors possess high sensitivity and excellent selectivity for the detection of many target molecules. With their potential market in medicine, food, agriculture, and environmental applications, the development of biosensors has seen sustained growth in both the research and application areas.[11-13] With the recent advent of DNA probe technology, a number of selective fluorescence biosensors that interact with specific DNA sequences have been identified and used to provide a new type of selective biorecognition information. Because of their functionality and molecular specificity, MBs have provided a variety of exciting opportunities in DNA, RNA, and protein studies both inside living cell specimens and in solution.[14-17] In recent years, various applications of surface-immobilized MB biosensors have been described, including the ultrasensitive fiber-optic MB biosensor,[18-20] the microwell MB biosensor,[21] the nanoparticle MB biosensor,[22,23] the MB array biosensor,[24,25] and the MB single molecular biosensor microarrays.[26] We will mainly discuss these biosensors based on MBs for gene analysis.

4.3 THE MOLECULAR BEACON DNA PROBES FOR BIOSENSORS

4.3.1 THE MOLECULAR BEACON AND ITS APPLICATIONS

Molecular beacons operate on the principle of DNA base pairing. They are synthetic DNA molecules that conform to a basic "stem-loop" or hairpin structure (Figure 4.1). DNA hybridization acts as the basis for target recognition and signal transduction in MB studies. The loop sequence (15–30 mer) is complementary to a target DNA (tDNA), whereas the stem is a 5–7-mer sequence complementary to it so that prior to binding tDNA sequences, the structure remains in the closed state. For signal transduction, fluorescence resonance energy transfer (FRET) is often employed.[27] A fluorescent moiety is covalently coupled to one end of the MB and a quenching moiety is coupled to the other end. There is a spectral overlap between the donor's emission spectrum and the acceptor's absorption spectrum. The fluorescent dye acts as an energy donor and the quencher acts as a nonfluorescent acceptor. When the stem hybridizes, the two moieties are kept near each other, causing the fluorescence of the donor to be quenched by FRET. In the presence of a target molecule, the loop region forms a hybrid with it that is longer and more stable than that of the stem. This forces the MB to undergo a spontaneous conformational change that forces the stem apart. With the quencher no longer positioned near the fluorophore, fluorescence is restored, thus signaling when the MB binds to its target (Figure 4.2).

Although the basic function of MBs is similar for most bioanalytical methods that employ them, each MB is individually tailored to meet the needs of the application. Engineering functional MBs is often a trial-and-error process; however, intuitive design plays a major role. A good understanding of how their fluorescence changes with temperature in the presence and absence of targets is critically important.

This built-in signaling property makes the MB a highly sensitive and selective DNA probe to report label-free targets. There are three primary advantages to

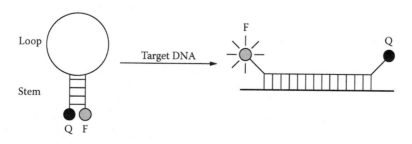

FIGURE 4.1 Mechanism of MB DNA probes. MB alone is nonfluorescent because the stem keeps the fluorophore close to the quencher and emits intense fluorescence when it hybridizes with the complementary DNA.

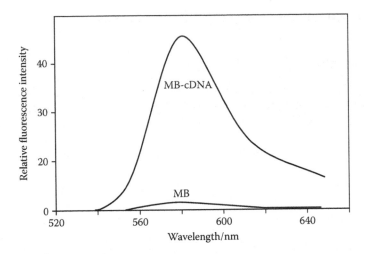

FIGURE 4.2 Fluorescence spectra of molecular beacon before (MB) and after (MB-cDNA) hybridization with its complementary DNA.

using MBs for the detection of specific gene targets over traditional fluorescent probes:

1. An inherent signal transduction mechanism for high sensitivity. The inherent fluorescent signal transduction mechanism enables an MB to function as a sensitive probe with a high signal:background ratio for real-time monitoring and without labeling target samples. Under optimal conditions, the fluorescence intensity may increase more than 200 times when the MB meets its target molecule.[15] This high fluorescence enhancement provides the MBs with a significant advantage over other fluorescent probes in the analysis of DNA, RNA, and proteins. With this inherent sensitivity, individual MB DNA molecules have been imaged and their hybridization process has been monitored on a single molecule basis.[26,28]

2. Detection without separation, that is, monitoring of target hybridization without separation of hybridized and nonhybridized probes. An unhybridized MB is in a closed state (not fluorescent), whereas the hybridized MB is in an open state (highly fluorescent). Though the unhybridized MBs are left behind in the reaction solution, they do not influence the fluorescence measurement produced by the hybridized MBs. MBs thus can be used in situations where it is not possible or desirable to isolate the probe–target hybrids from an excess of unhybridized probes, such as in the real-time monitoring of nucleic acid amplification (polymerase chain reaction [PCR] or nucleic acid sequence-based amplification [NASBA]) in sealed tubes[29,30] or in the detection of messenger RNA (mRNA) within living cells.[17] The usefulness of "detection without separation" for these applications cannot be overemphasized. This feature

allows the synthesis of nucleic acids to be monitored as it is occurring without additional manipulations.

3. Enhanced specificity over traditional linear DNA probes, that is, the ability to distinguish single base pair mismatches. The degree of molecular specificity of a fluorescence probe is very important in many bioanalytical applications, such as real-time DNA amplification,[31,32] DNA array development,[21,32,33] and DNA analysis.[16,19,34–36] Current techniques for routine detection of single base pair DNA mutations are often labor intensive and time consuming.[37] MBs, on the other hand, provide a simple and promising tool for the diagnosis of genetic disease and for gene therapy study. The stem-loop structure and the design of the MB make this discrimination possible. Experiments have shown that the range of temperatures within which perfectly complementary DNA targets form hybrids, but mismatched DNA targets do not, is significantly wider for MBs than for the corresponding range of conventional linear oligonucleotide probes.[38] Therefore, MBs can easily identify DNA targets that differ from one another by a single nucleotide. Thermodynamic studies reveal that the enhanced specificity is a general feature of structurally constrained DNA probes. Single base pair mismatch determinations may be reached by changing the GC content, melting temperature, or experimental conditions.

4.3.2 Synthesis and Characterization of MBs

Although the common MB may only be used in homogeneous solution, MBs immobilizable on a surface are critical for the development of biosensors that can be used for the study of biomolecular recognition processes at an interface. The design and synthesis of immobilizable MBs is a crucial step in the development of biosensors. Just as is the design and synthesis of regular MBs that are used in homogeneous solution, the design and synthesis of immobilizable MBs is also a trial-and-error process where intuitive design plays a major role. For most regular MBs, consideration is mainly needed regarding the sequence. Studies[14,15,37,39–43] indicate that a 15–25 base sequence, together with a 5 base pair stem, is an excellent balance. The sequence of the loop is chosen to be complementary with the target and the length of the loop sequence (15–40 nucleotides) is chosen to form a stable probe–target hybrid at the temperature of probing. The stem sequences (5–7 nucleotides) are chosen to be strong enough to form the hairpin structure for efficient fluorescence quenching and weak enough to be dissociated when the loop hybridizes with a complementary form of DNA. In addition, the stem sequence must be designed so as not to interfere with the probe sequences. Generally, the longer the loop sequence and the higher its GC content, the higher the melting temperature (more stable) of the probe–target hybrid. Consequently, the sensitivity of the probe for its target may be optimized for the application by adjusting the sequence to be more GC rich or deficient (see www.molecular-beacons.org). When designing probes for single base mismatch determinations, the stem acts as a counterweight to the loop, so shorter loop sequences are employed to effectively distinguish the single base difference in the loop at the cost of probe stability. If, on the other hand, single-nucleotide discrimination is not desired, longer and more stable probes may be chosen.

The fluorophore and quencher may be easily chosen for the desired MB. Different fluorescent dyes have been used in molecular studies, such as fluorescein, TMR, Cy3, or Texas Red, which emit fluorescence at different wavelengths. Dimethylaminophenylazobenzoic acid (DABCYL) may serve as a universal quencher for many fluorophores,[15] and Biosearch Technologies' (Novato, CA) Black Hole Quencher (BHQ) is much more specific and efficient for dyes that fluoresce in a certain wavelength range. Depending on the fluorophore, several quenchers (BHQ-0, BHQ-1, BHQ-2, and BHQ-3) are available: BHQ-1 has a maximal absorption at 534 nm and may be used as the quencher of dyes fluorescing at about 480 to 580 nm, such as FAM, HEX, TET, JOE, and Oregon Green. BHQ-2 has a maximal absorption at 579 nm and may be used as the quencher of dyes fluorescing at about 550 to 650 nm, such as Cy3, Cy3.5, TAMRA, ROX, and Texas Red.

However, some other important considerations must be taken into account regarding the design of the MB to be immobilized[44]:

1. The selection of a functional group for surface immobilization, such as the biotin functional group, amino group, carboxyl group, or thiol group.
2. The position for the functional group. Different positions may be used: the loop sequence, the second base pair position of the fluorophore side of the stem, and the same position on the quencher side of the stem. In order to minimize any effects that the functional group might have on fluorescence, quenching, and hybridization of the MB, the quencher side of the stem was chosen to link the functional group.
3. A spacer between the functional group used for attachment and the sequence. A poly-T oligos may provide easy access for target DNA molecules to efficiently interact with the loop sequence and an adequate separation to minimize potential interactions between the solid substrate surface and the DNA sequence.
4. The fluorophore. Although fluorescein has commonly been used in MBs, it has less quantum yield and is less photostable than some other dyes such as rhodamine. This may not be a critical problem in larger samples but will be important when an ultratrace amount of the MB is used or the MBs are immobilized on a surface. In addition, the fluorescence intensity of fluorescein is highly dependent on the pH of the sample matrix. However, it is also worth noting that the overlap between the emission spectrum of rhodamine dyes and the absorption spectrum of DABCYL is not as good as that for fluorescein.

There are also approximately 10 commercial companies specializing in the custom synthesis of MBs, such as TriLink BioTechnologies, Qiagen, Sigma-Genosys, Integrated DNA Technologies, and Gene Link. Today, MBs with specific sequences are also readily available at affordable prices and without tedious synthesis by individual investigators. A detailed protocol for MB synthesis may be found at www.molecular-beacons.org. The synthesis of MBs is similar to that of dual labeling a short oligomer with two dyes. MBs are generally synthesized from 3′ terminal to 5′ terminal, starting from a controlled pore glass (CPG) substrate that has been derivatized with the quencher group.

Once the MBs are synthesized, two important methods are used to evaluate the quality of the product. A matrix-assisted laser desorption/ionization time-of-flight mass spectrometer (MALDI-TOF MS) has been used to confirm the molecular weight of the designed MB.[45] Two important peaks were shown on the mass spectrum. The main peak appeared at the position corresponding to the molecular weight, and a small peak appeared near a position corresponding to half of the molecular weight of the MB, due to the molecules that lost two electrons in the desorption process during the measurement. The newly synthesized MBs are also tested by hybridization to evaluate their activity in the DNA/RNA reaction. The MB will usually incubate in the hybridization buffer, either with a fivefold molar excess of its complementary DNA or with a fivefold molar excess of a noncomplementary DNA. The fluorescence spectra and intensity are then recorded. The hybridization of the MB inside the solution will show a strong fluorescence signal when its loop reacts with the complementary DNA molecule. Incubation with the noncomplementary DNA has no enhancement under the same conditions. A high fluorescence enhancement confirms that the MB synthesized is what has been designed and whether it may be used for DNA/RNA studies.

4.3.3 Surface Treatment for Biosensor Preparation

Efficient immobilization of an MB is necessary in order to make a useful biosensor. Three well-known conjugation reactions may be used. The avidin–biotin binding mechanism is the first among these reactions used for the surface immobilization of MBs.[44,46,47] In this reaction, a biotinylated capture DNA is immobilized on an avidin-coated substrate (Figure 4.3). The second reaction is carbodiimide chemistry, successfully adopted in antibody immobilization, using either a 5′-amino terminal end or a carboxyl terminal end.[48,49] The third reaction is disulfide bonding, where a 3-mercaptopropyltrimethoxysilane (MPTS)-treated substrate is allowed to react with 5′-thiol-modified DNA.[48]

Among these three approaches, biotin–avidin binding is the most common and effective method for biomolecule immobilization onto a solid surface and was first tested with MB immobilization.[44] Biotinylated MBs were linked on the avidin-coated surface through biotin–avidin interaction. The binding process is fast and efficient, and equilibrium coverage can be reached within a few minutes. Immobilized MBs stay with the surface even after immersion in a buffer solution for a few days. Two methods—chemical linking and physical adsorption—may be used to modify the substrate surface to start MB immobilization.[18,19,44] The following sections describe the two approaches for MB immobilization via biotin–avidin binding.

4.3.3.1 Method 1 (Chemical Binding)

A schematic of the immobilization of an MB through chemical binding is shown in Figure 4.4. After being sufficiently washed with water, the substrate was first cleaned by immersion in a 1:1 (v/v) concentrated hydrogen chloride/methanol (HCl/MeOH)

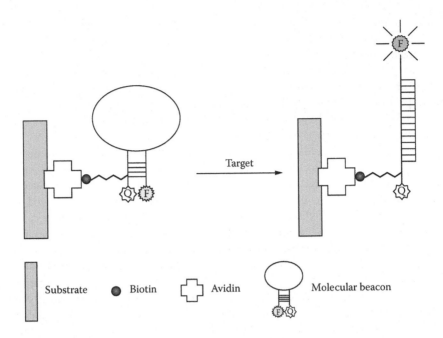

FIGURE 4.3 A schematic of the surface-immobilized MB and its hybridization with cDNA.

mixture for 30 min, rinsed in water, and submerged in concentrated sulfuric acid for 30 min. Further rinsing and then boiling in water for 8 to 10 min followed. Silanization of the surface was performed by immersing it in a freshly prepared 1% (v/v) solution of diethylenetriamine (DETA) in 1 mM acetic acid for 20 min at room temperature (23°C). The DETA-modified surface was thoroughly rinsed with water to remove excess DETA, dried under nitrogen, and fixed by heating in a 120°C oven for 5 min. Then the silanized substrate was immersed in 0.5 mg/ml N-hydroxysuccinimide (NHS)-biotin linker in a 0.1 M bicarbonate buffer (pH 8.5) for 3 h at room temperature. Streptavidin or avidin was bound to the substrate surface by incubating the biotinylated substrate overnight at 4°C in a solution containing 1.0 mg/ml of streptavidin or avidin.

4.3.3.2 Method 2 (Physical Absorption)

The substrate was first immersed into a 10-M sodium hydroxide (NaOH) solution overnight and thoroughly rinsed with deionized water. The treated surface was incubated in avidin solution (1 mg/ml, 10 mM phosphate buffer, pH 7.0) for 12 h at 4°C. The avidin layer was stabilized by cross-linking it with glutaraldehyde (1% in 100 mM phosphate buffer)[50] for 1 h at room temperature, followed by incubation in 1 M Tris-HCl (pH 6.5) for 3 h at 4°C to remove the unbound avidin.

   The hybridization properties of MBs immobilized on the solid surface are similar to those in solution. The immobilized MB probe was tested with different concentrations of complementary DNA molecules ranging from 5 to 600 nM. Fluorescence signal monitoring was achieved with a highly efficient setup for ultrasensitive

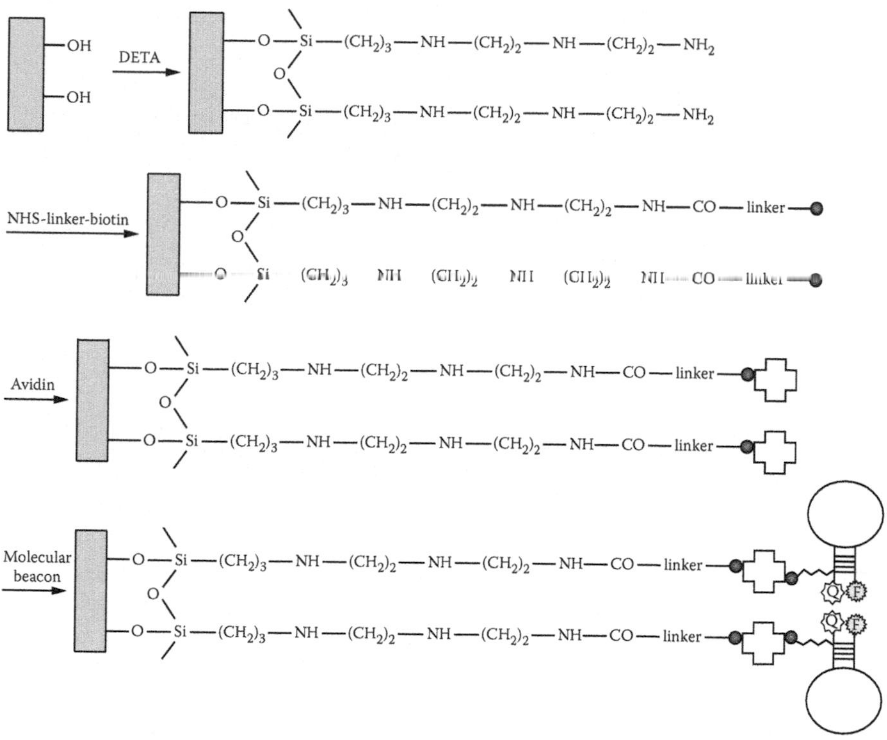

FIGURE 4.4 A schematic of MB immobilization through chemical binding.

optical detection. Optical fiber was used to conduct the excitation laser beam to a prism. The prism was put on the stage of the microscope and was sandwiched with the immobilized MB silica plate glass. An evanescent field was generated on the surface of the prism and was used to excite the immobilized MB. Fluorescent signals thus produced were collected by an objective and directed to an intensified charge-coupled device (ICCD). The results indicate that an MB is highly efficient in DNA hybridization after it is immobilized on the solid surface. Furthermore, the immobilized MB substrate may be used to detect target DNA molecules in the subnanomolar range (Figure 4.5). In addition, experiments have shown that the surface-immobilized DNA molecules can be regenerated after hybridization, so as to be reusable multiple times in DNA detection and interaction studies.

4.4 MBS USED FOR BIOSENSOR DEVELOPMENT

4.4.1 MB Optical Fiber Biosensor

With the results we obtained from the immobilization method, we moved forward to develop a real MB biosensor. Optical fiber biosensors have several advantages over other biosensors, including small size, flexible geometry, remote working capability,

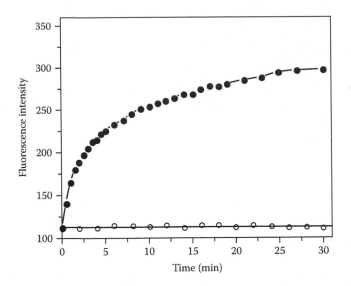

FIGURE 4.5 Immobilized MB hybridization kinetics study. Real-time measurements of the hybridization dynamics of immobilized MBs were obtained with target DNA molecules (solid circle) and noncomplementary DNA molecules (hollow circle).

and simplicity,[51-55] which means that they are ideal for clinical applications, environmental monitoring, and process control.

Optical fiber biosensors are based on total internal reflection, a fundamental characteristic of an optical fiber (Figure 4.6). When a laser beam travels in the core of the optical fiber, a small portion of the light penetrates into the surrounding medium and builds an evanescent field. The intensity of the evanescent field decays exponentially from the fiber core surface into the medium, and only those fluorophores immobilized on the core surface may be excited by the evanescent wave. This means that evanescent wave biosensors have the advantages of low background and low noise and are capable of monitoring surface interactions that are important in biological and clinical studies. Fluorescence produced from the fluorophores will then be coupled back into the fiber core for transmission to an optical detection system. Optical fiber biosensors are thus convenient and effective in practical applications because they avoid laborious optical adjustment and have the potential to be adapted to other measuring devices.

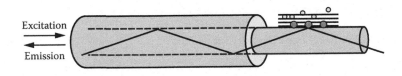

FIGURE 4.6 Configuration of the evanescent wave fiber-optical DNA biosensor.

Many optical fiber biosensors have been developed for sensitive DNA detection and biomolecular interaction studies.[52–54,56] Although these biosensors were remarkable in detection capability and useful for some interesting applications, most of them were based on one of three signal transduction mechanisms—labeled target DNA molecules,[52,53] intercalation reagents,[54] or competitive assay[52]—so that they required either labeled targets, a fluorescent DNA stain reagent, or a competitive assay. Those critical limitations made it difficult to carry out real-time hybridization studies and to quantitatively monitor hybridization kinetics on the surface. It also made it very difficult to study biological processes in real time and *in vivo*.

4.4.1.1 Point Sensor

During the design of fiber-optic evanescent wave sensors based on fluorescence measurement, the fiber core geometry in the sensing region has a large impact on the collection efficiency of the fluorescence light. A tapered fiber core geometry shows better enhancement in signal acquisition compared to a cylindrical fiber core geometry.[57] Our group has taken the advantages of the MB and made MB optical fiber biosensors, among which the point sensor is one application.[18] MBs are immobilized on the surface of the core of the optical fiber through biotin–avidin binding. The optical fiber cladding is stripped away from the core by chemical etching (such as with 49% hydrofluoric acid solution) or standard pulling procedures (such as using a P-2000 puller) at the end of the fiber probe. With the chemical modification method discussed above, the surface of the fiber core is coated with a layer of streptavidin. The streptavidin-immobilized optical fibers are then immersed with a biotinylated MB solution (1 µM in 10 mM pH 7.0 phosphate buffer) for as long as 20 min or overnight at 4°C to allow the biotinylated MB to be immobilized on the surface. The fiber probes are stored in a 10 mM phosphate buffer at 4°C for future use.

Results show that MB optical fiber biosensors may be used to directly detect, in real time, target DNA/RNA molecules with high sensitivity and one base mismatch selectivity without using competitive assays. Those sensors are stable, reproducible, and have remote detection capability. As shown in Figure 4.7, there is a linear relationship between the initial hybridization reaction rate and the concentration of the complementary DNA for the optical fiber MB biosensor. The concentration detection limit may be as low as 1.1 nM.[18] Those MB optical fiber biosensors have been applied to the quantitative detection of mRNA sequences that were amplified by the PCR as well as to study DNA hybridization kinetics.[18] They also hold the potential of direct detection of DNA/RNA targets in living cells without DNA/RNA amplification.

4.4.1.2 Optical Fiber Bundle Biosensor

With the success obtained in the preparation of the point biosensor described above, we moved forward to prepare MB biosensors for an optical fiber bundle. An ultrasmall (submicrometer) optical fiber MB biosensor was first prepared. These miniaturized biosensors also have higher sensitivity and selectivity. The concentration detection limits and mass detection limits are 0.3 nM and 15 amol for a 105-µm biosensor and 10 nM and 0.27 amol for a submicrometer biosensor.[19] They can be

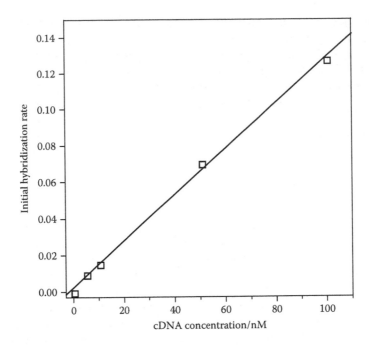

FIGURE 4.7 Relationship between the hybridization rate and the concentration of the target DNA for optical fiber MB biosensor.

easily regenerated by a 1-min rinse with a 90% formamide solution.[19] Since MBs with different loop sequences may be immobilized onto the different optical fiber probes, precise identification of multiple targets may be obtained simultaneously.[19,58]

We made an optical fiber bundle MB biosensor with four fiber probes and two different MBs—MB1 and MB2 (Figure 4.8). The four fibers were divided into

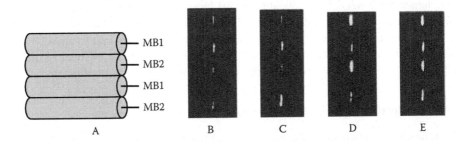

FIGURE 4.8 Optical fiber bundle biosensor: (A) schematic of the biosensor; (B) fluorescence image before hybridization; (C) fluorescence image after hybridization with cDNA of MB1 (100 nM); (D) fluorescence image after hybridization with cDNA of MB2 (100 nM); (E) fluorescence image after hybridization with cDNA of both MB1 and MB2 (100 nM each). The biosensor was regenerated after each hybridization.

two groups, and each group was immobilized with one kind of MB. Each optical fiber biosensor had high sensitivity toward its own target DNA molecules in the sample solution. As shown in Figure 4.8, when the complementary DNA of MB1 was added into the testing sample, only the MB1 biosensor became brighter, whereas the MB2 biosensor had no response, and vice versa. If a mixture of complementary DNA molecules for both MB1 and MB2 is added into the testing sample, all four biosensors become brighter. This result clearly demonstrates that the hybridization between DNA molecules may be monitored in real time and different nonlabeled target DNA molecules may be detected selectively and simultaneously.

4.4.2 MB MICROWELL BIOSENSOR

In biomedical diagnosis, drug development, and forensic investigation, the problems caused by extremely small sample sizes have been difficult to deal with. Our group developed an MB microwell biosensor using laser-induced fluorescence imaging for the detection of DNA/RNA molecules (Figure 4.9).[21] Lithographic and wet chemical etching techniques were used to fabricate the microwells. Each biosensor comprises 400 microwells; each well is 6 μm in diameter and 1 μm deep with a volume of about 28 fL (10^{-15} L). Each well is an independent reactor of femtoliter volume. This ultrasmall volume reduces the costs associated with reagents and samples as well as analysis time and will be highly useful for single-cell gene profiling and for multiple gene determination in disease diagnosis.

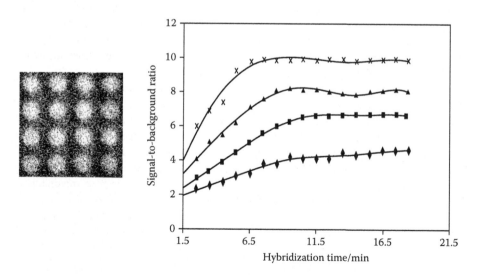

FIGURE 4.9 Microwell biosensor. (Left) Subframe fluorescence images (4 × 4 wells) of 1.0 μM MB and 0.5 μM cDNA in 20 mM Tris-HCl, 0.5 mM MgCl$_2$, 5.0 mM KCl, pH 8.0 buffer. Exposure time was 1 sec and hybridization time was 10 min. (Right) Hybridization kinetics of 1.0 μM MB with different concentrations of complementary oligonucleotide monitored by an ICCD: (a) [MB]:[cDNA] = 1:2; (b) [MB]:[cDNA] = 1:1; (c) [MB]:[cDNA] = 1:0.5; (d) [MB]:[cDNA] = 1:0.2. All experiments were performed in a hybridization buffer of 20 mM Tris-HCl, 0.5 mM MgCl$_2$, 5.0 mM KCl, pH 8.0.

Our microwell biosensor had a detection limit of nine rhodamine 6G molecules in each well. After the microwells were filled with the MB solution, the hybridization kinetics of the MB and the complementary DNA targets were monitored with a concentration detection limit of 3.0 nM, which means as few as 50 target DNA copies could be detected in each well. Differentiation of one base mismatched target DNA was also acquired in the microwell. This MB microwell biosensor was applied to measure rat γ-actin mRNA fragments (bases 782–985) amplified by a reverse transcription polymerase chain reaction (RT-PCR), and 34 nM, or 600 copies, of mRNA could be detected in each microwell.

4.4.3 MB NANOPARTICLE BIOSENSOR

In the areas of disease diagnostics, gene expression studies, and biotechnology development, the separation and collection of rare DNA and RNA molecules with one base mismatch in complex matrices are very important. Commonly used methods such as reversed-phase high-performance liquid chromatography (HPLC), ion exchange HPLC,[59] and gel electrophoresis[60] can easily resolve DNA molecules based on their length but have little sequence specificity. Bioconjugated nanoparticles have demonstrated many unique advantages in bioanalysis.[61–63] Due to their higher separation efficiency in a magnetic field, magnetic nanoparticles have proved to be excellent molecular carriers for DNA separation.[64,65] Meanwhile, besides the high sequence selectivity and excellent detection sensitivity, MBs can induce a wide range of melting temperature differences between the hybrids of its perfectly complementary DNA and the complex of the single base mismatched complementary DNA (Figure 4.10). As a result, an MB-based nanoparticle biosensor or genomagnetic nanocapturer (GMNC) was developed to collect, separate, and detect trace amounts of DNA/RNA molecules with one single base difference.[23]

The GMNC was constructed by bioconjugating MB DNA probes onto magnetic nanoparticle surfaces through avidin–biotin interactions. The excellent ability to differentiate single base mismatched DNA/mRNA samples was accomplished by combining the exceptional specificity of MBs with the separation power of magnetic nanoparticles as well as real-time monitoring of the collected gene products. As shown in Figure 4.10, the GMNC was first incubated at temperature T1 with a sample solution that contained perfectly complementary target DNA1, one base mismatched DNA2, an excess amount of random DNA, and an excess amount of proteins. The target molecules—DNA1 and DNA2—bind with immobilized molecular beacons, whereas random DNA and proteins will not. These target DNA molecules are carried on the GMNC and are separated from the mixture when the sample is exposed to a magnet. After incubating those collected from the GMNC at temperature T2 in a buffer solution, DNA2 completely dissociates from the GMNC and separates from DNA1, which is still bound to the surface-immobilized MB. Further incubating the DNA1-bound GMNC at temperature T3 in a buffer solution results in the complete dissociation of DNA1 from the GMNC as well.

The GMNC biosensor is a unit for separation, collection, and detection. It is highly efficient and selective and can monitor the separation process in real time. Results have shown that this biosensor can capture complementary DNA molecules

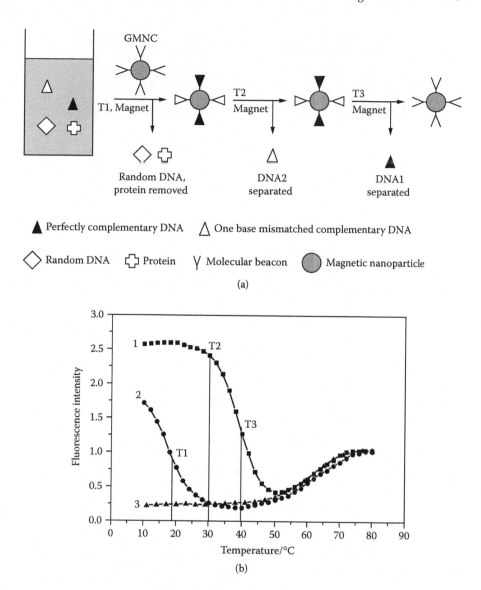

FIGURE 4.10 Isolation and detection of two single-base mismatched DNA molecules using an MB nanoparticle biosensor. (a) Schematic diagram of the working principle of the. (b) The melting curves of the MB and its target duplexes in buffer solutions. The solution contained 0.6 μM DNA1 and 0.1 μM MB for curve 1 and 0.6 μM DNA2 and 0.1 μM MB for curve 2. Curve 3 is for MB only in solution.

at a concentration of 3 fM (10^{-15} M) and capture one base mismatched DNA at a concentration of 9 fM. The collection efficiency is 90% or more. The GMNC biosensor has been demonstrated to work well in collecting trace amounts of mRNA targets from artificial and cancer cell samples. This newly developed

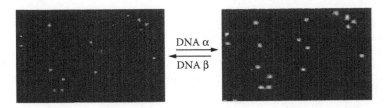

FIGURE 4.11 Fluorescence images of nanomotor-based nanoparticle biosensor switched between the shrinking and extending states on exposure to different DNA solutions. Nanomotor-bearing nanoparticles were immobilized on the surface of a substrate.

technique may be useful for a variety of sample sources in forensics, medicine, and biotechnology.

Another kind of MB nanoparticle biosensor is the DNA nanomotor-immobilized nanoparticle (Figure 4.11).[22] DNA nanomotors are often constructed through self-assembly of several DNA strands and produce twisting or opening–closing movements. Our nanomotor was built with a single-stranded DNA, with the fluorophore and quencher attached at the end of each terminal to report the motion of the nanomotor. This single-stranded DNA has a structure similar to a regular MB and can change between the closed and open state. The nanomotor was switched between the two conformations by alternating DNA hybridization and strand exchange reactions with two different DNAs (DNAα and DNAβ), which enabled the nanomotor to perform an inchwormlike extending–shrinking motion. The motion causes a change in the energy transfer between the fluorophore and the quencher, and it can be viewed in real time by monitoring the fluorescent signal. The nanomotor may be immobilized on the surface of the nanoparticle through biotin–actin binding. The nanomotor-bearing nanoparticles may be further used as biosensors to track the presence of DNAα and DNAβ. Results show that those nanoparticles lit up when exposed to DNAα, but they turn dark when exposed to DNAβ (Figure 4.11).

4.4.4 MB Array Biosensor

In the past decade, the DNA array has become one of the leading methods for the investigation of biomolecules, the diagnosis of disease, and the discovery of genes.[66–74] The major advantages of the DNA array over conventional methods are its ability to simultaneously detect different targets, its capacity for virtual automation, and its functional integration for high-throughput screening,[69] which potentially allows rapid and cost-effective screens for all possible mutations and sequence variations in genomic DNA.

The major tasks of DNA array technology include the manufacture of the array, fluorescent labeling of the complementary DNA (cDNA) probes, hybridization of the probes to the immobilized target DNA, and subsequent analysis of the hybridization results. The DNA array is prepared through either the on-chip synthesis of nucleic acids or the attachment of presynthesized oligonucleotides.[66,75] The on-chip method was developed in the early 1990s by Fodor et al.[76] and is often referred to as the Affymetrix method (Affymetrix Inc., Santa Clara, CA). In this method, DNA oligonucleotides are synthesized directly onto the array using photoprotecting groups and masks to direct the selective addition of nucleotides. The second method was achieved by spotting presynthesized oligonucleotides onto a substrate with a precision robot.[77] Although the first method presents an elegant approach to chip fabrication, it requires resources and expertise that may limit facile implementation, so that the on-chip method is difficult to perform and may cause failure sequences, which can affect the accuracy and sensitivity of further analysis. Therefore, the use of presynthesized probes modified with an appropriate linking group is common.[66]

The signal mechanism of most DNA arrays is based on the intensity change of the fluorescent dye. Because of its intrinsic advantages, an MB may also be used in a DNA array. After we immobilized MBs on a solid surface through biotin–avidin interaction, we obtained DNA biosensors with rapid response and high reproducibility.[18,19] Steemers et al.[58] immobilized MBs on a randomly ordered optical fiber and detected unlabeled DNA targets at subnanomolar concentrations. These results made it possible to build a DNA array to detect different label-free DNA targets simultaneously on a large scale and with high throughput.

However, the MB array has an unsolved bottleneck. Although an MB may have a fluorescence enhancement of tens to hundreds in homogeneous solution, MB arrays exhibit an enhancement of only one to two times upon hybridizing with the cDNA. This low enhancement factor prevented us from exploring the MB's full potential in DNA array and biosensor applications.

This weak enhancement arises from the interaction between immobilized MB and the surface of the substrate. First, this interaction partially destroys the hairpin structure of unhybridized MB, which in turn causes a high fluorescence background that decreases the fluorescence enhancement after hybridization. Second, this interaction changes the electrostatic properties and local environment of the immobilized MB, so as to hinder the hybridization of the MB with its target DNA if the MB is on or near the surface.

In order to minimize the interfacial effect between MBs and the substrate, we designed new MB probes with different linker lengths to enable a larger separation between the surface and the solid support surface.[24] The rationale is to use poly-T as a way to obtain various separation distances. It is believed that poly-T is quite rigid[78] and, thus, the linker length is increased with the number of Ts in the strand. Each poly-T linker is labeled with a biotin molecule and can be immobilized on an avidin film on the substrate through a biotin–avidin interaction. Results showed that the linker length affects the performance of the MB array. As shown in Figure 4.12, if the linker length is shorter than the optimal length (25 bases), the MB array sensitivity (fluorescence enhancement) increases with linker length, whereas further increases in linker length result in a decrease in sensitivity. Results also revealed that the pH and ionic strength of the hybridization buffer have an important effect on the MB array. With optimal linker length, pH, and ionic strength, the MB array exhibits good sensitivity and selectivity (Figure 4.13), with an enhancement of 5.5-fold, much better than in the previous study (about one- to twofold), and may be

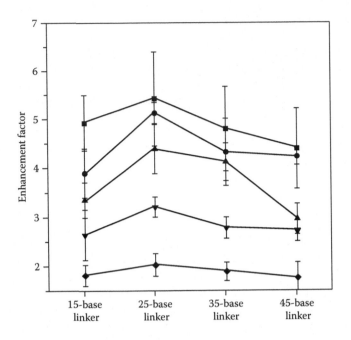

FIGURE 4.12 Effect of linker length on fluorescence enhancement for MB array at different pH levels (from top to bottom: pH 6.0, 7.2, 8.2, 9.1, 9.9).

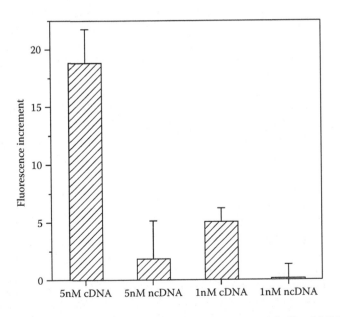

FIGURE 4.13 Sensitivity and selectivity of the MB array in pH 9.9, 20 mM Tris-HCl, 100 mM $MgCl_2$ buffer.

used for 0.2 nM DNA detection with excellent selectivity as well as with multiple DNA targets.

Improvements were also made by immobilizing MBs on an agarose film-coated slide.[25] Three-dimensional functionalized hydrophilic gel films, such as polyacrylamide and agarose, have the advantages of a porous structure and a planar surface and are attracting more and more interest as substrates. They can provide high binding capacity and a solution-like environment where hybridization and other processes resemble a homogeneous liquid phase reaction rather than a heterogeneous liquid–solid reaction.[25] Results demonstrated that an MB array on an agarose film had a low fluorescence background and an excellent discrimination ratio for single nucleotide mismatches.

4.4.5 MB SINGLE MOLECULE BIOSENSOR

In the past decade tremendous strides have been made in single molecule techniques by several groups, both in ultrasensitive detection and in practical applications such as gene chips and fundamental molecular mechanism studies.[79–85] It is a technique that directly observes each target molecule individually and explicitly removes the average effect of the population, so that a single molecule technique can study and characterize detailed physical and chemical properties at a fundamental level and can lead to methodological and technological improvements, with applications in medical science and biotechnology. Fluorescent probes have been playing a greater role in the study of single molecules. However, the challenge is to extract the fluorescence signals from single molecules from among the background noise. One approach is to improve the single molecule fluorescence imaging system by using the confocal technique or the total internal reflection technique or by fluorescence correlation spectroscopy. Another approach is to use probes with a higher signal:noise ratio.

Because the MB has a high fluorescence enhancement upon hybridization, it has the potential to be used in a biosensor for single molecule research. Despite the fact that it has proven to be a useful probe in the study of DNA–DNA, DNA–RNA, and DNA–protein interactions in solution, its complex behavior in a traditional bulk state cannot provide any information on whether all the molecules share a common distribution or whether each molecule gives its own specific contribution to the distribution seen for many molecules.

Our group has used MBs as biosensors to monitor the reaction dynamics of the DNA hybridization process on a liquid–solid interface at the single molecule level with total internal reflection fluorescence microscopy in real time.[26] A biotinylated MB was immobilized on the surface of a quartz slide through biotin–avidin binding. The MB molecules are excited with an evanescent wave field produced by a quartz prism. Time-lapse fluorescence images of the surface hybridization progression were obtained by a fluorescence microscope equipped with an ICCD (Figure 4.14). Results showed that the interactions between DNA molecules are not of the same kind.

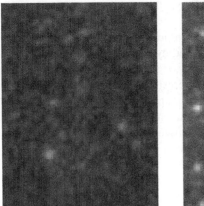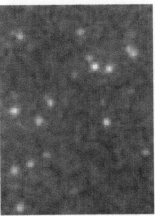

FIGURE 4.14 Single molecule fluorescence images of MB before (left) and after (right) hybridization with cDNA in a 15.62 μm × 21.34 μm area.

The surface-immobilized MBs have two major types of kinetics during their hybridization: fast dynamics with an abrupt fluorescence increase for 87.5% of the MB probes (Figure 4.15a) and slow dynamics with a gradual fluorescence increase for 12.5% of the MB probes (Figure 4.15b). Statistical data also reveal that DNA hybridization is not a single-step process, which means hybrid intermediates might exist in the DNA hybridization process. It has been demonstrated that using an MB as a biosensor in a single molecule study will not only expand the research means to better understand the DNA interaction mechanism at the solid–liquid interface but will also improve biosensor development where an immobilized MB can provide its unique advantages in signal transduction.

4.5   CONCLUSION

The MB has proven to be an excellent probe for homogeneous solutions in biological, biochemical, and clinic studies. The review here further confirms that MBs may be applied successfully to optical fiber biosensors, microwell biosensors, nanoparticle biosensors, MB array biosensors, and single molecule biosensors. The MB biosensors inherit the advantages of MBs, such as high sensitivity, nonlabeled target, detection without separation, and one base match specificity. They have also demonstrated good target detection and monitoring ability.

Considering that the commercialization of new devices is the aim of biosensor development, efforts are needed to optimize the performance of MB biosensors to obtain better analytical results, especially in the areas of decreasing nonspecific binding, increasing fluorescence enhancement, and multiple target detection.

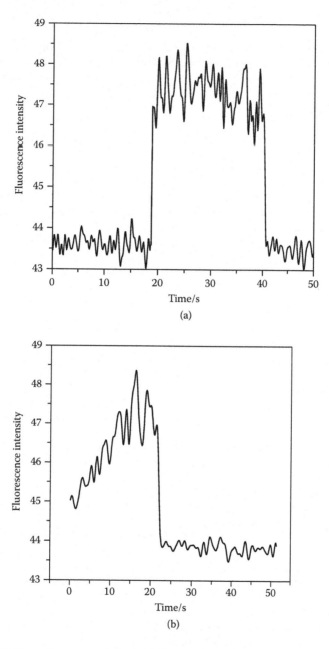

FIGURE 4.15 Dynamic curves of the hybridization of a single MB and its cDNA: (a) an abrupt increase of fluorescent intensity and (b) a gradual increase in fluorescence intensity.

ACKNOWLEDGMENTS

We thank our colleagues at the University of Florida for their help with the work reported here. This work was supported by NIH CA92581 and GM66137.

REFERENCES

1. Wolfbeis, O.S., *Fiber Optic Chemical Sensors and Biosensors,* vol. 1. Boca Raton, FL: CRC Press, 1991.
2. Prasad, P.N., *Introduction to Biophotonics.* New York: John Wiley & Sons, 2003.
3. Wang, J., Rivas, G., Cai, X., Palecek, E., Nielsen, P., Shiraishi, H., Dontha, N., Luo, D., Parrado, C., Chicharro, M., Farias, P.A.M., Valera, F.S., Grant, D.H., Ozsoz, M., and Flair, M.N., DNA electrochemical biosensors for environmental monitoring: a review. *Anal Chim Acta,* 347, 1–8, 1997.
4. Shah, J. and Wilkins, E., Electrochemical biosensors for detection of biological warfare agents. *Electroanalysis,* 15, 157–167, 2003.
5. Marose, S., Lindemann, C., Ulber, R., and Scheper, T., Optical sensor systems for bioprocess monitoring. *Trends Biotechnol,* 17, 30–34, 1999.
6. Baird, C.L. and Myszka, D.G., Current and emerging commercial optical biosensors. *J Mol Recognit,* 14, 261–268, 2001.
7. Epstein, J.R., Biran, I., and Walt, D.R., Fluorescence-based nucleic acid detection and microarrays. *Anal Chim Acta,* 469, 3–36, 2002.
8. Leonard, P., Hearty, S., Brennan, J., Dunne, L., Quinn, J., Chakraborty, T., and O'Kennedy, R., Advances in biosensors for detection of pathogens in food and water. *Enzyme Microb Technol,* 32, 3–13, 2003.
9. Tamayo, J., Alvarez, M., and Lechuga, L.M., Digital tuning of the quality factor of micromechanical resonant biological detectors. *Sensors Actuator B Chem,* 89, 33–39, 2003.
10. Ramanathan, K., Rank, M., Svitel, J., Dzgoev, A., and Danielsson, B., The development and applications of thermal biosensors for bioprocess monitoring. *Trends Biotechnol,* 17, 499–505, 1999.
11. Rodriguez-Mozaz, S., Marco, M.P., de Alda, M.J.L., and Barcelo, D., Biosensors for environmental monitoring of endocrine disruptors: a review article. *Anal Bioanal Chem,* 378, 588–598, 2004.
12. Nakamura, H. and Karube, I., Current research activity in biosensors. *Anal Bioanal Chem,* 377, 446–468, 2003.
13. Mello, L.D. and Kubota, L.T., Review of the use of biosensors as analytical tools in the food and drink industries. *Food Chem,* 77, 237–256, 2002.
14. Tyagi, S. and Kramer, F.R., Molecular beacons: probes that fluoresce upon hybridization. *Nat Biotechnol,* 14, 303–308, 1996.
15. Tyagi, S., Bratu, D.P., and Kramer, F.R., Multicolor molecular beacons for allele discrimination. *Nat Biotechnol,* 16, 49–53, 1998.
16. Fang, X., Li, J.J., and Tan, W., Using molecular beacons to probe molecular interactions between lactate dehydrogenase and single-stranded DNA. *Anal Chem,* 72, 3280–3285, 2000.
17. Perlette, J. and Tan, W., Real-time monitoring of intracellular mRNA hybridization inside single living cells. *Anal Chem,* 73, 5544–5550, 2001.
18. Liu, X. and Tan, W., A fiber-optic evanescent wave DNA biosensor based on novel molecular beacons. *Anal Chem,* 71, 5054–5059, 1999.

19. Liu, X., Farmerie, W., Schuster, S., and Tan, W., Molecular beacons for DNA biosensors with micrometer to submicrometer dimensions. *Anal Biochem,* 283, 56–63, 2000.
20. Epstein, J.R., Leung, A.P.K., Lee, K.H., and Walt, D.R., High-density, microsphere-based fiber optic DNA microarrays. *Biosens Bioelectron,* 18, 541–546, 2003.
21. Lou, H.J. and Tan, W., Femtoliter microarray wells for ultrasensitive DNA/mRNA detection. *Instrum Sci Technol,* 30, 465–476, 2002.
22. Li, J.J. and Tan, W., A single DNA molecule nanomotor. *Nano Lett,* 2, 315–318, 2002.
23. Zhao, X., Tapec-Dytioco, R., Wang. K., and Tan, W., Collection of trace amounts of DNA/mRNA molecules using genomagnetic nanocapturers. *Anal Chem,* 75, 3476–3483, 2003.
24. Yao, G. and Tan, W., A molecular beacon based array for sensitive DNA analysis. *Anal Biochem,* 331, 216–223, 2004.
25. Wang, H., Li, J., Liu, H., Liu, Q., Mei, Q., Wang, Y., Zhu, J., He, N., and Lu, Z., Label-free hybridization detection of a single nucleotide mismatch by immobilization of molecular beacons on an agarose film. *Nucleic Acids Res,* 30, e61, 2002.
26. Yao, G., Fang, X., Yokota, H., Yanagida, T., and Tan, W., Monitoring molecular beacon DNA probe hybridization at the single-molecule level. *Chemistry,* 9, 5686–5692, 2003.
27. Haugland, R.P., *Handbook of Fluorescent Probes and Research Chemicals.* Eugene, OR: Molecular Probes, 1994.
28. Fang, X. and Tan, W., Imaging single fluorescent molecules at the interface of an optical fiber probe by evanescent wave excitation. *Anal Chem,* 71, 3101–3105, 1999.
29. Li, Q.G., Liang, J.X., Luan, G.Y., Zhang, Y., and Wang, K., Molecular beacon-based homogeneous fluorescence PCR assay for the diagnosis of infectious diseases. *Anal Sci,* 16, 245–248, 2000.
30. Chen, W., Martinez, G., and Mulchandani, A., Molecular beacons: a real-time polymerase chain reaction assay for detecting salmonella. *Anal Biochem,* 280, 166–172, 2000.
31. Manganelli, R., Tyagi, S., and Smith, I., Real-time PCR using molecular beacons: a new tool to identify point mutations and to analyze gene expression in *Mycobacterium tuberculosis.* In: *Mycobacterium tuberculosis Protocols,* Parish, T. and Stoker, N.G., eds. Totowa, NJ: Humana Press, 2001:295–310.
32. Vet, J.A.M., Van der Rijt, B.J.M., and Blom, H., Molecular beacons: colorful analysis of nucleic acids. *Exp Rev Mol Diagn,* 2, 77–86, 2002.
33. Broude, N.E., Stem-loop oligonucleotides: a robust tool for molecular biology and biotechnology. *Trends Biotechnol,* 20, 249–256, 2002.
34. Kota, R., Holton, T.A., and Henry, R.J., Detection of transgenes in crop plants using molecular beacon assays. *Plant Mol Biol Rep,* 17, 363–370, 1999.
35. Kuhn, H., Demidov, V.V., Gildea, B.D., Fiandaca, M.J., Coull, J.C., Frank-Kamenetskii, M.D., PNA beacons for duplex DNA. *Antisense Nucleic Acid Drug Dev,* 11, 265–270, 2001.
36. Yates, S., Penning, M., Goudsmit, J., Frantzen, I., De Weijer B., Van Strijp, D., and Van Gemen, B., Quantitative detection of hepatitis B virus DNA by real-time nucleic acid sequence-based amplification with molecular beacon detection. *J Clin Microbiol,* 39, 3656–3665, 2001.
37. Kostrikis, L.G., Tyagi, S., Mhlanga, M.M., Ho, D.D., and Kramer, F.R. Molecular beacons—spectral genotyping of human alleles. *Science,* 279, 1228–1229, 1998.
38. Bonnet, G., Tyagi, S., Libchaber, A., and Kramer, F.R., Thermodynamic basis of the enhanced specificity of structured DNA probes. *Proc Natl Acad Sci USA,* 96, 6171–6176, 1999.

39. Piatek, A.S., Tyagi, S., Pol, A.C., Telenti, A., Miller, L.P., Kramer, F.R., and Alland, D., Molecular beacon sequence analysis for detecting drug resistance in *Mycobacterium tuberculosis. Nat Biotechnol,* 16, 359–363, 1998.

40. Kostrikis, L.G., Huang, Y., Moore, J.P., Wolinsky, S.M., Zhang, L.Q., Guo, Y., Deutsch, L., Phair, J., Neumann, A.U., and Ho, D.D., A chemokine receptor CCR2 allele delays HIV-1 disease progression and is associated with a CCR5 promoter mutation. *Nat Med,* 4, 350–353, 1998.

41. Giesendorf, B.A.J., Vet, J.A.M., Tyagi, S., Mensink, E.J.M.G., Trijbels, F.J.M., and Blom, H.J., Molecular beacons: a new approach for semiautomated mutation analysis. *Clin Chem,* 44, 482–486, 1998.

42. Ehricht, R., Kirner, T., Ellinger, T., Foerster, P., and McCaskill, J.S., Monitoring the amplification of CATCH, a 3SR based cooperatively coupled isothermal amplification system, by fluorimetric methods. *Nucleic Acids Res,* 25, 4697–4699, 1997.

43. Gao, W., Tyagi, S., Kramer, F.R., and Goldman, E., Messenger RNA release from ribosomes during 5-translational blockage by consecutive low-usage arginine but not leucine codons in *Escherichia coli. Mol Microbiol,* 25, 707–716, 1997.

44. Fang, X., Liu, X., Schuster, S., and Tan, W., Designing a novel molecular beacon for surface-immobilized DNA hybridization studies. *J Am Chem Soc,* 121, 2921–2922, 1999.

45. Matsuo, T., *In situ* visualization of mRNA for basic fibroblast growth factor in living cells. *Biochim Biophys Acta,* 1379, 178–184, 1998.

46. Anzai, J., Hoshi, T., and Osa, T., Avidin–biotin complexation for enzyme sensor applications. *Trends Anal Chem,* 13, 205–210, 1994.

47. Dontha N., Nowall, W.B., and Kuhr, W.G., Generation of biotin/avidin/enzyme nanostructures with maskless photolithography. *Anal Chem,* 69, 2619–2625, 1997.

48. Hilliard, L.R., Zhao, X.J., and Tan, W., Immobilization of oligonucleotides onto silica nanoparticles for DNA hybridization studies. *Anal Chim Acta,* 470, 51–56, 2002.

49. Tapec, R., Development and characterization of nanomaterials for biotechnology and bioapplications. Ph.D. dissertation, University of Florida, Gainesville, 2002.

50. Cordek, J., Wang, X., and Tan, W., Direct immobilization of glutamate dehydrogenase on optical fiber probes for ultrasensitive glutamate detection. *Anal Chem,* 71, 1529–1533, 1999.

51. Henke, L., Piunno, P.A.E., McClure, A.C., and Krull, U.J., Covalent immobilization of single-stranded DNA onto optical fibers using various linkers. *Anal Chim Acta,* 344, 201–213, 1997.

52. Abel, A.P., Weller, M.G., Duveneck, G.L., Ehrat, M., and Wildmer, H.M., Fiber-optic evanescent wave biosensor for the detection of oligonucleotides. *Anal Chem,* 68, 2905–2912, 1996.

53. Pilevar, S., Davis, C.C., and Portugal, F., Tapered optical fiber sensor using near-infrared fluorophores to assay hybridization. *Anal Chem,* 70, 2031–2037, 1998.

54. Piunno, P.A., Krull, U.J., Hudson, R.H.E., Damha, M.J., and Cohen, H., Fiberoptic DNA sensor for fluorometric nuclei acid determination. *Anal Chem,* 67, 2635–2643, 1995.

55. Kleinjung, F., Bier, F.F., Warsinker, A., and Scheer, F.W., Fibre-optic genosensor for specific determination of femtomolar DNA oligomers. *Anal Chim Acta,* 350, 51–58, 1997.

56. Hutchinson, A.M., Evanescent-wave biosensors—real-time analysis of biomolecular interactions. *Mol Biotechnol* 3, 47–54, 1995.

57. Nath, N. and Anand, S., Evanescent wave fiber optic fluorosensor: effect of tapering configuration on the signal acquisition. *Optic Eng,* 37, 220–228, 1998.

58. Steemers, F.J., Ferguson, J.A., and Walt, D.R., Screening unlabeled DNA targets with randomly ordered fiber-optic gene arrays. *Nat Biotechnol,* 18, 91–94, 2000.
59. Levison, P.R., Mumford, C., Streater, M., Brandt-Nielsen, A., Pathiramna, N.D., and Badger, S.E., Performance comparison of low-pressure ion-exchange chromatography media for protein separation. *J Chromatogr A,* 760, 151–158, 1997.
60. Behne, D., Kyriakopoeulos, A., Weiss-Nowak, C., Kalckloesch, M., Westpal, C., and Gesssner, H. Newly found selenium-containing proteins in the tissues of the rat. *Biol Trace Elem Res,* 55, 99–110, 1996.
61. Santra, S., Wang, K.M., Tapec, R., and Tan, W., Development of novel dye-doped silica nanoparticles for biomarker application. *J Biomed Opt,* 6, 160–166, 2001.
62. Santra, S., Zhang, P., Wang, K.M., Tapec, R., and Tan, W., Conjugation of biomolecules with luminophore-doped silica nanoparticles for photostable biomarkers. *Anal Chem,* 73, 4988–4993, 2001.
63. Kolodny, L.A., Willard, D.M., Carillo, L.L., Nelson, M.W., and Van Orden, A., Spatially correlated fluorescence/AFM of individual nanosized particles and biomolecules. *Anal Chem,* 73, 1959–1966, 2001.
64. Hawkins, T.L., McKernan, K.J., Jacotot, L.B., MacKenzie, B., Richardson, P.M., and Lander, E.S., DNA sequencing—a magnetic attraction to high-throughput genomics. *Science,* 276, 1887, 1997.
65. Chemla, Y.R., Grossman, H.L., Poon, Y., McDermott, R., Stevens, R., Alper, M.D., and Clarke, J., Ultrasensitive magnetic biosensor for homogeneous immunoassay. *Proc Natl Acad Sci USA,* 97, 14268–14272, 2000.
66. Steel, A.B., Levicky, R.L., Herne, T.M., and Tarlov, M.J., Immobilization of nucleic acids at solid surfaces: effect of oligonucleotide length on layer assembly. *Biophys J,* 79, 975–981, 2000.
67. Broude, N.E., Woodward, K., Cavallo, R., Cantor, C.R., and Englert, D., DNA microarrays with stem-loop DNA probes: preparation and applications. *Nucleic Acids Res,* 29, e92, 2001.
68. Stillman, B.A. and Tonkinson, J.L., Expression microarray hybridization kinetics depend on length of the immobilized DNA but are independent of immobilization substrate. *Anal Biochem,* 295, 149–157, 2001.
69. Preininger, C. and Chiarelli, P., Immobilization of oligonucleotides on crosslinked poly(vinylalcohol) for application in DNA chips. *Talanta,* 55, 973–980, 2001.
70. Peterson, A.W., Heaton, R.J., and Georgiadis, R.M., The effect of surface probe density on DNA hybridization, *Nucleic Acids Res,* 29, 5163–5168, 2001.
71. Walsh, M.K., Wang, X., and Weimer, B.C., Optimizing the immobilization of single-stranded DNA onto glass beads. *J Biochem Biophys Methods,* 47, 221–231, 2001.
72. Zhao, X., Nampalli, S., Serino, A.J., and Kumar, S., Immobilization of oligodeoxyribonucleotides with multiple anchors to microchips. *Nucleic Acids Res,* 29, 955–959, 2001.
73. Beier, M. and Hoheisel, J.D., Versatile derivatisation of solid support media for covalent bonding on DNA-microchips. *Nucleic Acids Res,* 27, 1990–1977, 1999.
74. Southern, E., Mir, K., and Shchepinov, M., Molecular interactions on microarrays. *Nat Genet,* 21(suppl), 5–9, 1999.
75. Dolan, P.L., Wu, Y., Ista, L.K., Metzenberg, R.L., Nelson, M.A., and Lopez, G.P., Robust and efficient synthetic method for forming DNA microarrays. *Nucleic Acids Res,* 29, e107, 2001.
76. Fodor, S.P.A., Read, J.L., Pirrung, M.C., Stryer, L., Lu, A.T., and Solas, D., Light-directed, spatially addressable parallel chemical synthesis. *Science,* 251, 767–773, 1991.

77. Schena, M., Shalon, D., Davis, R.W., and Brown, P.O., Quantitative monitoring of gene expression patterns with a complementary DNA microarray. *Science,* 270, 467–470, 1995.

78. Devlin, T.M., ed., *Textbook of Biochemistry with Clinical Correlations,* 5th ed. New York: John Wiley & Sons, 2002: chap. 2.

79. Moerner, W.E. and Kador, L., Optical detection and spectroscopy of single molecules in a solid. *Phys Rev Lett,* 62, 2535–2538, 1989.

80. Funatsu, T., Harada, Y., Tokunaga, M., Saito, K., and Yanagida, T., Imaging of single fluorescent molecules and individual ATP turnovers by single myosin molecules in aqueous solution. *Nature,* 374, 555–559, 1995.

81. Nie, S. and Zare, R.N., Optical detection of single molecules. *Annu Rev Biophys Biomol Struct,* 26, 567–596, 1997.

82. Xie, X.S. and Trautman, J.K., Optical studies of single molecules at room temperature. *Annu Rev Phys Chem,* 49, 441–480, 1998.

83. Dickson, R.M., Norris, D.J., Tzeng, Y.L., and Moerner, W.E., Three-dimensional imaging of single molecules solvated in pores of poly(acrylamide) gels. *Science,* 274, 966–969, 1996.

84. Xu, X. and Yeung, E.S., Direct measurement of single-molecule diffusion and photodecomposition in free solution. *Science,* 275, 1106–1109, 1997.

85. Zhang, P. and Tan, W., Direct observation of single-molecule generation at a solid-liquid interface. *Chemistry,* 6, 1087–1092, 2000.

5 Fluorescence Resonance Energy Transfer-Based Sensors for Bioanalysis

Gabriela Blagoi, B.S., Nitsa Rosenzweig, Ph.D., and Zeev Rosenzweig, Ph.D.

CONTENTS

5.1   INTRODUCTION

Biosensors have been valuable analytical tools for several decades, with glucose sensing being the most impressive success story of biosensor technology.[1,2] Recent developments in the field of biosensors enable their employment in new applications primarily focusing on real-time monitoring of bimolecular interactions. For example, biosensors have been used to monitor processes such as ligand fishing,[3] signal transduction in cells,[4,5] cell adhesion,[6] enzymatic reactions,[7] and protein conformation changes and aggregation.[8,9] Biosensors have also been used for primary screening of drug candidates.[10,11]

Several recently published reviews have focused on the application of biosensors in drug discovery.[12-15] The present review discusses the theory of fluorescence resonance energy transfer (FRET) and the application of FRET as a unique optical signaling strategy in fluorescence sensors. Because FRET signals depend largely on the distance between the donor and acceptor molecules, FRET-based sensors provide valuable information in systems involving ligand–receptor binding or interactions between donor-labeled drugs and acceptor-labeled cells. The mechanisms and parameters that determine the success or failure of FRET-based sensors are discussed. This review also summarizes recent studies in our group in which FRET-based sensors were used to monitor binding interactions between carbohydrates and glycoproteins and particles labeled with carbohydrate–binding protein molecules.

5.2 FRET

Fluorescence resonance energy transfer is a nonradiative energy transfer between the excited states of a fluorescent donor (D) and an acceptor (A). This energy transfer process is followed by photon emission of the acceptor molecules. FRET occurs when the emission spectrum of a fluorescent donor overlaps with the absorption spectrum of an acceptor (Figure 5.1). As a result, the donor lifetime is shortened and the acceptor fluorescence is sensitized.[16–18] The FRET phenomenon was first described by Forster in 1946.[19,20] The main finding of Forster was that FRET between a donor-acceptor pair could occur over distances of up to 100 Å. For effective FRET between a donor and acceptor molecule, the donor and acceptor must be spatially separated to prevent interactions between their electronic clouds that could alter their electronic spectra. However, as shown in Figure 5.1, for FRET to occur, there must be a spectral overlap between the donor emission and acceptor absorption.

The resonance energy transfer phenomenon between donor and acceptor molecules may be described as a resonance between two mechanical pendulums. Considering the wave function a product of the electronic wave function, the vibrational function, and the normalized spectrum of the donor, the rate constant of FRET, k_t, is expressed as follows:

$$k_t = \frac{9000(\ln 10)Q_D K^2}{128\pi^5 N n^4 \tau_D R^6} \int_0^\infty F_D(v)\varepsilon_a(v)\frac{dv}{v^4} \qquad (5.1)$$

where K^2 is an orientation factor, $F_D(v)$ is the spectral distribution of the donor fluorescence (normalized to frequency), $\varepsilon_a(v)$ is the molar extinction coefficient of

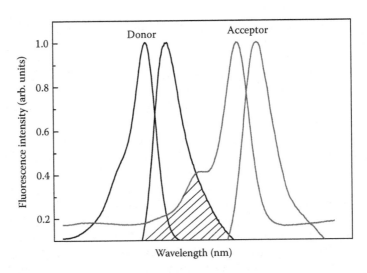

FIGURE 5.1 A spectral overlap (marked in diagonal lines) between the donor emission and the acceptor absorption is required for FRET in a donor–acceptor pair.

the acceptor, N is Avogadro's number, n is the refractive index of the solvent, τ_D is the intrinsic lifetime of the donor in the absence of quenchers, Q_D is the quantum yield of the donor in the presence of the acceptor, and R is the distance between the donor and acceptor molecules.

The orientation factor K^2 depends on the relative orientations of the emission transition dipole of the donor and the absorption transition dipole of the acceptor. Values of K^2 range between zero and four. For randomly oriented donors and acceptors, K^2 is equal to 2/3. There is an extensive discussion in the literature about the effect of specific orientation geometries resulting from rotational and transitional diffusion during the excited lifetime of the donor on FRET efficiency.[21-27] These considerations are relevant when the interacting molecules are well defined. In most biological applications of FRET, as in our experiments, labeling of the interacting biomolecules with donor and acceptor molecules introduces considerable heterogeneity. Under these conditions, the orientation factor is less important.[16,21]

To simplify the description of FRET, Forster wrote Equation 5.1 in the following reduced form:

$$k_t = \tau_D^{-1} \left(\frac{R_0}{R} \right)^6 \tag{5.2}$$

where τ_D is the measured lifetime of the donor in the absence of the acceptor, R_0 is the critical radius of the transfer, or the Forster distance, which is the distance at which the energy transfer efficiency is 50%. The Forster distance, R_0, depends on the spectral characteristics of the donor–acceptor pair and is expressed in angstroms (Å).

As can be seen in Equation 5.2, the rate of energy transfer is highly dependent on the distance between the donor and acceptor molecules. For a donor–acceptor pair that is covalently bound together, the energy transfer efficiency, E, is expressed as

$$E = \frac{R_0^6}{R_0^6 + R^6} \tag{5.3}$$

On the other hand, for randomly distributed donor and acceptor molecules in solution, R_0 depends on the acceptor concentration and is expressed as

$$R_0 = \left(\frac{3000}{4\pi N |A|_{1/2}} \right)^{1/3} \tag{5.4}$$

where N is Avogadro's number and $|A|_{1/2}$ is the concentration of the acceptor at which the energy transfer efficiency E is 50%. Equation 5.4 reveals the most important feature of energy transfer between donor and acceptor molecules that are randomly distributed in solution; that is, the concentration of acceptor has to be rather high, often in the millimolar level, to produce measurable FRET signals. The high acceptor concentration that is required complicates FRET measurements in solution due to competing inner filter effects, which result from absorption of the excitation light as well as donor emission by the highly concentrated acceptor molecules.

The decrease in donor fluorescence due to FRET depends on the dimensionality of the acceptor distribution. Acceptor molecules may be distributed randomly in solution (three-dimensionally), in a plane (two-dimensionally), or even in a linear dimension (one-dimensionally). For example, a planar distribution occurs when the donor and acceptor molecules are incorporated into planar phospholipid bilayers. A linear distribution occurs when the donor and acceptor molecules are bound to a DNA helix or elongated protein such as collagen. For a random distribution of donors and acceptors in solution, assuming no diffusion and no excluded volume, the fluorescence intensity of the donor molecules in the presence of acceptors is expressed as[19,20,28–30]

$$I_{DA} = I_D^0 \exp\left[-\frac{t}{\tau_D} - 2\gamma \left(\frac{t}{\tau_D} \right)^{d/6} \right] \qquad (5.5)$$

where d is 3, 2, or 1, for three-, two-, or one-dimensional distributions, respectively; τ_D is the donor decay time in the absence of acceptor molecules; and γ is a function related to the concentration of the acceptor and the dimensionality of the system.

In our laboratory we explored the use of FRET to monitor the permeation of fluorescent molecules into single cells. We also developed FRET-sensing particles and used them for carbohydrate and glycoprotein screening. The results of these studies are briefly summarized in the following sections.

5.3 DELIVERY OF FLUOROPHORES INTO SINGLE CELLS

Since FRET efficiency depends strongly on the distance between the donor and acceptor, it may be used as a highly specific molecular ruler.[18a] For example, FRET measurements have been employed for the detection of colocalization of proteins in cellular membranes,[31–33] hybridization of nucleic acids,[34,35] intracellular signaling,[36,37] and protein–protein interaction.[38,39] The popularity of FRET imaging microscopy increased significantly following the recent development of the green fluorescent protein (GPF) and its mutants.[40] Fluorescence microscopy techniques have often been used to monitor the delivery of fluorescent molecules into cells.[41,42] However, using ordinary fluorescence microscopy techniques, it is difficult to determine whether the fluorescent molecules truly permeate the cell membrane or only adsorb to the cell surface.

We hypothesized that FRET measurements between fluorescent molecules pre-loaded into cells and permeating fluorescent molecules (donors or acceptors) would provide direct proof that the fluorescent molecules truly permeate cells. Fluorescein and rhodamine derivatives were utilized as donor and acceptor molecules in our cellular studies. We investigated two strategies. The first involved the delivery of donor molecules into acceptor-labeled cells. The second strategy involved the delivery of acceptor molecules into donor-labeled cells. To monitor the delivery of donor molecules into acceptor-labeled cells, J774 murine macrophages were preincubated with 5 µM sulforhodamine (acceptor). Then the cell-permeating dye calcein-AM (donor) was added to the cultures. Fluorescence images of the cells were collected 15 min after the addition of calcein-AM to the cultures. The ratio of the fluorescence intensity of cells that were obtained through the acceptor channel ($\lambda_{ex} = 480$ nm, $\lambda_{em} > 590$ nm), FRET, and the donor channel ($\lambda_{ex} = 480$ nm, 520 nm $>$ $\lambda_{em} > 550$ nm), Fd, was calculated after background subtraction. Figure 5.2 depicts the FRET:Fd ratio plotted as a function of increasing donor concentrations. Curves (a) and (b) describe the concentration dependence of the FRET:Fd ratio in the presence and absence of sulforhodamine. It can be seen that the FRET:Fd ratio of acceptor-labeled macrophages was twofold larger than the FRET:Fd ratio of cells in the absence of acceptor molecules. The increase in the FRET/Fd ratio in the absence of acceptor molecules was attributed to red-tail emission of donor molecules. This phenomenon is often described in the literature as "bleeding" and should be minimized to obtain quantitative FRET measurements. In spite of the bleeding effect, it is fair to conclude that, although limited in dynamic range, FRET between permeating

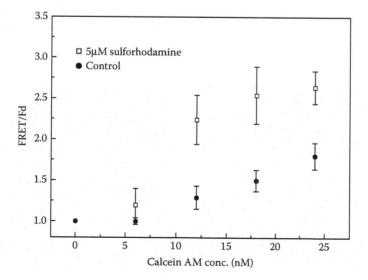

FIGURE 5.2 FRET efficiency between sulforhodamine (5 µM)-labeled macrophages and calcein-AM. The FRET magnitude was determined by recording the increase in the ratio between the sensitized acceptor signal FRET and donor fluorescence Fd in the presence (a) and absence (b) of sulforhodamine.

donor molecules and acceptor-labeled cells may be used to confirm the delivery of donor fluorescent molecules into cells.

The high emission quantum yield of calcein-AM combined with the low loading efficiency of sulforhodamine has been a major shortcoming of this FRET-sensing geometry. It resulted in a large red-tail emission of donor molecules in the FRET channel. To overcome this problem, we investigated a FRET system in which acceptor molecules permeated macrophages that were preloaded with calcein-AM donor molecules. Furthermore, we replaced sulforhodamine with CellTracker Orange, another acceptor dye that, unlike sulforhodamine, did not show a tendency to be sequestered in endosomes. The FRET efficiency was determined based on the fluorescence decrease of the donor molecules (calcein-AM) as acceptor molecules (CellTracker Orange) permeated the cells. It was expressed as the ratio between the donor fluorescence in the presence and absence of acceptor molecules in the cells. Complete delivery of CellTracker Orange into the cells was realized in about 30 min at 37°C.

Fluorescence images of murine macrophage cells loaded with the donor calcein-AM (5 μM) before and after delivery of CellTracker Orange (500 μM) are shown in Figure 5.3. The decrease in calcein-AM-labeled cell emission resulted from energy transfer between the acceptor molecules delivered into the observed macrophages and the donor-labeled cells. A 40% decrease in donor fluorescence was observed as a result of acceptor delivery into the cells. The use of acceptor molecule permeation into donor-labeled cells covered a wider range of acceptor concentrations. However, a relatively high acceptor concentration in the submillimolar range was needed to obtain measurable FRET signals in these three-dimensional systems. The results underscore the need to reduce the dimensionality of FRET systems in order to increase FRET efficiency between donor and acceptor molecules. This led to the development of FRET-sensing particles, which are described in the following section.

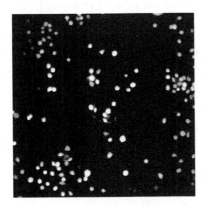 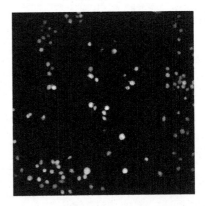

FIGURE 5.3 Fluorescence images of the calcein-AM (5 μM)-labeled murine macrophages before (a, left) and after incubation with 500 μM CellTracker Orange for 30 min (b, right). Fluorescence images were taken through the donor channel (λ_{ex} = 480 nm, 520 nm > λ_{em} > 550 nm).

5.4 FRET-BASED BIOSENSORS

Fluorescence resonance energy transfer-based sensors were recently developed and have found uses in various applications. For example, FRET lifetime-based sensors have been used to determine inorganic analytes such as pH, carbon dioxide, and metal cations with high sensitivity and selectivity.[43–45] FRET sensors have also been developed for DNA and RNA hybridization studies. Research in this area has focused on improving the donor and acceptor pairs in order to reduce false-positive signals.[46–48] For example, the use of lanthanides as donors in FRET-based sensors has increased the sensitivity of DNA and RNA analysis.[46] Several research groups have recently developed highly sensitive and selective FRET-based sensors for monosaccharide quantification. Specific lectins were used as biorecognition components in these sensors to increase the specificity toward carbohydrate analytes.[47–49] FRET sensors have also been used in sensing arrays designed for high-throughput drug screening.[50]

The guiding principle in the design of our particle-based sensors was the need to decrease system dimensionality to increase FRET efficiency. The underlying idea was to immobilize donor fluorophores to micrometric-size particles along with a biomolecular recognition component. We predicted that binding of fluorescent acceptor molecules or molecules that are labeled with acceptor molecules to the particles would result in FRET between the donor–acceptor pair. Fluorescein was chosen as the donor fluorophore and concanavalin A (Con A), a carbohydrate binding protein, was selected as the biorecognition component. Con A is a tetrameric lectin that is specific to mannose and glucose moieties.[51,52] To simplify the system, we attached fluorescein-labeled Con A directly to the particle surface. The ability of the FRET-sensing particles to monitor binding interactions with carbohydrates was tested by incubating particle suspensions with dextran, a polymeric sugar, which was labeled with the fluorescent acceptor Texas Red. The optical signal resulted from FRET between the fluorescein (donor)-labeled Con A and the Texas Red (acceptor)-labeled dextran (dextran-TR). A digital fluorescence imaging microscopy system consisting of an inverted fluorescence microscope and a high-performance charge-coupled device (CCD) camera was used to measure the FRET between the donor and acceptor molecules.

Figure 5.4 shows the temporal dependence of the fluorescence spectra of the FRET-sensing particles in the presence of 0.12 μM dextran-TR (molecular weight 10,000). Curve (a) illustrates the fluorescence spectrum of the FRET-sensing particles prior to the addition of dextran-TR. The fluorescence spectrum of the FRET-sensing particles 15 min after the addition of 0.12 μM dextran-TR to the suspension is shown in curve (b). The fluorescence intensity of fluorescein at 525 nm decreased slightly immediately upon adding dextran-TR to the sample. Curve (b) shows an approximately 40% decrease in the fluorescence intensity of fluorescein (donor) at 525 nm and a similar increase in the fluorescence intensity of Texas Red (acceptor) at 615 nm when the fluorescence spectrum was recorded 15 min after the addition of dextran-TR to the sample. Longer incubation times did not increase the donor quenching or acceptor fluorescence increase, suggesting that the system reached equilibrium in less than 15 min.

The FRET efficiency between the FRET-sensing particles and dextran-TR increased with increasing dextran-TR concentrations and reached a maximum at about

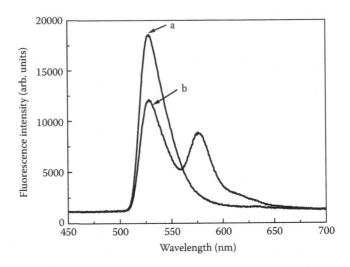

FIGURE 5.4 Digital fluorescence spectra of FRET-sensing particles ($\lambda_{ex} = 470$ nm, $\lambda_{em} > 520$ nm) (a) in the absence of dextran-TR and (b) when dextran-TR was bound to the FITC-ConA-labeled particles and the system reached equilibrium (15 min).

1 μM, where the donor fluorescence decreased by about 60%. The FRET-sensing particles were used in inhibition assays to screen unlabeled carbohydrates and glycoproteins based on their affinity to the particles. These substrates competed with dextran-TR for the Con A binding sites on the surface of the particles. The inhibitor concentration required for 50% inhibition (IC_{50}) of the energy transfer between the FRET-sensing particles and dextran-TR was calculated for several inhibitors. We found that glycoproteins such as glucose oxidase, which is characterized by a large number of glycosidic residues, competed more effectively with dextran-TR for the lectin-binding sites and therefore reduced the FRET efficiency between the fluorescein-labeled particles and dextran-TR more effectively. Glycoproteins such as ovalbumin and avidin have higher IC_{50} values. These glycoproteins do not inhibit as effectively as glucose oxidase because they contain a smaller number of glycosidic residues compared to glucose oxidase. Figure 5.5 shows IC_{50} values on a logarithmic scale for several inhibitors. The calculated IC_{50} values for glucose oxidase, ovalbumin, and avidin were found to be 0.5 μM, 4.5 μM, and 75 μM, respectively. Furthermore, mannose was able to displace dextran-TR from the particles at millimolar concentrations. Mannose was the weakest inhibitor screened, with an IC_{50} value of 1.7 mM. Another monomeric carbohydrate, galactose, was found to be an ineffective FRET inhibitor. It is likely that replacing Con A with a galactose-selective lectin would result in higher binding affinity of the FRET-sensing particles to galactose.

5.5 CONCLUSION

Fluorescence resonance energy transfer has often been used in cellular biology applications. In our laboratory, we have focused on the use of FRET as a quantitative probe in real-time monitoring of the permeation of fluorescent molecules

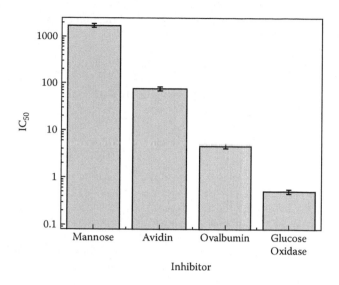

FIGURE 5.5 The calculated IC_{50} values for mannose, avidin, ovalbumin, and glucose oxidase from the inhibition of dextran-TR binding to FRET-sensing particles.

into single cells. We discovered that FRET measurements in three-dimensional systems, such as cellular cytoplasm or any other solution between randomly distributed donors and acceptors, are quite limited in quantitative power. FRET measurements following the delivery of donor molecules into cells preloaded with acceptor molecules showed nanomolar sensitivity. However, the dynamic range of the technique was limited to the nanomolar range due to inner filter effects. FRET measurements of the delivery of acceptor molecules into cells preloaded with donor molecules were less sensitive and required acceptor levels in the range of hundreds of micromoles to generate measurable FRET signals. It became clear that the dimensionality of the FRET system must be reduced to enable quantitative measurements in the micromolar range. We then synthesized FRET-sensing particles by immobilizing lectins and donor molecules on the surface of micrometric particles. The particles exhibited high binding affinity to polymeric sugars, with apparent binding constants of about 1×10^{-7} M. This binding constant is four orders of magnitude higher than the binding constant of mannose to Con A, which was used as a biorecognition component in our sensing particles. The FRET-sensing particles were able to quantify and discriminate between several inhibitors that decreased the binding efficiency of dextran to the FRET-sensing particles. Highly glycated glycoproteins were more effective inhibitors than other glycoproteins with smaller numbers of glycosidic moieties.

The newly developed FRET-sensing particles expand the scope of analytes that can be detected with fluorescence sensing. Glycoproteins and carbohydrates are integral parts of many drugs. The ability to quantify and discriminate between glycoproteins and carbohydrates has the potential to strongly impact the field of drug discovery, where high–throughput screening techniques to monitor drug-target interactions are in great demand. Future studies will focus on the fabrication of FRET-sensing arrays for high-throughput sensing applications.

ACKNOWLEDGMENTS

This work was supported by NSF Grant CHE-0134027 and DoD/DARPA Grant HR0011-04-C-0068.

REFERENCES

1. Ballerstadt, R. and Schultz, J.S., A fluorescence affinity hollow fiber sensor for continuous transdermal glucose monitoring. *Anal Chem,* 72, 4185–4192, 2000.
2. Badugu, R., Lakowicz, J.R., and Geddes, C.D., Noninvasive continuous monitoring of physiological glucose using a monosaccharide-sensing contact lens. *Anal Chem,* 76, 610–618, 2004.
3. Catimel, B., Weinstock, J., Nerrie, M., Domagala, T., and Nice, E.C., Micropreparative ligand fishing with a cuvette-based optical mirror resonance biosensor. *J Chromatogr A,* 869, 261–273, 2000.
4. Chan, S., Horner, S.R., Fauchet, P.M., and Miller, B.L., Identification of gram-negative bacteria using nanoscale silicon microcavities. *J Am Chem Soc,* 123, 11797–11798, 2001.
5. Yoshida, T., Sato, M., Ozawa, T., and Umezawa, Y., An SPR-based screening method for agonist selectivity for insulin signaling pathways based on the binding of phosphotyrosine to its specific binding protein. *Anal Chem,* 72, 6–11, 2000.
6. Svetlicic, V., Ivosevic, N., Kovac, S., and Zutic, V., Charge displacement by adhesion and spreading of a cell: amperometric signals of living cells. *Langmuir,* 16, 8217–8220, 2000.
7. Mei, S.H.J., Liu, Z., Brennan, J.D., and Li, Y., An efficient RNA-cleaving DNA enzyme that synchronizes catalysis with fluorescence signaling. *J Am Chem Soc,* 125, 412–420, 2003.
8. Thanh, N.T.K. and Rosenzweig, Z., Development of an aggregation-based immunoassay for anti-protein a using gold nanoparticles. *Anal Chem,* 74, 1624–1628, 2002.
9. Ho, H.-A., and Leclerc, M., Optical sensors based on hybrid aptamer/conjugated polymer complexes. *J Am Chem Soc,* 126, 1384–1387, 2004.
10. Cooper, M.A., Optical biosensors in drug discovery. *Nat Rev Drug Discov,* 1, 515–528, 2002.
11. Frostell-Karlsson, A., Remaeus, A., Roos, H., Andersson, K., Borg, P., Hamalainen, M., and Karlsson, R., Biosensor analysis of the interaction between immobilized human serum albumin and drug compounds for prediction of human serum albumin binding levels. *J Med Chem,* 43, 1986–1992, 2000.
12. Numann, R. and Negulescu, P.A., High-throughput screening strategies for cardiac ion channels. *Trends Cardiovasc Med,* 11, 54–59, 2001.
13. Turner, A.P.F., Biochemistry: biosensors—sense and sensitivity. *Science,* 290, 1315–1317, 2000.
14. Rich, R.L. and Myszka, D.G., Advances in surface plasmon resonance biosensor analysis. *Curr Opin Biotechnol,* 11, 54–61, 2000.
15. Zlokarnik, G., Fluorescent molecular sensor for drug discovery. *Anal Chem,* 71, 290A–291A, 1999.
16. Lakowicz, J.R., *Principles of Fluorescence Spectroscopy.* New York: Plenum Press, 1999:367–424.

17. Vekshin, N., *Energy Transfer in Macromolecules*. Bellingham, WA: SPIE Optical Engineering Press, 1997:8–26.
18. (a) Stryer, L., Fluorescence energy transfer as a spectroscopic ruler. *Annu Rev Biochem*, 47, 819–846, 1978; (b) Stryer, L., Thomas, D.D., and Meares, C.F., Diffusion-enhanced fluorescence energy transfer. *Annu Rev Biophys Biol*, 11, 203–222, 1982; (c) Wu, P. and Brand, L., Resonance energy transfer: methods and applications. *Anal Biochem*, 218, 1–13, 1994.
19. Forster, T., Energy migration and fluorescence. *Naturwissenschaften*, 33, 166–175, 1946.
20. Forster, T., Transfer mechanisms of electronic excitation. *Disc Faraday Soc*, 27, 7–17, 1959.
21. Srinivas, G. and Bagchi, B., Effect of orientational motion of mobile chromophores on the dynamics of Forster energy transfer in polymers. *J Phys Chem B*, 105, 9370–9374, 2001.
22. Sacca, B., Fiori, S., and Moroder, L., Studies of the local conformational properties of the cell-adhesion domain of collagen type IV in synthetic heterotrimeric peptides. *Biochemistry*, 42, 3429–3436, 2003.
23. Domanov, Y.A. and Gorbenko, G.P., Analysis of resonance energy transfer in model membranes: role of orientational effects. *Biophys Chem*, 99, 143–154, 2002.
24. Selvin, P.R., Rana, T.M., and Hearst, J.E., Luminescence resonance energy transfer. *J Am Chem Soc*, 116, 6029–6030, 1994.
25. Wu, P. and Brand, L., Orientation factor in steady-state and time-resolved resonance energy transfer measurements. *Biochemistry*, 31, 7939–7947, 1992.
26. Eisinger, J. and Dale, R.E., Interpretation of intramolecular energy transfer experiments. *Biophys Chem*, 84, 643–647, 1974.
27. dos Remedios, C.G. and Moens, P.D.J., Fluorescence resonance energy transfer spectroscopy is a reliable "ruler" for measuring structural changes in proteins: dispelling the problem of the unknown orientation factor. *J Struct Biol*, 115, 175–185, 1995.
28. Koppel, D.E., Fleming, P.J, and Strittmatter, P., Intramembrane positions of membrane-bound chromophores determined by excitation energy transfer. *Biochemistry*, 24, 5450–5457, 1979.
29. (a) Tamai, N., Yamazaki, T., and Yamazaki, I., Excitation-energy transfer between dye molecules adsorbed on a vesicle surface. *J Phys Chem*, 91, 3503–3508, 1987; (b) Kwok-Keung Fung, B. and Stryer, L., Surface density determination in membranes by fluorescence energy transfer. *Biochemistry* 24, 5421–5248, 1978.
30. Maliwal, B.P., Kusba, J., and Lakowicz, J.R., Fluorescence energy transfer in one dimension-frequency-domain fluorescence study of DNA–fluorophore complexes. *Biopolymers* 35, 245–255, 1995.
31. Kenworthy, A.K. and Edidin, M., Imaging fluorescence resonance energy transfer as probe of membrane organization and molecular associations of GPI-anchored proteins. *Methods Mol Biol*, 116, 37–49, 1999.
32. Jovin, T.M. and Ardnt-Jovin, D.J., Luminescence digital imaging microscopy. *Annu Rev Biophys Biophys Chem*, 18, 271–308, 1989.
33. Oliveria, S.F., Gomez, L.L., and Dell'Acqua, M.L., Imaging kinase–AKAP79–phosphatase scaffold complexes at the plasma membrane in living cells using FRET microscopy. *J Cell Biol* 160, 101–112, 2003
34. Dirks, R.W., Molenaar, C., and Tanke, H.J., Methods for visualizing RNA processing and transport pathways in living cells. *Histochem Cell Biol*, 115, 3–11, 2001.

35. Tsuji, A., Sato, Y., Hirano, M., Suga, T., Koshimoto, H., Taguchi, T., and Ohsuka, S., Development of a time-resolved fluorometric method for observing hybridization in living cells using fluorescence resonance energy transfer. *Biophys J,* 81, 501–515, 2001.

36. Jobin, C.M., Chen, H., Lin, A.J., Yacono, P.W., Igarashi, J., Michel, T., and Golan, D.E., Receptor-regulated dynamic interaction between endothelial nitric oxide synthase and calmodulin revealed by fluorescence resonance energy transfer in living cells. *Biochemistry,* 42, 11716–11725, 2003.

37. Miyakawa-Naito, A., Uhlen. P., Lal, M., Aizman, O., Mikoshiba, K., Brismar, H., Zelenin, S., and Aperia, A., Cell signaling microdomain with Na,K-ATPase and inositol 1,4,5-trisphosphate receptor generates calcium oscillations. *J Biol Chem,* 278, 50355–50361, 2003.

38. Niethammer, P., Bastiaens, P., and Karsenti, E., Stathmin–tubulin interaction gradients in motile and mitotic cells. *Science,* 19, 303, 1862–1866, 2004.

39. Szczesna-Skorupa, E., Mallah, B., and Kemper, B., Fluorescence resonance energy transfer analysis of cytochromes P450 2C2 and 2E1 molecular interactions in living cells. *J Biol Chem,* 278, 31269–31276, 2003.

40. Tsien, R.Y., Bacskai, B.J., and Adams, S.R., FRET for studying intracellular signaling. *Trends Cell Biol,* 3, 242–245, 1993.

41. Liu, J., Zhang, Q., Remsen, E.E., and Wooley, K.L., Nanostructured materials designed for cell binding and transduction. *Biomacromolecules,* 2, 362–368, 2001.

42. Luedtke, N.W., Carmichael, P., and Tor, Y., Cellular uptake of aminoglycosides, guanidinoglycosides, and poly-arginine. *J Am Chem Soc* 125, 12374–12375, 2003.

43. Bambot, S.B., Sipior, J., Lakowicz, J.R., and Rao, G., Lifetime-based optical sensing of pH using resonance energy-transfer in sol-gel films. *Sensor Actuator B Chem,* 22, 181–188, 1994.

44. Neurauter, G., Klimant, I., and Wolfbeis, O.S., Microsecond lifetime-based optical carbon dioxide sensor using luminescence resonance energy transfer. *Anal Chim Acta* 382, 67–75, 1999.

45. (a) Thompson, R.B., Cramer, M.L., Bozym, R., and Fierke, C.A., Excitation ratiometric fluorescent biosensor for zinc ion at picomolar levels. *J Biomed Opt,* 7, 555–560, 2002; (b) Birch, D.J.S., Rolinski, O.J., and Hatrick, D., Fluorescence lifetime sensor of copper ions in water. *Rev Sci Instrum,* 67, 2732–2737, 1996.

46. Tsourkas, A., Behlke, M.A., Xu, Y., and Bao, G., Spectroscopic features of dual fluorescence/luminescence resonance energy-transfer molecular beacons. *Anal Chem,* 75, 3697–3703, 2003.

47. Ueberfeld, J. and Walt, D.R., Reversible ratiometric probe for quantitative DNA measurements. *Anal Chem,* 76, 947–952, 2004.

48. Wang, S., Gaylord, B.S., and Bazan, G.C., Fluorescein provides a resonance gate for FRET from conjugated polymers to DNA intercalated dyes. *J Am Chem Soc,* 126, 5446–5451, 2004.

49. Medintz, I.L., Goldman, E.R., Lassman, M.E., and Mauro, J.M., A fluorescence resonance energy transfer sensor based on maltose binding protein. *Bioconjugate Chem,* 14, 909–918, 2003.

50. Rodems, S.M., Hamman, B.D., Lin, C., Zhao, J., Shah, S., Heidary, D., Makings, L., Stack, J.H., and Pollok, B.A., A fret-based assay platform for ultra-high density drug screening of protein kinases and phosphatases. *Assay Drug Dev Technol,* 1, 9–19, 2002.

51. Lis, H. and Sharon, N., Lectins: carbohydrate-specific proteins that mediate cellular recognition. *Chem Rev,* 98, 637–674, 1998.
52. Goldstein, I.J., Hollerman, C.E., and Smith, E.E., Protein-carbohydrate interaction. II. Inhibition studies on the interaction of concanavalin A with polysaccharides. *Biochemistry,* 4, 876–883, 1965.

6 Carbonic Anhydrase-Based Biosensing of Metal Ions: Issues and Future Prospects

Richard B. Thompson, Ph.D., Rebecca A. Bozym, B.S., Michele L. Cramer, M.S., Andrea K. Stoddard, B.S., Nissa M. Westerberg, Ph.D., and Carol A. Fierke, Ph.D.

CONTENTS

6.1 INTRODUCTION

Over the past decade, fluorescence-based determinations of free metal ions in solution using variants of human apocarbonic anhydrase (apoCA) have been demonstrated to be highly sensitive, selective, and versatile. In particular, free divalent copper, zinc, cobalt, cadmium, and nickel have been determined at concentrations down to the picomolar range[1-3] by changes in fluorescence emission[4] and excitation wavelength ratios,[5] lifetimes,[6,7] and anisotropy (polarization).[8,9] A unique advantage of biosensing systems that use biomolecules as transducers over small molecule indicators was first demonstrated for this system, namely that the sensitivity,[10] selectivity,[11-13] analyte binding kinetics,[14-16] and potentially stability[17] may all be improved by subtle modification of the protein structure using standard molecular biology methods. Moreover, this reengineering may be done either by directed mutagenesis based on one's understanding of the protein structure and function or by quasi-random mutagenesis followed by screening for the desired properties.[16]

This development has resulted in fluorescence-based sensors for zinc and copper ions of unequalled sensitivity and selectivity. These attributes have permitted application of these sensors for problems otherwise beyond the state of the art. In particular, a carbonic anhydrase-based sensor was the first to obtain real-time measurements of free Cu(II) at picomolar levels in sea water[18]; importantly, the sensor was a fiber-optic sensor that permitted *in situ* measurements (see below). The immunity to interference from other divalent cations present at millimolar concentrations such as calcium or magnesium has permitted real-time measurement of extracellular free Zn(II) in the brains of living animals by dialysis probe (Frederickson et al., submitted) and fiber-optic sensor (Thompson et al., unpublished observations). Finally, the same attributes have also permitted for the first time the determination of free zinc ion at picomolar levels in ordinary cultured cells using a quantitative (excitation ratiometric) fluorescence microscopy method.[19]

Notwithstanding these achievements, there are still many outstanding questions of metallobiochemistry and analysis that could be addressed by improving or adapting the carbonic anhydrase-based sensor platform. Below is a series of issues that need to be addressed in the coming years to fully exploit this unique approach.

6.2   INTRACELLULAR IMAGING

We have overcome a key issue in using protein-based fluorescence sensors to measure intracellular analytes, namely the difficulty of getting something water soluble and as large as a protein across the cell membrane. Cells obviously can do this, and we adapted the discovery by Schwarze et al.[20] of the human immunodeficiency virus (HIV) TAT peptide to promote transmembrane importation of the protein; the success of this approach may be seen in Figure 6.1, which is a fluorescence micrograph of Alexa Fluor-labeled H36C apoCA in a PC-12 cell. Although this approach works well in isolated cultured cells, it clearly will be problematic in cultured tissues and whole organisms, particularly when one wishes to observe only a subset of the cells. Chalfie et al.[21] brilliantly initiated an approach to this problem by showing that the *Aequorea* green fluorescent protein (GFP) could be expressed in a transfected cell line or live organism to produce protein that emitted. Miyawaki et al.[22] showed that the fluorescent proteins could be fused at the DNA level with calmodulin to produce a calcium sensor transducer that could be expressed within a cell. This extraordinary achievement inspired other groups[23,24] to take the same approach to measure zinc, with only modest success in terms of sensitivity and signal change.

We have devised a somewhat different approach for expressing a zinc-sensitive fluorescent indicator, a modification of our excitation ratiometric approach[5] that is illustrated in Figure 6.2. Unlike calmodulin, carbonic anhydrase changes conformation almost imperceptibly upon binding zinc or other ions, so an approach predicated on structural change seemed unlikely. Similarly, the generally well-protected nature of the fluorescent moiety within the beta-barrel structure of

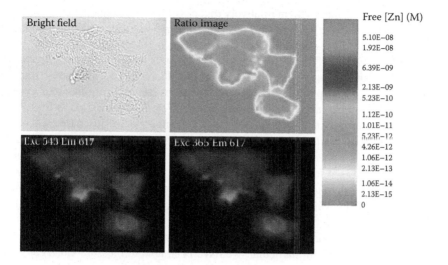

FIGURE 6.1 (See color insert following page 142.) Excitation ratiometric imaging of intracellular free zinc ion. Upper left is a bright field image of PC-12 cells; lower left is a fluorescence image of the same field with excitation at 543 nm and emission at 617 nm; lower right is a fluorescence image with excitation at 365 nm and emission at 617 nm; and upper right is a false color image (scale on right) of the ratio of the two lower images.

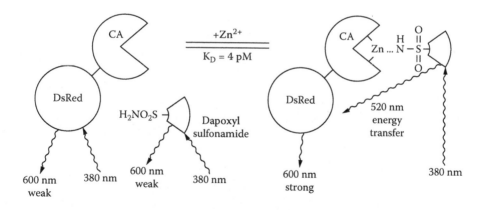

FIGURE 6.2 Scheme for expressible excitation ratiometric quantitation of free zinc. In the absence of zinc, Dapoxyl sulfonamide does not bind to DsRed-apoCA, and one observes weak emission at 600 nm with excitation at 380 nm from both the Dapoxyl and DsRed. Binding of Zn(II) to the apoCA promotes binding of Dapoxyl to the protein, which is accompanied by a large blue shift and quantum yield increase; because of its proximity to DsRed, energy transfer from Dapoxyl to DsRed occurs and one now observes strong emission at 600 nm. The signal is normalized to the emission of DsRed directly excited at 570 nm.

fluorescent proteins makes them relatively inaccessible to most quenchers (as exemplified by the quenching of the beta-barrel protein apoazurin[25]), which accounts for their high fluorescence efficiency, but makes them poor transducers for the most part. This is particularly true for zinc ion, which is often innocuous from the standpoint of quenching or otherwise perturbing fluorescence; the energy transfer approach (Figure 6.2) exploits the propensity of fluorescent aryl sulfonamides to bind to carbonic anhydrase (and thus be held near the protein) only if zinc is present in the active site, an approach we have used frequently.[3–5,9] In this case, when the fluorescent aryl sulfonamide Dapoxyl sulfonamide is bound (which ordinarily shifts its emission and increases its quantum yield 20-fold), it efficiently transfers its energy to the DsRed2[26] fused to the apoCA, and thus zinc binding is accompanied by an increase in the ultraviolet (UV)-excitable red fluorescence of the protein. In fact, this approach works well *in vitro* (Figure 6.3), and we expect to try it in bacteria and yeast in the near future.

The carbonic anhydrase-based sensing platform is also extremely sensitive for measurements of Cu(II). In particular, we have been able to measure Cu(II) at picomolar levels by changes in fluorescence intensity, lifetime, and anisotropy. In principle, any of these modes may be used to image copper in the cell because microscopes employing all three have been constructed and (except for anisotropy) are available commercially. Cu(II) is of interest in biology because of its ubiquity and high toxicity,[27] but it is of particular interest in biomedicine because of its central role in Menkes' and Wilson's diseases and potentially in Alzheimer's disease and the prion-mediated encephalopathies such as Creutzfeldt-Jakob and "Mad Cow" disease.[28,29] O'Halloran's group predicted that intracellular free copper levels are extremely low (less than one

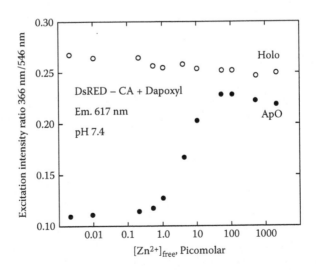

FIGURE 6.3 Calibration curve for zinc determination by the expressible fluorescence excitation ratiometric biosensor method. The ratio of emission at 617 nm excited at 365 nm to that excited at 543 nm is depicted as a function of free zinc concentration for apoCA (filled circles) and holocarbonic anhydrase (control, open circles).

ion per cell on average),[30] but this proposal remains to be tested. Indeed, although measures of total copper in many tissues and bodily fluids exist, there seem to be few measurements of free (and therefore biologically active) copper ion in any of these systems, particularly *in vivo*.

6.3 REMOTE MEASUREMENT

For some time we have pursued integrating the carbonic anhydrase based sensor transducer into a fiber-optic sensor. In a fiber-optic sensor, the fiber optic conducts exciting light to and fluorescence emission from the sensor transducer to the rest of the instrument. Such sensors permit analysis *in situ* of matrices that are remote, difficult to sample, or present in toxic or hazardous environments. In a very real sense, one is bringing the analysis to the sample rather than bringing a sample to the analytical instrument.[31,32] An illustration of such a sensor is shown in Figure 6.4; other optical configurations have been used by others, but we find the advantages of the single fiber system are substantial and compelling.[33] We have used such a sensor (measuring intensities and lifetimes) to quantitate free Cu(II) in sea water, *in situ*, at picomolar levels in real time.[18] This experiment showed that such

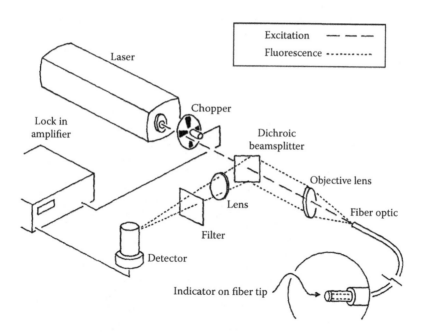

FIGURE 6.4 Schematic of fluorescence-based fiber-optic biosensor. Excitation from the laser (dashed line) passes through the dichroic beam splitter and is launched into the optical fiber, where it excites fluorescence in the distal tip. Fluorescence is coupled back into the fiber, is reflected off the dichroic beam splitter through the emission filter, and is collected by the detector. For emission ratiometric determinations, the filter is mounted in a wheel for rapid changing.

measurements were feasible, but several issues remain for application of this technology to other problems.

The measurements in Zeng et al.[18] were made in near-surface waters with only 25 m of fiber. For oceanographic applications especially, being able to measure through much longer lengths of fiber (kilometers) is of compelling interest, for two reasons. First, retrieving discrete samples from depths greater than a few meters for shipboard analysis is slow and costly, provides few data points per day, and requires great care to avoid contamination. Using a string of fiber sensors at varying depths would permit continuous (or quasi-continuous) monitoring of an analyte (kinetics permitting) during a transect (Figure 6.5). Clearly, for many analytes one would like to measure at depths measured in kilometers. Also, for coastal applications where, in addition to natural hazards, a buoy-mounted sensor may be hit by marine traffic or vandalized, it would be desirable to keep the expensive, fragile portion of the sensor safely on shore and expose only a length of fiber and inexpensive transducer to these hazards.

Measurements through much longer lengths of fiber are principally limited by two factors: optical attenuation and modal dispersion of the fiber.[32,34] Optical attenuation of the fiber is simply the amount of light transmitted by the fiber, which is strongly wavelength dependent. Fiber attenuation at visible to near-infrared wavelengths is due to scattering and absorption processes and for silica fibers is at a minimum (<1 dB/km, where $dB = -10\log I_{in}/I_{out}$) in the so-called telecommunications window, about 1330 nm and 1550 nm, which makes long-haul fiber-optic telecommunications possible.

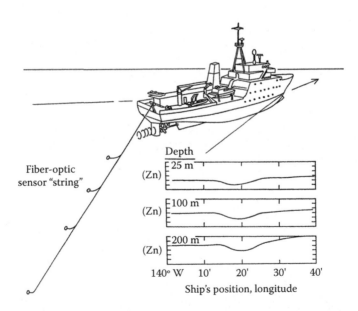

FIGURE 6.5 Fiber-optic sensor array for marine survey. The ship tows an array of fiber-optic sensors behind it as it moves, with the sensing tips at different depths in the water column; the inset (lower right) shows notional zinc levels measured at a series of depths as the ship transects the ocean.

At wavelengths in the red to near-infrared range (say 600 nm to 900 nm), attenuation is still low enough to permit measurements (<10 dB/km). The measurements of Zeng et al.[18] were done at 650 nm to 700 nm with 0.5 mW of excitation power. Increasing the laser power by a factor of 50 (using a commercially available source) and working at 780 nm to 830 nm with longer wavelength fluorescent labels (such as those offered commercially by LiCor) should also improve attenuation 100-fold, so that we can anticipate measuring through multi-kilometer lengths of fiber.

The other limiting factor is modal dispersion, which expresses the tendency of multimode fiber to conduct high-frequency optical signals (like those used for lifetime determination) without demodulating them. As expected, this is a big issue for high-frequency (gigabit per second) telecommunications, which is circumvented in most cases by the use of single-mode fibers that have no modal dispersion. The gradient index multimode fiber used in Zeng et al.[18] was specified to have a 3-dB bandwidth distance product of 100 MHz/km at 850 nm. For multikilometer measurements at 200 MHz, this is probably inadequate and single-mode fiber will be necessary; unfortunately, launching light with good efficiency into the small core of a single-mode fiber is more difficult, and the effective size of the transducer is reduced, limiting signal. Nevertheless, we hope to make such multikilometer measurements in the near future.

6.4 TRANSDERMAL IMAGING

Although we have demonstrated *in vivo* determination of free zinc in the brain of an animal using an optical fiber carbonic anhydrase-based sensor, the fiber sensor perforce "sees" only a relatively small voxel near the fiber tip and provides essentially no spatial information. By comparison, the intracellular measurements depicted in Figure 6.1 indicate the spatial distribution within the cultured cells, but this circumstance is not a true simulacrum of tissue. What one would prefer to do is image zinc levels in tissues *in situ* through the skin of the animal and, ideally, deep within the body. The high scattering coefficient of tissues such as skin and brain[35] as well of the absorbance of naturally occurring tissue components such as hemoglobin, flavins, and melanin make this technically quite challenging.

Nevertheless, some initial steps have been taken toward imaging fluorescence from sites within organisms such as nude mice.[36] It was recognized early on that longer wavelengths in the near-infrared range, above 600 nm, suffered substantially less absorbance and scattering, and this fueled interest in fluorophores absorbing and emitting at longer wavelengths.[37] The introduction of multiphoton excitation to fluorescence microscopy made several workers realize that multiphoton excitation (wherein a focused infrared laser beam consisting preferably of very brief pulses is able to excite fluorophores that only absorb at substantially shorter wavelengths[38]) may also permit transdermal fluorescence measurements by bringing the beam to a focus beneath the skin. These early results indicate the promise of the technique it will require substantial development before the technique can be adapted to carbonic anhydrase-based sensing.

6.5 LOWER DETECTION LIMITS

We can improve the detection limit for determination of various metal ions by different means. We have already shown that CA variants with higher affinity than the wild-type protein for zinc[39] and copper,[11,13] and potentially other metal ions, can be constructed. In the case of copper, variants are in hand with affinities as tight as 100 attomolar, which seems as tight as will be necessary for a practical sensor. In particular, the dissociation rate constant of such a binding site would, of necessity, be so slow (even with association as fast as that of diffusion-controlled association) that it would take some days for the sensor to equilibrate in the absence of catalysis. Although such a high-affinity site might be useful for capture-based assays, it would be of little use for continuous monitoring.

Whereas the variants with high affinity for copper can most likely be incorporated into sensor transducers with modest effort, the high-affinity zinc variant achieves its tight binding by adding a fourth ligand to the three provided by the wild-type binding site; with four coordination sites occupied, sulfonamide-based fluorescence transduction is infeasible. Moreover, the histidinyl ligands of the wild type appear almost ideally disposed for tetrahedral binding of zinc, so that it would appear difficult to engineer tighter binding with these three ligands. More exotic options include replacement of the imidazoles with other (perhaps abiotic) ligands. Such attempts also beg the question of the utility of metal ion sensors with much higher sensitivities, because even with diffusion-controlled association constants, their off-rates perforce must be very slow and call into question their utility as continuous sensors (vs. as alarms that are able to only detect an increase in zinc concentration).

6.6 INEXPENSIVE SENSORS

Although substantial progress has been made in developing fluorescence instrumentation (especially time-resolved instrumentation), which has greatly reduced costs,[40,41] fluorescence-based determinations still require fairly sophisticated hardware to get quantitative results, and especially to realize the sensitivity potential of the technique. However, there are somewhat less demanding applications that might benefit from a very cheap sensor or a test kit for discrete samples. In particular, home test kits that indicate some threshold level of analyte (such as human chorionic gonadotropin for pregnancy) by the appearance of a colored symbol have shown that even hormones at nanomolar levels and below may be detected without instruments. The simplest optical techniques for chemical determination have employed color changes, with litmus or pH paper being the canonical example. Although there are colorimetric indicators for Zn(II) and other metal ions such as pyridylazoresorcinol, they are neither very specific nor very sensitive: under ideal conditions, an extinction coefficient change from zero to 10^5/mol/cm at some wavelength would result in an appreciable change in absorbance only if the indicator were present at concentrations of perhaps 50 nM or more. The extinction coefficient changes some sulfonamides undergo[42] upon binding to holocarbonic anhydrase, or the d–d absorbance band shifts exhibited by Cu(II), Co(II), and Ni(II) upon binding to apoCA,[43] are far more modest by comparison and consequently less useful in this regard. Other investigators[44] have described quantitative fluorescence polarization assays done without instruments by simply using visual comparisons by the user. Overall, we view simpler assays as an important, yet underserved, objective for carbonic anhydrase-based determinations in particular and biosensors in general.

6.7 STABILITY

For many applications of carbonic anhydrase-based sensors, the stability of the transducer is a key issue.[43] In particular, we found that our copper ion sensor in sea water lost half its response within 4 h. Although we have improved upon this substantially since then (H. Zeng and R. Thompson, unpublished data), others have also attempted to improve the stability of carbonic anhydrase by engineering in a disulfide cross-link17,46 or including additives in media.[47]. It remains beyond the state of the art to design a carbonic anhydrase *de novo* with high stability, but adaptation of structural motifs by directed evolution may offer improvements. An important shortcoming of most classical carbonic anhydrases is their instability and reduced metal ion affinity at pH levels less than 6, which may limit their sensor applications. Directed evolution methods (repetitive cycles of mutagenesis and selection steps) have proven quite successful at preparing mutants with altered stability under a variety of applications. Therefore, it is likely that a carbonic anhydrase variant that is stable at low pH could be selected for, although at present there are no applications that demand this performance.

6.8 DETERMINATION OF OTHER ANALYTES

Heretofore we have focused on divalent cation analytes known to bind to the wild-type apoenzyme with appreciable affinity: copper, zinc, nickel, cobalt, and cadmium. However, there seems to be no *ab initio* reason why the carbonic anhydrase binding site could not be further modified to accept other cations. Analytes of particular interest include Pb(II), Mn(II), and Fe(II); the last may already bind, but it is troublesome to measure because of its lability in oxygenated solution. Although we demonstrated carbonic anhydrase-based determination of certain anions some time ago,[48] neither the affinity nor the response to these analytes was especially striking, and other approaches seem more appealing. Trivially, carbonic anhydrase may also act as a sensor for its other inhibitors, such as aryl sulfonamides, but these are of modest interest analytically.

6.9 CONCLUSION

Carbonic anhydrase-based sensing has been shown to be a viable approach for determining certain divalent cations and anions in some important applications, and we believe it can serve as an archetype for other fluorescence-based sensors. The recent invention by Looger et al.[49] of what appears to be a general approach for developing ligands and sensors for a wide range of potential (small molecule) analytes suggests that protein-based sensors can be devised for those many analytes for which classical fluorescent indicators do not exist. Particularly in biological systems, the host of metabolites, signaling molecules, biosynthetic intermediates, and breakdown products of potential interest from both research and therapeutic standpoints numbers at least several hundred. Inasmuch as many metabolic changes and disease processes need not be accompanied by distinguishing changes in gene expression, the new tools for studying such expression may be less informative or useful for pinpointing therapeutic targets. This is likely to be especially true for rapid responses to trauma and other overt insults, when there is no time for protein synthesis. Although we have found several ways to transduce the binding of analytes as a change in a fluorescence observable, these approaches are not necessarily general, and a reliable means of transduction for an arbitrary analyte remains to be devised.

ACKNOWLEDGMENTS

The authors wish to thank the Office of Naval Research, National Science Foundation, and National Institute of Biomedical Imaging and Bioengineering for their support.

REFERENCES

1. Fierke, C.A. and Thompson, R.B., Fluorescence-based biosensing of zinc using carbonic anhydrase. *BioMetals,* 14, 205–222, 2001.
2. Thompson, R.B., Maliwal, B.P., Feliccia, V.L., Fierke, C.A., and McCall, K., Determination of picomolar concentrations of metal ions using fluorescence anisotropy: biosensing with a "reagentless" enzyme transducer. *Anal Chem,* 70, 4717–4723, 1998.
3. Thompson, R.B. and Jones, E.R., Enzyme-based fiber optic zinc biosensor. *Anal Chem,* 65, 730–734, 1993.
4. Thompson, R.B., Whetsell, W.O. Jr., Maliwal, B.P., Fierke, C.A., and Frederickson, C.J., Fluorescence microscopy of stimulated Zn(II) release from organotypic cultures of mammalian hippocampus using a carbonic anhydrase-based biosensor system. *J Neurosci Methods,* 96, 35–45, 2000.
5. Thompson, R.B., Cramer, M.L., Bozym, R., and Fierke, C.A., Excitation ratiometric fluorescent biosensor for zinc ion at picomolar levels. *J Biomed Opt,* 7, 555–560, 2002.
6. Thompson, R.B. and Patchan, M.W., Lifetime-based fluorescence energy transfer biosensing of zinc. *Anal Biochem,* 227, 123–128, 1995.
7. Thompson, R.B., Ge, Z., Patchan, M.W., Huang, C.-C., and Fierke, C.A., Fiber optic biosensor for Co(II) and Cu(II) based on fluorescence energy transfer with an enzyme transducer. *Biosens Bioelectron,* 11, 557–564, 1996.
8. Elbaum, D., Nair, S.K., Patchan, M.W., Thompson, R.B., and Christianson, D.W., Structure-based design of a sulfonamide probe for fluorescence anisotropy detection of zinc with a carbonic anhydrase-based biosensor. *J Am Chem Soc,* 118, 8381–8387, 1996.
9. Thompson, R.B., Maliwal, B.P., and Zeng, H.H., Zinc biosensing with multiphoton excitation using carbonic anhydrase and improved fluorophores. *J Biomed Opt,* 5, 17–22, 2000.

10. Kiefer, L.L., Paterno, S.A., and Fierke, C.A., Hydrogen bond network in the metal binding site of carbonic anhydrase enhances zinc affinity and catalytic efficiency. *J Am Chem Soc,* 117, 6831–6837, 1995.

11. Hunt, J.A., Ahmed, M., and Fierke, C.A., Metal binding specificity in carbonic anhydrase is influenced by conserved hydrophobic amino acids. *Biochemistry,* 38, 9054–9060, 1999.

12. DiTusa, C.A., McCall, K.A., Chritensen, T., Mahapatro, M., Fierke, C.A., and Toone, E.J., Thermodynamics of metal ion binding. 2. Metal ion binding by carbonic anhydrase variants. *Biochemistry,* 40, 5345–5351, 2001.

13. McCall, K.A. and Fierke, C.A., Probing determinants of the metal ion selectivity in carbonic anhydrase using mutagenesis. *Biochemistry,* 43, 3979–3986, 2004.

14. Kiefer, L.L., Paterno, S.A., and Fierke, C.A., Second shell hydrogen bonds to histidine ligands enhance zinc affinity and catalytic efficiency. *J Am Chem Soc,* 117, 6831–6837, 1995.

15. Huang, C.-C., Lesburg, C.A., Kiefer, L.L., Fierke, C.A., and Christianson, D.W., Reversal of the hydrogen bond to zinc ligand histidine-119 dramatically diminishes catalysis and enhances metal equilibration kinetics in carbonic anhydrase II. *Biochemistry,* 35, 3439–3446, 1996.

16. Hunt, J.A. and Fierke, C.A., Selection of carbonic anhydrase variants displayed on phage: aromatic residues in zinc binding site enhance metal affinity and equilibration kinetics. *J Biol Chem,* 272, 20364–20372, 1997.

17. Burton, R.E., Hunt, J.A., Fierke, C.A., and Oas, T.G., Novel disulfide engineering in human carbonic anhydrase II using the PAIRWISE side-chain geometry database. *Protein Sci,* 9, 776–785, 2000.

18. Zeng, H.H., Thompson, R.B., Maliwal, B.P., Fones, G.R., Moffett, J.W., and Fierke, C.A., Real-time determination of picomolar free Cu(II) in seawater using a fluorescence-based fiber optic biosensor. *Anal Chem,* 75, 6807–6812, 2003.

19. Bozym, R.A., Zeng, H.H., Cramer, M., Stoddard, A., Fierke, C.A., and Thompson, R.B., *In vivo* and intracellular sensing and imaging of free zinc ion. In: *Proceedings of the SPIE Conference on Advanced Biomedical and Clinical Diagnostic Systems II,* Cohn, G.E., Grundfest, W.S., Benaron, D.A., and Vo-Dinh, T., eds. Bellingham, WA: SPIE, 2004.

20. Schwarze, S.R., Ho, A., Vocero-Akbani, A., and Dowdy, S.F., *In vivo* protein transduction: delivery of a biologically active protein into the mouse. *Science,* 285, 1569–1572, 1999.

21. Chalfie, M., Prasher, D., and Ward, W., GFP as expression probe. *Science,* 263, 802–805, 1994.

22. Miyawaki, A., Llopis, J., Heim, R., McCaffery, J.M., Adams, J.A., Ikura, M., and Tsien, R.Y., Fluorescent indicators for Ca^{2+} based on green fluorescent proteins and calmodulin. *Nature,* 388, 882–887, 1997.

23. Pearce, L.L., Gandley, R.E., Han, W., Wasserloos, K., Stitt, M., Kanai, A.J., McLaughlin, M.K., Pitt, B.R., and Levitan, E.S., Role of metallothionein in nitric oxide signaling as revealed by a green fluorescent fusion protein. *Proc Natl Acad Sci USA,* 97, 477–482, 2000.

24. Jensen, K.K., Martini, L., and Schwartz, T.W., Enhanced fluorescence resonance energy transfer between spectral variants of green fluorescent protein through zinc-site engineering. *Biochemistry,* 40, 938–945, 2001.

25. Eftink, M.R., Fluorescence quenching: theory and applications. In: *Topics in Fluorescence Spectroscopy,* vol. 2, *Principles,* Lakowicz, J.R., ed. New York: Plenum Press, 1991:53–126.

26. Campbell, R.E., Tour, O., Palmer, A.E., Steinbach, P.A., Baird, G.S., Zacharias, D.A., and Tsien, R.Y., A monomeric red fluorescent protein. *Proc Natl Acad Sci USA,* 99, 7877–7882, 2002.

27. Linder, M.C., *Biochemistry of Copper.* New York: Plenum Press, 1991.

28. Prince, R.C. and Gunson, D.E., Prions are copper-binding proteins. *Trends Biochem Sci,* 23, 197–198, 1998.

29. Cherny, R.A., Atwood, C.S., Xilinas, M.E., Gray, D.N., Jones, W.D., McLean, C.A., Barnham, K.J., Volitakis, I., Fraser, F.W., Kim, Y.S., et al., Treatment with a copper–zinc chelator markedly and rapidly inhibits beta-amyloid accumulation in Alzheimer's disease transgenic mice. *Neuron,* 30, 665–676, 2001.

30. Rae, T.D., Schmidt, P.J., Pufahl, R.A., Culotta, V.C., and O'Halloran, T.V., Undetectable intracellular free copper: the requirement of a copper chaperone for superoxide dismutase. *Science,* 284, 805–808, 1999.

31. Wolfbeis, O.S., ed., *Fiber Optic Chemical Sensors and Biosensors.* Boca Raton, FL: CRC Press, 1991.

32. Thompson, R.B., Fiber optic chemical sensors. *IEEE Proc Circuits Devices,* CD-10, 14–21, 1994.

33. Thompson, R.B., Levine, M., and Kondracki, L., Component selection for fiber optic fluorometry. *Appl Spectrosc,* 44, 117–122, 1990.

34. Snyder, A.W. and Love, J.D., *Optical Waveguide Theory.* New York: Chapman & Hall, 1983.

35. Cheong, W.F., Prahl, S.A., and Welch, A.J., A review of the optical properties of biological tissues. *IEEE J Quantum Electron,* QE-26, 2166–2185, 1990.

36. Fisher, G., Ballou, B., Srivastava, M., and Farkas, D.L., Far-red fluorescence-based high specificity tumor imaging *in vivo. Biophys J,* 70, 212A, 1996.

37. Thompson, R.B., Red and near-infrared fluorometry. In: *Topics in Fluorescence Spectroscopy,* vol. 4, *Probe Design and Chemical Sensing,* Lakowicz, J.R., ed. New York: Plenum Press, 1994:151–181.

38. Denk, W., Strickler, J.H., and Webb, W.W., Two-photon laser scanning fluorescence microscopy. *Science,* 248, 73–76, 1990.

39. Ippolito, J.A., Baird, T.T., McGee, S.A., Christianson, D.W., and Fierke, C.A., Structure-assisted redesign of a protein–zinc binding site with femtomolar affinity. *Proc Natl Acad Sci USA,* 92, 5017–5021, 1995.

40. Sipior, J., Carter, G.M., Lakowicz, J.R., and Rao, G., Single quantum well light emitting diodes demonstrated as excitation sources for nanosecond phase-modulation fluorescence lifetime measurements. *Rev Sci Instrum,* 67, 3795–3798, 1996.

41. Harms, P., and Rao, G., Low-cost phase-modulation fluorometer for measuring nanosecond lifetimes using a lock-in amplifier. In: *SPIE Conference on Advances in Fluorescence Sensing Technology IV,* Lakowicz, J.R., Soper, S., and Thompson, R.B., eds. San Jose, CA: SPIE, 1999:52–59.

42. Einarsson, R. and Zeppezauer, M., Visible absorption and circular dichroism spectra of 5(4-sufamylphenylazo)-8-hydroxyquinoline bound to carbonic anhydrase and alcohol dehydrogenase. *Acta Chem Scand,* 24, 1098–1102, 1970.

43. Lindskog, S. and Nyman, P.O., Metal-binding properties of human erythrocyte carbonic anhydrases. *Biochim Biophys Acta* 85, 462–474, 1964.

44. Lakowicz, J.R., Gryczynski, I., Gryczynski, Z., Tolosa, L., Dattelbaum, J.D., and Rao, G., Polarization-based sensing with a self-referenced sample. *Appl Spectrosc,* 53, 1149–1157, 1999.

45. Thompson, R.B., Zeng, H.H., Loetz, M., and Fierke, C., Issues in enzyme-based metal ion biosensing in complex media. In: *In-vitro Diagnostic Instrumentation,* Cohn, G.E., ed. San Jose, CA: SPIE, 2000:120–127.

46. Martensson, L.G., Karlsson, M., and Carlsson, U., Dramatic stabilization of the native state of human carbonic anhydrase by an engineered disulfide bond. *Biochemistry,* 41, 15867–15875, 2002.

47. Cleland, J.L., Hedgepeth, C., and Wang, D.I.C., Polyethylene glycol enhanced refolding of bovine carbonic anhydrase. *J Biol Chem,* 267, 13327–13334, 1992.

48. Thompson, R.B., Lin, H.-J., Ge, Z., Johnson, K., and Fierke, C., Fluorescence lifetime-based determination of anions using a site-directed mutant enzyme transducer. In: *Advances in Fluorescence Sensing Technology III,* Thompson, R.B., ed. San Jose, CA: SPIE, 1997:247–257.

49. Looger, L.L., Dwyer, M.A., Smith, J.J., and Hellinga, H.W., Computational design of receptor and sensor proteins with novel functions. *Nature,* 423, 185–190, 2003.

7 Metal-Enhanced Fluorescence Sensing

Chris D. Geddes, Ph.D., Kadir Aslan, Ph.D.,
Ignacy Gryczynski, Ph.D., and Joseph R.
Lakowicz, Ph.D.

CONTENTS

7.1 INTRODUCTION

Over the last 15 years, fluorescence has become the dominant detection/sensing technology in medical diagnostics and biotechnology. Although fluorescence provides high sensitivity, there exists the need for ever-reduced detection limits and small copy number detection. Fluorophore detectability is usually limited by autofluorescence of the samples and the photostability of the fluorophores. In an effort to obtain increased fluorescence sensitivity, we have recently investigated the use of metallic surfaces or nanoparticles to favorably modify the spectral properties of fluorophores.[1–4] The presence of a nearby metal surface (m) can favorably increase the radiative decay rate by addition of a new rate Γ_m (Figure 7.1, right). The metallic surface can cause Forster-like quenching with a rate k_m, can concentrate the incident field E_m, and can, importantly for sensing applications, increase the radiative decay rate Γ_m. These new phenomena typically occur at different distances from the metal surface, as depicted in Figure 7.2. As the value of Γ_m, that is, the spontaneous rate at which a fluorophore emits photons, increases, the quantum yield increases while the lifetime decreases. This is unusual to most fluorescence spectroscopists, as the quantum yield and lifetime usually change in unison.

To illustrate this point, we calculated the lifetime and quantum yield for fluorophores with an assumed natural lifetime, $\tau_N = 10$ ns, $\Gamma = 10^8$/sec, and various values for the nonradiative decay rates and quantum yields. The values of k_{nr} ranged from 0 to 9.9×10^7/sec, resulting in quantum yields from 1.0 to 0.01. Suppose now the metal results in increasing values of Γ_m. Since Γ_m is a rate process returning the fluorophore to the ground state, the lifetime decreases as Γ_m becomes comparable and larger than Γ (Figure 7.3, left). In contrast, as Γ_m/Γ increases, Q_m increases, but no change is observed for fluorophores where $Q_0 = 1$ (Figure 7.3, right).

As a result of these calculations, we predicted that the metallic surfaces could create new unique fluorophores with increased quantum yields and shorter lifetimes. Figure 7.4 illustrates that the presence of a metal surface near a fluorophore with low quantum yield ($Q_0 = 0.01$) increases its quantum yield approximately

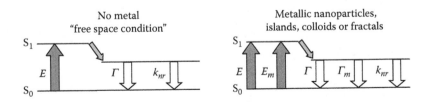

FIGURE 7.1 Classic Jablonski diagram for the free-space condition and the modified form in the presence of metallic particles, islands, colloids, or silver nanostructures. E, excitation; E_m, metal-enhanced excitation rate; Γ_m, radiative rate in the presence of metal. For our studies, we do not consider the effects of metals on k_{nr}.

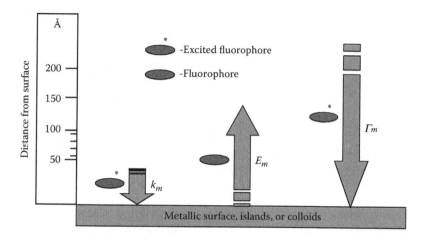

FIGURE 7.2 Predicted distance dependencies for a metallic surface on the transitions of a fluorophore. Γ_m modifications typically occur at distances greater than 50 Å from the metal surface.

10-fold, resulting in brighter emission, while reducing its lifetime 10-fold, resulting in an enhanced photostability of the fluorophore due to spending less time in an excited state, that is, less time for oxidation and other processes.

We speculated that the properties of the fluorophores with different quantum yields would be affected differently by the presence of nearby metal due to the difference in radiative rate modifications, with a view to understanding metal–fluorophore interactions for sensing applications. We tested this hypothesis by investigating the changes in emission properties of two fluorophores with similar

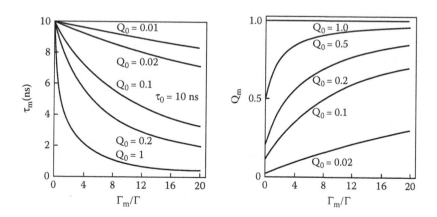

FIGURE 7.3 The effect of an increase in radiative decay rate (Γ_m/Γ) on the lifetime and quantum yield of a fluorophore.

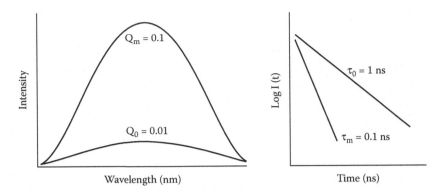

FIGURE 7.4 Metallic surfaces can create unique fluorophores with high quantum yields and short lifetimes. These are likely to find multifarious applications in MEF sensing.

absorption/emission spectra but with different quantum yields (rhodamine B, quantum yields $Q_0 = 0.48$; Rose Bengal, $Q_0 = 0.02$). Figure 7.5 summarizes our observations with these fluorophores. The emission from rhodamine B on silver island films (SiFs) was approximately 20% more than the emission from the glass side. On the other hand, emission from Rose Bengal on SiFs was approximately five times that on the glass side. In these experiments, the fluorophores were placed between two quartz plates (approximately half of the plates were coated with SiFs). We estimated the distance between the plates to be about 1 μm, and such a configuration of the samples results in only a small fraction of the fluorophores being present within the distance over which metallic surfaces can exert effects (cf. Figure 7.2). The region of varying photonic mode density (PMD) is expected to extend about 200 Å into the solution. Hence, only about 4% of the liquid volume between the plates is estimated to be within this active volume. This suggests that the fluorescence intensity of Rose Bengal within 200 Å of the islands is effectively increased 125-fold.

The emission spectra in Figure 7.5 do not solely demonstrate an increase in the radiative decay rate. However, the demonstration of such an increase can be done by additional intensity decay (lifetime) measurements. Figure 7.6 shows that the intensity decay is more rapid near SiFs. We interpret this decrease in lifetime to be due to the fluorophore's interacting with the metal surface. Hence, an increase in intensity accompanied by a reduction in lifetime may be explained only by a radiative rate modification.

The results from rhodamine B and Rose Bengal (Figures 7.5 and 7.6) were consistent with our expectations that the presence of metal increases the emission intensity (quantum yield) and decreases the lifetime of the fluorophores. Nonetheless, one could be concerned with possible artifacts due to dye binding to the surfaces or other unknown effects. For this reason, we examined a number of additional fluorophores between uncoated quartz plates and between SiFs. In all cases, the

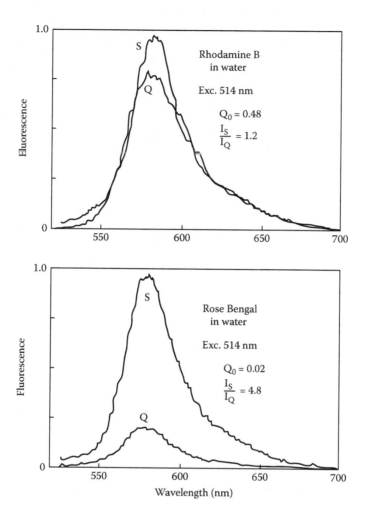

FIGURE 7.5 The effect of SiFs on the emission spectra of rhodamine B (top) and Rose Bengal (bottom). These probes were chosen to demonstrate MEF, as they both have similar spectral characteristics but different quantum yields. (Adapted from Lakowicz et al.[2])

emission was more intense for the solution between the silver islands. For example, [Ru(bpy)$_3$] and [Ru(phen)$_2$dppz] have quantum yields near 0.02 and 0.001, respectively. A larger enhancement was found for [Ru(phen)$_2$dppz] than for [Ru(bpy)$_3$] (data not shown). The enhancements for 10 different fluorophore solutions are shown in Figure 7.7. In all cases, lower bulk-phase quantum yields result in larger enhancements for samples between SiFs. We rationalized that the enhancement factor is approximately $1/Q_0$, where Q_0 is the quantum yield (intensity) in the absence of metal.

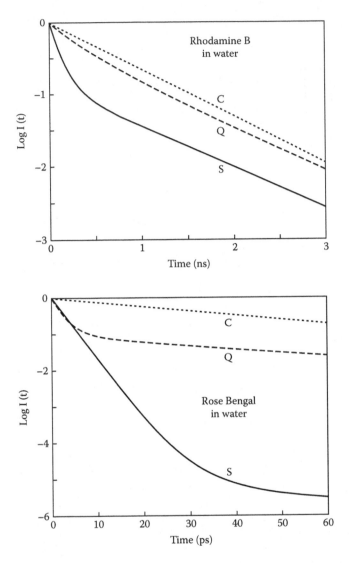

FIGURE 7.6 Intensity decays for rhodamine B (top) and Rose Bengal (bottom) in a cuvette (C), between quartz slides (Q), and between silvered slides (S). (Adapted from Lakowicz et al.[2])

7.2 NOBLE-METAL SUBSTRATES FOR METAL-ENHANCED FLUORESCENCE WITH APPLICATION TO THE ENHANCED SPECTRAL PROPERTIES OF INDOCYANINE GREEN AND FLUORESCEIN

We have shown that the proximity of fluorophores to metallic silver particles results in increased intensities, quantum yields, and decreased lifetimes (Figures 7.1–7.7). These results are consistent with increased radiative decay rates of the fluorophores, Γ, induced

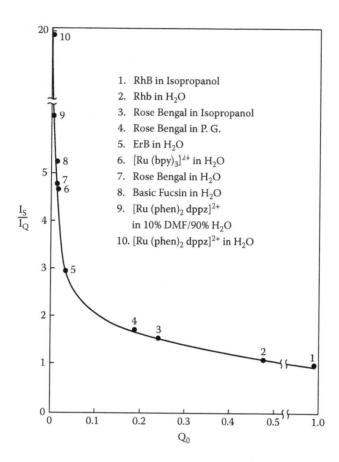

FIGURE 7.7 The effect of SiFs on the quantum yields of both high and low quantum yield fluorophores. I_S, intensity between silvered plates; I_Q, intensity between quartz plates (control sample). (Adapted from Lakowicz et al.[2])

by an interaction with these metallic surfaces. These effects have been predicted theoretically[5-7] and are analogous to the increases in Raman signals observed for surface-enhanced Raman scattering (SERS), except that the effects on fluorescence are due to through-space interactions, which do not require molecular contact between the fluorophores and the metal. For use in medical and biotechnology applications, such as diagnostic or microfluidic devices, it would be useful to obtain metal-enhanced fluorescence (MEF) at desired locations in the measurement device, that is, MEF on demand. We subsequently investigated various methods for the preparation of silver nanostructures and used indocyanine green (ICG) and fluorescein-labeled proteins and oligonucleotides (Figure 7.8) to test our new potential-sensing platforms; we chose these fluors because of their widespread use in medical applications and their U.S. Food and Drug Administration (FDA) approval for use in humans.

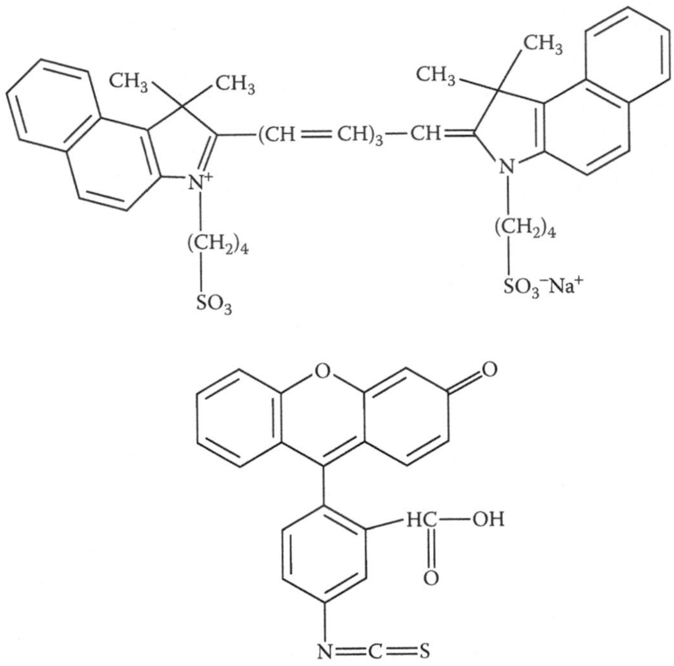

FIGURE 7.8 Molecular structure of ICG (top) and fluorescein (bottom). These probes were chosen to demonstrate MEF because they are FDA approved for use in humans.

7.2.1 SiFs for MEF

In recent years we have been reporting our observations on the favorable effects of silver nanoparticles deposited randomly on glass substrates (SiFs) for increasing the intensity and photostability of fluorophores, particularly those with low quantum yields. A typical absorption spectrum of SiFs, which display an absorption maximum near 430 nm, is given in Figure 7.9 (top). Figure 7.9 (top) also shows a schematic for a fluorophore between two SiFs. We have investigated the effects of SiFs on the properties of ICG when bound to human serum albumin (HSA). The emission spectra of ICG-HSA bound to quartz and SiFs are shown in Figure 7.9 (bottom). The intensity of ICG is increased approximately 10-fold on the SiFs when compared with quartz. The spectral shape is similar on quartz and SiFs. We found the same amount of increase in the emission of ICG whether the surfaces were coated with HSA, which already contains bound ICG, or if the surfaces were first coated with HSA followed by exposure to a dilute solution of ICG. From ongoing studies of albumin-coated surfaces, we estimated that the same amount of HSA binds to each surface, with the difference in binding being less than a factor of two. Hence, the observed increase in the intensity on SiFs is not due to increased ICG–HSA binding but, rather, is due to a change in the quantum yield and/or rate of excitation of ICG near metallic silver nanoparticles.[8]

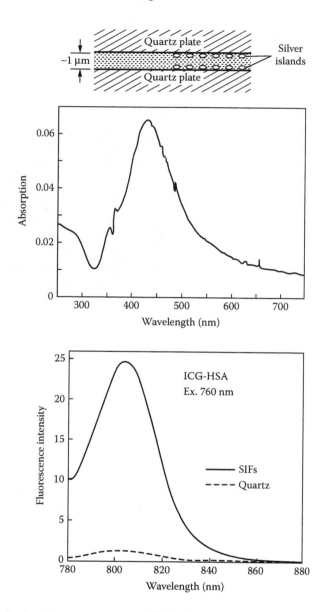

FIGURE 7.9 Typical absorption-spectrum SiFs (top) and schematic for a fluorophore solution between two SiFs (inset). Fluorescence spectra of ICG–HSA on SiFs and on quartz (bottom). (Adapted from Malicka et al.[8])

7.2.2 Silver Colloid Films for MEF

In contrast to SiFs, where there is little control on the size of the silver structures deposited on glass slides, the preparation of colloidal suspensions of silver is rather standard and easily controlled to yield homogeneously sized spherical silver particles.

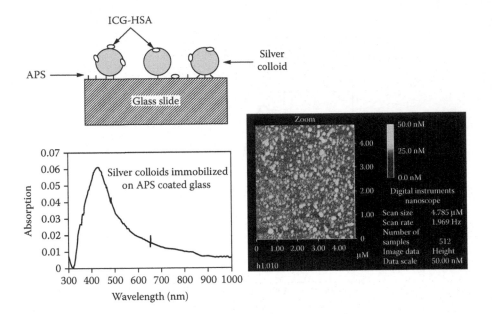

FIGURE 7.10 Glass surface geometry (top). APS is used to functionalize the surface of the glass with amine groups that readily bind silver colloids. Absorption spectrum of silver colloids on APS-coated glass (bottom left) and AFM image of a silver colloid-coated glass (bottom right). APS, 1-amino-3-propylethoxysilane. (Adapted from Geddes et al.[9])

An advantage of a colloidal suspension is the possibility of injection for medical imaging, such as in retinal angiography, which could widely benefit from an enhanced ICG and fluorescein quantum yield and photostability. Thus, we also investigated the effects of spherical silver particles on the emission of ICG. The sample geometry was similar to that of SiFs, except the spherical silver particles were prepared separately and were immobilized to aminopropylsilane (APS)-coated glass slides by immersing the glass in a solution of the particles (Figure 7.10, top). Figure 7.10 shows an absorption spectrum typical of our colloid-coated APS glass slides. The absorption peak centered near 430 nm is typical of colloidal silver particles with subwavelength dimensions. An atomic force microscopy (AFM) image of silver colloid-coated glass slides shows that the size of the silver colloids was smaller than 50 nm with partly aggregated sections (Figure 7.10, bottom). The surfaces were incubated with ICG–HSA to obtain a monolayer surface coating. The emission spectra showed an approximately by 30-fold larger intensity on the surfaces coated with silver colloids but only a minor shift (Figure 7.11).[9] The fluorescence intensity typically increased with the concentration of APS used to treat the cleaned surfaces (Figure 7.12), which also appeared to correlate with the optical density of silver colloids at 430 nm (not shown). We also measured the lifetimes of ICG on both surfaces and observed a significant reduction in the lifetimes on the silver colloids (Figure 7.13), providing additional evidence that the increase in intensity is, in fact, due to a modification of the radiative decay rate, Γ_m, by the proximity to silver colloids.[9]

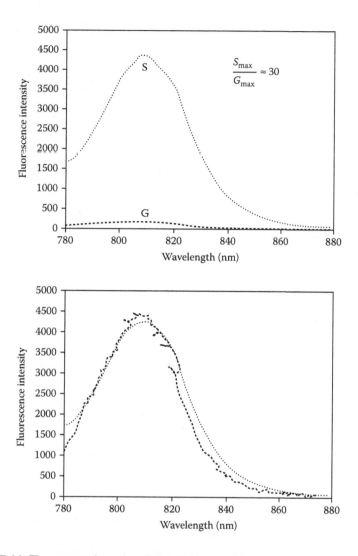

FIGURE 7.11 Fluorescence intensity of ICG–HSA-coated glass, G, and above silver colloids, S; excitation = 760 nm (top). Fluorescence intensities normalized to the intensity on silver colloids (bottom). (Adapted from Geddes et al.[9])

7.2.3 NONSPHERICAL SILVER COLLOIDS
AND NANOSTRUCTURES FOR MEF

In addition to the interest in spherical silver colloids, the optical properties of nonspherical particles have been attracting many researchers since the early 20th century. Typical absorption spectra of silver colloids are shown in Figure 7.14. The long wavelength absorption is called the surface plasmon absorption, which is due to electron

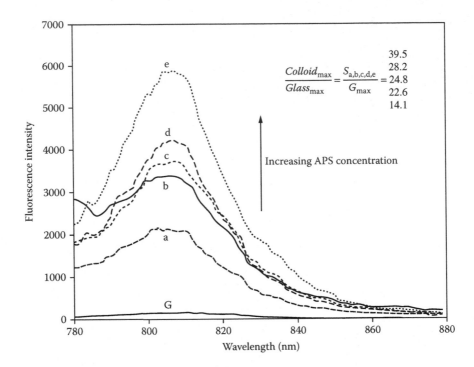

FIGURE 7.12 Fluorescence intensity of ICG–HSA on silver colloids as a function of increased APS use. Cleaned glass slides were initially soaked in (a) 0.1%, (b) 0.25%, (c) 0.5%, (d) 1.0%, and (e) 1.25% (v/v) APS solution for 4 h and then washed and soaked in colloid solution for 4 days. G, fluorescence intensity of ICG–HSA deposited on 0.5% APS-covered glass. ICG coverage for the glass controls (i.e., for the different-percentage APS-coated glass slides) was approximately constant. (Adapted from Geddes et al.[9])

oscillations on the metal surface. These spectra may be calculated for the small particle limit ($r \ll \lambda$) from the complex dielectric constant of the metal.[10,11] Larger particles display longer wavelength absorption. The absorption spectra are also dependent on the shape of the particles, with prolate spheroids displaying longer absorption wavelengths. Most studies of surface effects on fluorescence have been performed using silver particles to avoid the longer wavelength absorption of gold.

Several groups have considered the effects of metallic spheroids on the spectral properties of nearby fluorophores.[5,6,12–14] A typical model is shown in Figure 7.14 (bottom left) for a prolate spheroid with an aspect ratio of a/b. The particle is assumed to be a metallic ellipsoid with a fluorophore positioned near the particle. The fluorophore is located outside the particle at a distance r from the center of the spheroid and a distance d from the surface. The fluorophore is located on the major axis and may be oriented parallel or perpendicular to the metallic surface. The presence of a metallic particle may have dramatic effects on the radiative decay rate of nearby fluorophores. Figure 7.14 (bottom right) shows the radiative rates expected for a fluorophore

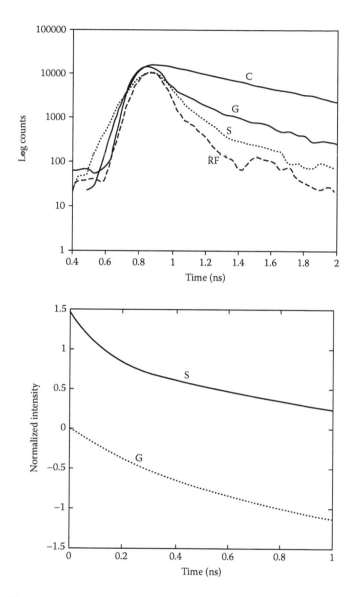

FIGURE 7.13 Complex intensity decays of ICG–HSA in a cuvette (buffer), *C*; on glass slides, *G*; and silver colloids, *S*. *RF*, instrumental response function (top). Data from the convolution process normalized to steady-state intensity, that is, the area under $S = 39.5 \times$ area under *G* (bottom). (Adapted from Geddes et al.[9])

at various distances from the surface of a silver particle and for different orientations of the fluorophore transition moment. The most remarkable effect is for a fluorophore perpendicular to the surface of a spheroid with $a/b = 1.75$. In this case the radiative rate may be enhanced by a factor of 1000 or greater. The effect is much smaller for

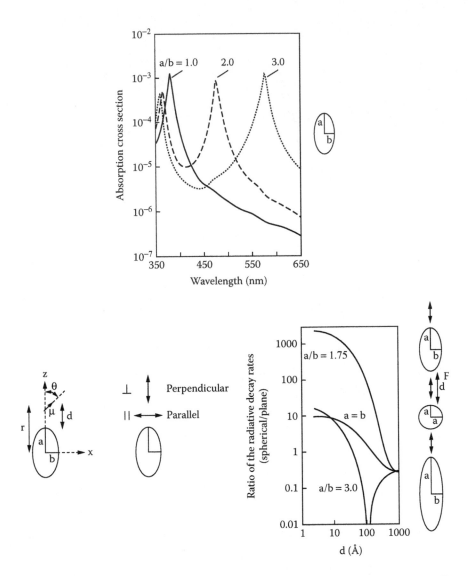

FIGURE 7.14 Calculated absorption cross section of silver nanostructures (top) and the effect of a metal spheroid on the radiative decay rate (bottom). (Adapted from Lakowicz[1].)

a sphere ($a/b = 1.0$) and much smaller for a more elongated spheroid ($a/b = 3.0$) when the optical transition is not in resonance.

Recently we developed a methodology for depositing silver nanorods with controlled sizes and loadings onto glass substrates. The absorption spectra of silver nanorods deposited on glass substrates are shown in Figure 7.15 (top). Silver nanorods display two distinct surface plasmon peaks—transverse and longitudinal—that typically appear at approximately 420 nm and 650 nm, respectively. In our experiments, the longitudinal surface plasmon peak shifted and increased in absorbance as more

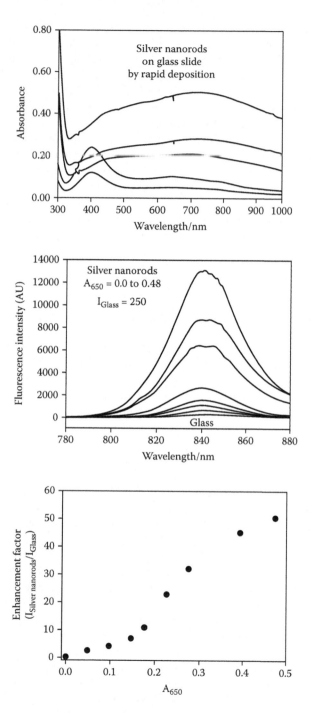

FIGURE 7.15 Absorption spectra of silver nanorods (top), emission spectra of ICG–HSA on silver nanorods (middle), and enhancement factor vs. absorbance (bottom).

nanorods were deposited on the surface of the substrates. In parallel to these measurements, we observed an increase in the size of the nanorods (by AFM, 100 nm in width and several hundred nanometers in length; data not shown). In order to compare the extent of enhancement of fluorescence with respect to the extent of loading of silver nanorods deposited on the surface, we have arbitrarily chosen the value of absorbance of 650 nm as a means of loading of the nanorods on the surface. This is because the 650-nm band is solely attributed to the longitudinal absorbance of the nanorods.

Figure 7.15 (middle) shows the fluorescence emission intensity of ICG–HSA measured from both glass and silver nanorods and the enhancement factor vs. the loading density of silver nanorods. We have obtained up to a 50-fold enhancement in emission of ICG on silver nanorods compared to the emission on glass (Figure 7.15, bottom). We have also measured the lifetime of ICG on both glass and silver nanorods and observed a significant reduction in the lifetime of ICG on silver nanorods, providing evidence that increased emission is due to radiative rate modifications, Γ_m.

7.2.4 LASER-DEPOSITED SILVER FOR MEF

It would be useful to obtain MEF at desired locations in a measurement device for use in medical and biotechnology applications, such as diagnostic or microfluidic devices. Although a variety of methods may be used, we reasoned that the light-directed deposition of silver would be widely applicable. In a typical preparation of light-induced deposition of silver on glass slides, the silver colloid-forming solution was prepared by adding 4 ml of 1% trisodium citrate solution to a warmed 200 ml 10^{-3} M silver nitrate ($AgNO_3$) solution. This warmed solution already contains some silver colloids, as seen from a surface plasmon absorption optical density near 0.1. A 180-ml aliquot of this solution was injected between the glass microscope slide and the plastic cover slip, creating a microsample chamber 0.5 mm thick (Figure 7.16). For all experiments, a constant volume of 180 ml was used. Irradiation of the sample chamber was performed using a helium–cadmium (HeCd) laser with a power of about 8 mW, which was collimated and defocused using a 10× microscope objective, numerical aperture (NA) 0.40, to provide illumination over a 0.5-mm diameter spot.[15]

We examined the emission spectrum of ICG–HSA when bound to illuminated or nonilluminated regions of the APS-treated slides. For APS-treated slides (Figure 7.17), the intensity of ICG was increased about sevenfold in the regions with laser-deposited silver. We examined the photostability of ICG–HSA when bound to glass or laser-deposited silver. We reasoned that ICG molecules with shortened lifetimes should also be more photostable because there is less time for photochemical processes to occur. The intensity of ICG–HSA was recorded with continuous illumination at 760 nm. When excited with the same incident power, the fluorescence intensities, when considered on the same intensity scale, decreased somewhat more rapidly on the silver (Figure 7.18, top). However, the difference was minor. Because the observable intensity of the ICG molecules prior to photobleaching is given by the area under these curves, it is evident that at least 10 times more signal may be observed from ICG near silver as compared to glass.

Alternatively, one may consider the photostability of ICG when the incident intensity is adjusted to result in the same initial signal intensities on silver and

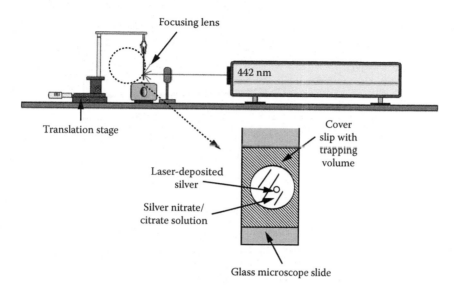

FIGURE 7.16 Experimental setup for light-induced deposition of silver on APS-coated glass microscope slides. (Adapted from Geddes et al.[15].)

glass. In this case (Figure 7.18, bottom), photobleaching is slower on the silver surfaces. The fact that the photobleaching is not accelerated for ICG on silver indicates that the increased intensity on silver is likely not due to an increased rate of excitation.

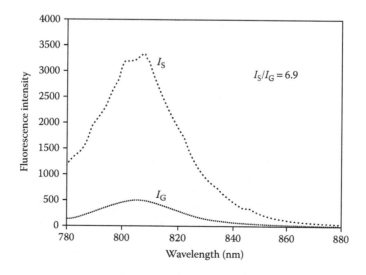

FIGURE 7.17 Fluorescence spectra of ICG–HSA on glass and on light-deposited silver. I_S, intensity on silver; I_G, intensity on glass. (Adapted from Geddes et al.[15])

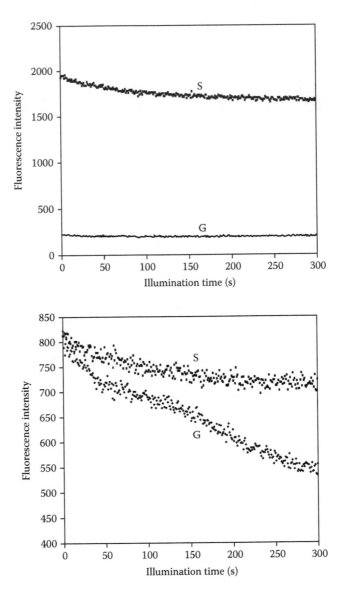

FIGURE 7.18 Photostability of ICG–HSA on glass (G) and laser-deposited silver (S) measured at an excitation power of 760 nm (top) and with the laser power at 760 nm adjusted for the same initial fluorescence intensity (bottom). Laser-deposited samples were made by focusing 442-nm laser light onto APS-coated glass slides immersed in an AgNO$_3$ citrate solution for 15 min. The optical density (OD) of the sample was approximately 0.3. (Adapted from Geddes et al.[15])

7.2.5 ELECTROCHEMICALLY DEPOSITED SILVER FOR MEF

One of our methods for silver deposition onto glass/quartz substrates was to pass a controlled current between two electrodes in pure water (Figure 7.19, top).[16] The two silver electrodes were mounted in a quartz cuvette containing deionized water (Millipore, Billerica, MA). The silver electrodes had dimensions of 9 mm × 35 mm × 0.1 mm

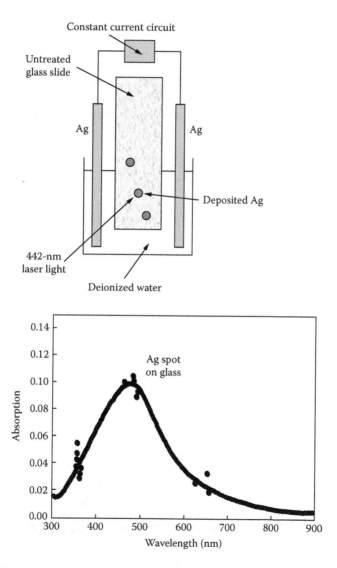

FIGURE 7.19 Light-deposited silver produced electrochemically. Constant current circuit (top) and absorbance spectrum of silver spot on glass (bottom). (Adapted from Geddes et al.[16])

separated by a distance of 10 mm. For the production of silver colloids, a simple constant current generator circuit (60 µA) was constructed and used. After 30 min of current flow, a clear glass microscope slide was positioned within the cuvette (no chemical glass surface modifications) and was illuminated (HeCd, 442 nm). We observed silver deposition on the glass microscope slide, the amount depending on the illumination time. Simultaneous electrolysis and 442-nm laser illumination resulted in the deposition of metallic silver in the targeted illuminated region, a 5-mm focused spot size (Figure 7.19, top). The absorption spectrum of the deposited silver is shown in Figure 7.19 (bottom). A single absorption band is present on glass, indicating that the silver particles are somewhat spherical. We examined ICG–HSA when coated on glass (G) and silver particles (S). The emission intensity increased about 18-fold on the silver particles (Figure 7.20). We found a dramatic increase in the photostability near the silver particles (Figure 7.21). This very encouraging result indicates that a much greater signal may be obtained from each fluorophore prior to photodestruction and that more photons may be obtained per fluorophore before the ICG on silver eventually degrades.[16]

7.2.6 ELECTROPLATING OF SILVER FOR MEF

This method is similar to the one employed in the previous section, except that the silver cathode electrode was replaced with an indium tin oxide (ITO)-coated glass electrode (Figure 7.22, top), or it may be replaced with any surface to be silvered. The current was again 60 µA. After a short period of time, silver readily deposited on the ITO surface (no laser illumination), the extent of which was again dependent on the exposure time. Two maxima were found on the ITO absorption spectrum, which eventually formed one large broad band. This suggests that the particles are

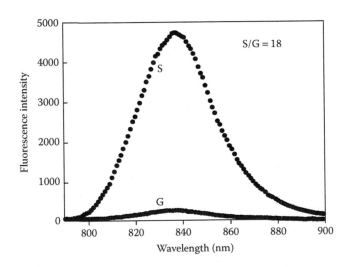

FIGURE 7.20 Fluorescence spectra of ICG–HSA on glass (G) and on light-deposited silver (S). (Adapted from Geddes et al.[16])

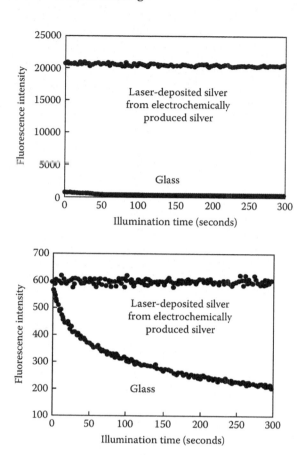

FIGURE 7.21 Photostability of ICG–HSA on glass and laser-deposited silver produced via electrolysis measured using the same excitation power at 760 nm (top) and with power adjusted to give the same initial fluorescence intensities (bottom). In all the measurements, vertically polarized excitation was used, whereas the fluorescence emission was observed at the magic angle (i.e., 54.7°). (Adapted from Geddes et al.[16])

elongated and display both transverse and longitudinal resonances (not shown). Enhanced fluorescence emission from ICG–HSA was also found for silver particles on ITO (Figure 7.22, bottom), and there appeared to be a small blue shift on silvered ITO.[16]

7.2.7 ROUGHENED SILVER ELECTRODES FOR MEF

In a typical preparation, commercially available silver electrodes (Sigma-Aldrich, St. Louis, MO) are placed in deionized water 10 mm apart.[17] A constant current of 60 μA was supplied across the two electrodes for 10 min by a constant-current generator (Figure 7.23). Figure 7.24 shows the time-dependent growth of the silver nanostructures on the silver cathode. In comparison, the anode was relatively unperturbed. Binding the ICG–HSA to both the silver anode and cathode after electrolysis

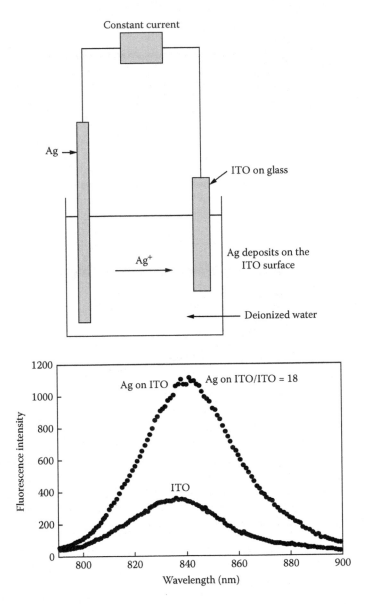

FIGURE 7.22 Electroplating of silver on substrates (top), and fluorescence emission spectra of ICG–HSA on ITO and silver deposited on ITO (bottom). (Adapted from Geddes et al.[16])

was accomplished by soaking the electrodes in a solution of ICG–HSA overnight, followed by rinsing with water to remove the unbound material. As a control sample, an unused silver electrode was coated with ICG–HSA. A roughened silver cathode was also dipped in 10^{-4} M sodium chloride (NaCl) for 1 h before washing and then coated with ICG–HSA.

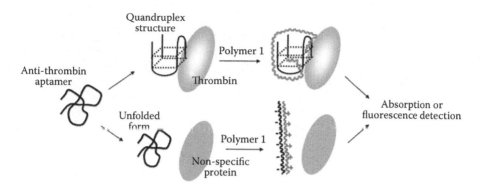

COLOR FIGURE 2.3

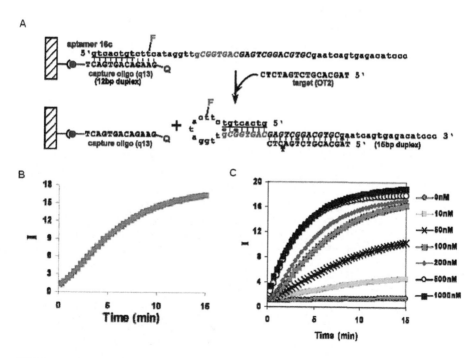

COLOR FIGURE 2.7

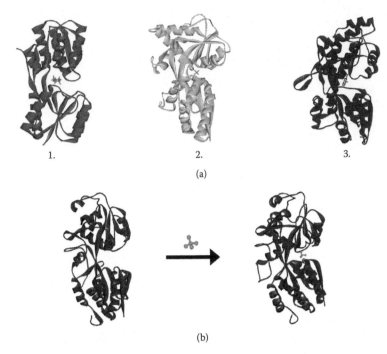

1.   2.   3.

(a)

(b)

COLOR FIGURE 3.1

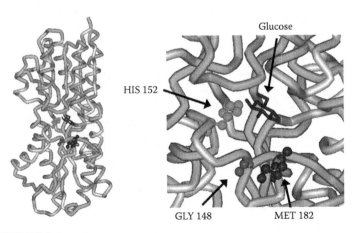

Glucose

HIS 152

GLY 148   MET 182

COLOR FIGURE 3.2

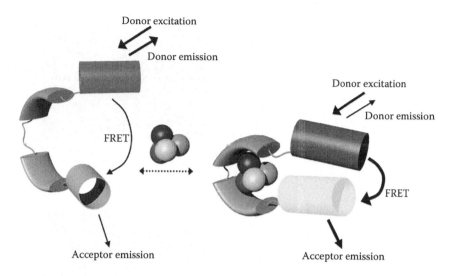

Donor excitation

Donor emission

FRET

Acceptor emission

Donor excitation

Donor emission

FRET

Acceptor emission

COLOR FIGURE 3.3

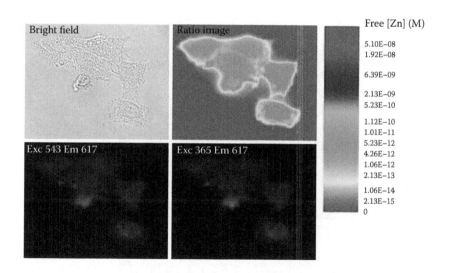

Bright field

Ratio image

Exc 543 Em 617

Exc 365 Em 617

Free [Zn] (M)

5.10E–08
1.92E–08

6.39E–09

2.13E–09
5.23E–10

1.12E–10
1.01E–11
5.23E–12
4.26E–12
1.06E–12
2.13E–13

1.06E–14
2.13E–15
0

COLOR FIGURE 6.1

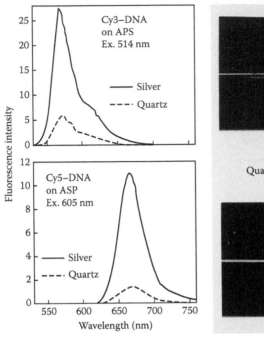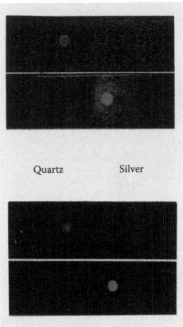

COLOR FIGURE 7.34

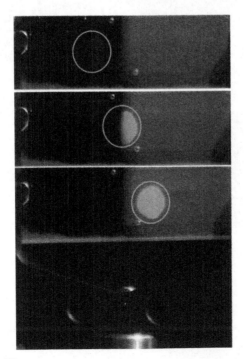

COLOR FIGURE 7.48

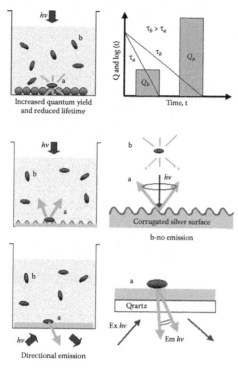

COLOR FIGURE 7.57

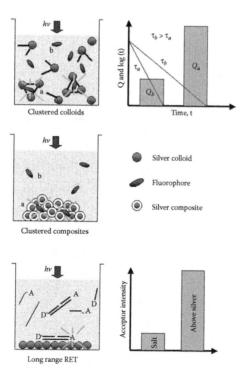

COLOR FIGURE 7.58

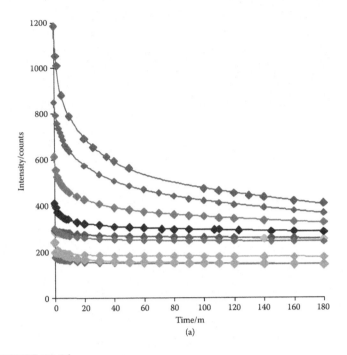

COLOR FIGURE 10.2A

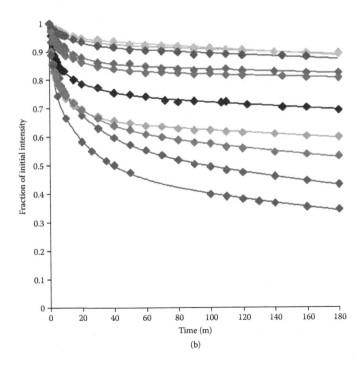

COLOR FIGURE 10.2B

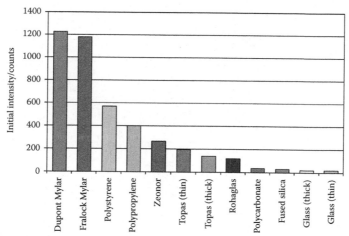

COLOR FIGURE 10.3

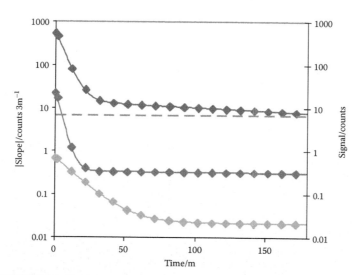

COLOR FIGURE 10.4

COLOR FIGURE 11.1

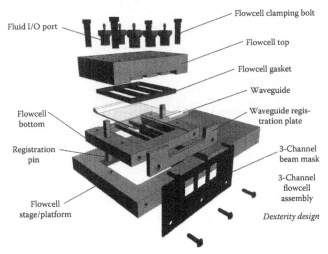

Fluid I/O port

Flowcell clamping bolt

Flowcell top

Flowcell gasket

Waveguide

Waveguide registration plate

Flowcell bottom

Registration pin

3-Channel beam mask

3-Channel flowcell assembly

Flowcell stage/platform

Dexterity design

COLOR FIGURE 12.3

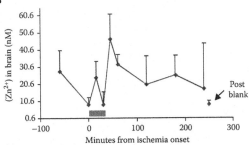

COLOR FIGURE 14.7

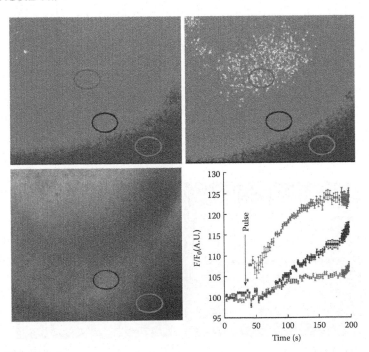

COLOR FIGURE 14.8

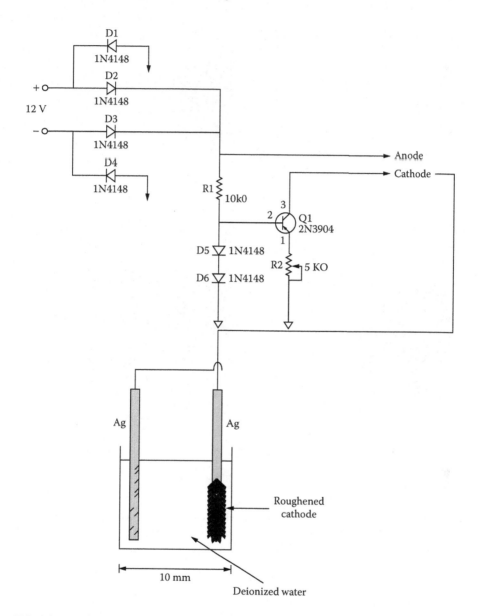

FIGURE 7.23 Experimental setup for the production of roughened silver electrodes. (Adapted from Geddes et al.[16])

Three electrodes were coated with ICG–HSA and studied: the roughened cathode, the anode, and an unroughened electrode. Essentially no emission was seen from ICG–HSA on an unroughened, bright silver surface (Ag in Figure 7.25 (top), the control). However, a dramatically larger signal was observed on the roughened cathode and a somewhat smaller signal was observed on the anode. In all our experiments, we typically found that the roughened silver cathode was 20- to 100-fold

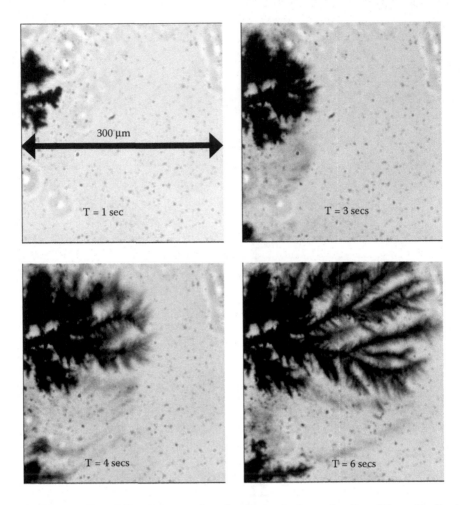

FIGURE 7.24 Fractal-like silver growth on the silver cathode as a function of time, visualized using transmitted light. This structure was characteristic of the whole electrode. (Adapted from Geddes et al.[17])

more fluorescent than the unroughened control electrode. In comparison, the silver anode was 5 to 50 times more fluorescent than the silver control. When we increased the time for roughening to more than 1 h, the intensities of both electrodes after coating with ICG–HSA were essentially the same but were still 50-fold more fluorescent than the unroughened silver control. The emission spectra on the two electrodes probably had the same emission maximum, where the slight shift seen in Figure 7.25 (bottom) is thought to be due to the filters used to reject the scattered light. It should be noted that the amount of material coated on both surfaces was approximately the same, and the effect was not due to the increased surface area, and therefore increased protein coverage, of the roughened surface.[17]

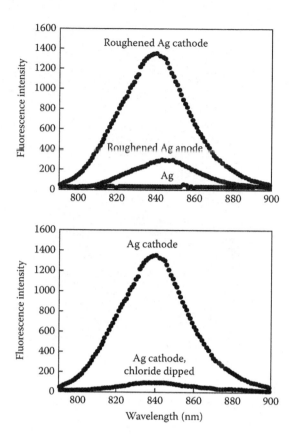

FIGURE 7.25 Fluorescence emission spectra of ICG–HSA on roughened silver electrodes (top) and silver cathode (bottom). (Adapted from Geddes et al.[17])

To place our findings in context with the enhanced Raman signals observed with chloride-dipped electrodes, we examined the silver cathode after dipping in 10^{-4} M NaCl for 1 h. In contrast to SERS electrodes, we observed a decrease in fluorescence intensity for ICG–HSA chloride-dipped electrodes, confirming our hypothesis of MEF being a through-space phenomenon as compared to SERS, which is due to contact interaction.

7.2.8 SILVER FRACTAL-LIKE STRUCTURES ON GLASS SUBSTRATES FOR MEF

Fractal-like silver structures were also generated on glass using two silver electrodes held between two glass microscope slides (Figure 7.26, top).[18] The electrodes were 10 mm × 35 mm × 0.1 mm, with about 20 mm between the two electrodes. Deionized water was placed between the slides. A direct current of 10 μA was passed between the electrodes for about 10 min, during which the voltage started near 5 V and

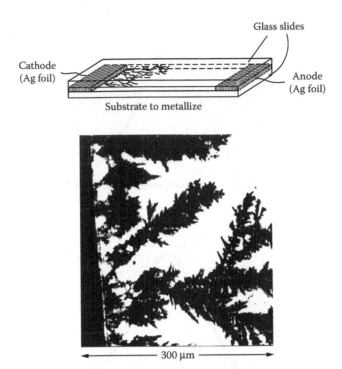

FIGURE 7.26 Configuration for creation of fractal-like silver surfaces on glass (top) and bright-field image of fractal-like silver surfaces on glass. (Adapted from Parfenov et al.[18])

decreased to 2 V. During the current flow, fractal silver structures grew on the cathode and then on the glass near the cathode (Figure 7.26), thus producing silver nanostructures on glass, as compared to those grown on silver electrodes as described in the previous section. Similar to the silver electrodes, the structures grew rapidly, but they appeared to twist as they grew. In addition, we found that dipping the slides in 0.001 mg/dl stannous chloride ($SnCl_2$) for 30 min before electrolysis resulted in structures that were firmly bound to the glass during working. Without the $SnCl_2$, similar structures were formed but were partially removed during washing. Following passage of the current, the silver structures on glass were soaked in 10 μM fluorescein isothiocyanate (FITC)–HSA overnight at 4°C, which is thought to result in a monolayer of surface-bound HSA. For the FITC–HSA-coated fractal silver surfaces on glass, we were able to measure a fluorescence image very similar to that of the fractal silver surface alone (bright field image) using the same apparatus (Figure 7.27). Interestingly, regions of high and low fluorescence intensity were observed (Figure 7.27, bottom).

This result is roughly consistent with recent SERS data, which showed the presence of intense signals that appeared to be located between clusters of particles.[19,20] Figure 7.28 represents the intensity of FITC–HSA on silvered glass. The emission intensities range from 100-fold (position 6) to 600-fold (position 1) greater than the

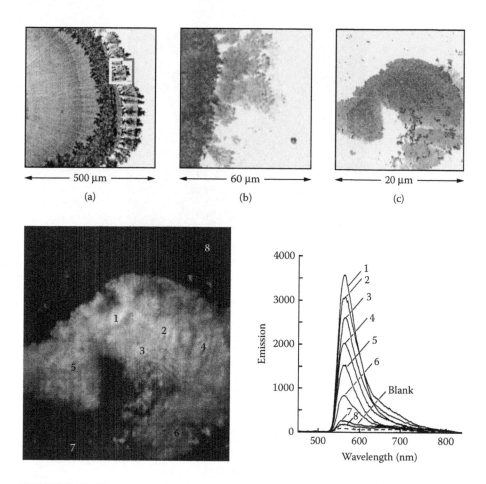

FIGURE 7.27 Silver nanostructures deposited on glass during electroplating (panel a). Panels b and c are consecutive magnifications of the marked area on panel a. Bright-field image (top). Fluorescence image of FITC–HSA deposited on the silver structure above (bottom left), and the emission spectra of the numbered areas shown on the right (bottom right). (Adapted from Parfenov et al.[18])

signal from FITC–HSA on unsilvered glass. We recognized that some of the increase in intensity of FITC–HSA could be due to binding of more FITC–HSA to silver structures with large surface areas. We note that the fluorescein is not quenched on the surface, probably because the size of an HSA molecule positions the fluorescein about 40 Å from the surface, which is near the distance for maximal radiative rate and therefore fluorescence enhancement (cf. Figure 7.2).

We tested the use of fractal silver for providing emission selectively from fluorophores near the surface. For this purpose we examined glass coated with FITC–HSA, to which we added a solution of Nile blue to yield a more-than-10-fold larger signal from Nile blue (Figure 7.29, top). In contrast to the glass surface, the signal was dominated by FITC–HSA emission for the silver surface

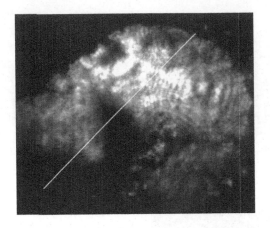

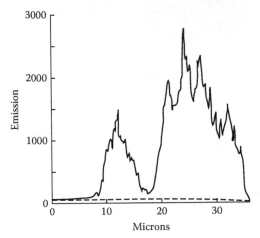

FIGURE 7.28 Diagonal scan of the emission intensity of FITC–HAS on silvered glass. The dashed line is the intensity observed across a line of equivalent length across unsilvered glass. (Adapted from Parfenov et al.[18])

(Figure 7.29, bottom). Intuitively, this result suggests that unwanted autofluorescence from biological samples can be avoided by MEF of surface-localized fluorophores.

We also studied the photostability of FITC on fractal silver, SiFs, and uncoated quartz. Although the relative photobleaching is higher on fractal silver, the increased rate of photobleaching is less than the increase in intensity (Figure 7.30). From the areas under these curves, we estimate approximately 16-fold and 160-fold more photons can be detected from the FITC–HSA on SiFs and fractal silver, respectively, relative to quartz, before photobleaching.[18]

As we have shown, metallic silver may be deposited on different surfaces by a variety of methods. The deposited silver is still useful for MEF, and indeed some deposits seem better for overall fluorescence intensity enhancements (e.g., silver

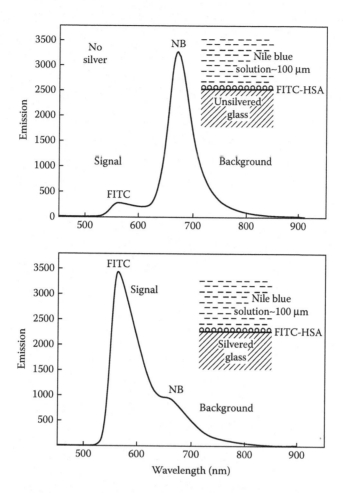

FIGURE 7.29 Emission spectra of a monolayer of FITC–HSA on glass (top) or silver (bottom) in the presence of 2×10^{-6} M Nile blue. The sample was 100 mM thick. (Adapted from Parfenov et al.[18])

fractals). Given the versatility of these deposition techniques, we may readily envisage their widespread use in MEF-sensing applications. In the following sections we show some actual enhanced fluorescence-sensing schemes that are a direct result of MEF and our knowledge gained from studying different silvered surfaces.

7.3 ENHANCED DNA DETECTION USING MEF

In DNA, each nucleotide residue contains an ultraviolet (UV) absorbing (approximately 260 nm) base that might be expected to display fluorescence. However, the intrinsic fluorescence of the bases is exceedingly weak in DNA and in the isolated bases.[21,22] The emission is so weak that there are no practical uses of intrinsic DNA emission. For this reason a vast array of fluorophores that bind

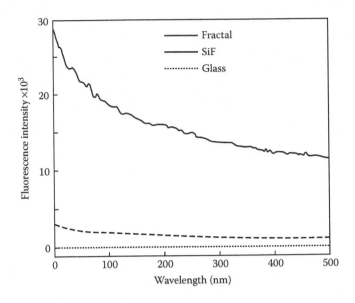

FIGURE 7.30 Photostability of FITC–HSA on silver fractals, SiFs, and glass. (Adapted from Parfenov et al.[18])

to DNA and display UV, visible, or near infrared fluorescence has been developed.[23–26] In fact, the renaissance of fluorescence some 15 years ago is attributed to the wide-spread use of fluorescent probes developed for DNA sequencing, compared to the now-declining use of radiolabels. The difficulty with DNA is not that the radiative rates are slow but that the nonradiative rates are exceedingly fast, so the excited bases return to the ground state prior to emission. That is, fluorescence is a competitive process between radiative and nonradiative decay. In the case of DNA, the low quantum yields are due to nonradiative rates that are much larger than the radiative rate. Suppose that the PMD near the DNA can be increased so that the radiative rate increases. This may be accomplished by bringing DNA into the proximity of silver particles that display plasmon resonance, suggesting usefully high intrinsic fluorescence from DNA.

7.3.1 Intrinsic DNA Detection: A New Opportunity for Sequencing

We examined the emission of DNA in micron-thick samples between quartz plates and between SiFs. When a solution of DNA is examined between two quartz slides, the emission is barely detectable (Figure 7.31). However, there is a dramatic increase in intrinsic emission for the DNA between SiFs.[3] In this experiment, the information is only qualitative because the DNA is not bound to the surfaces and only a small fraction is near the silver particles. Hence, the increase in quantum yield for those molecules near the silver particles is likely to be larger than that actually seen in Figure 7.31. The emission spectra in Figure 7.31 do not demonstrate an increase in

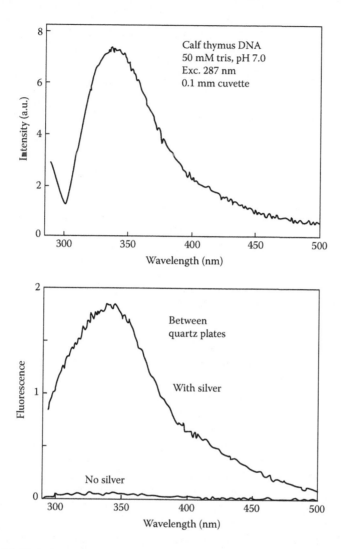

FIGURE 7.31 Emission spectra of DNA solution in a cuvette (top), between quartz plates (bottom, no silver), and between SiFs (bottom, with silver). (Adapted from Lakowicz et al.[3])

the radiative rate. However, such an increase may be demonstrated by the additional measurement of the intensity of decay. These results show that the intensity of decay is more rapid near the SiFs (Figure 7.32). We interpret this decrease in lifetime as being due to the increased PMD near the DNA bases. As previously mentioned, an increase in intensity accompanied by a reduction in lifetime can only be explained by a radiative rate modification.

Although the human genome and other organisms have been sequenced,[27,28] there is still a need for faster, less expensive, and more sensitive DNA sequences. Some groups are attempting to sequence a single DNA strand using fluorescence-based techniques.[29,30]

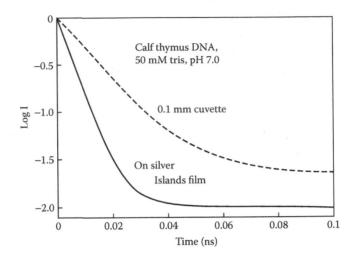

FIGURE 7.32 Time-dependent intensity decays of calf thymus DNA without metal (__) and between SiFs (_).(Adapted from Lakowicz et al.[3])

The basic idea is to allow an exonuclease to sequentially cleave single nucleotides from the strand, label the nucleotide with a fluorophore, and detect and identify the labeled nucleotide. This goal is more difficult than single molecule detection because every nucleotide must be detected and identified, not the simpler task of finding one fluorophore among many. Also, the labeling of nucleotides is likely to result in a larger number of fluorophores that have not reacted.

The use of MEF could allow base detection and identification without labeling. Suppose the released nucleotides pass through a specially designed flow chamber (Figure 7.33). The size and shape of the chamber could be such that the bases display intrinsic emission. Also, the surface would be shaped and periodic in a manner that directs the emission toward a detector.[1] The design would be such that the directed emission occurs wherever the unlabeled nucleotides flow through the laser beam. SERS of DNA bases has been observed using silver and gold colloids,[31] which suggests to us that MEF will also be observed for DNA and its bases. In addition, it has now become possible to create silver particles on surfaces with broadly tunable SPR spectra using nanolithography,[32] and a variety of methods are appearing for self-assembly of metallic nanoparticles.[33-35] If successful, such an apparatus would allow DNA sequencing using intrinsic nucleotide fluorescence.

7.3.2 ENHANCED DNA LABELS

We also investigated the emission spectral properties of fluorophore-labeled DNA using the geometry as shown in Figure 7.9 (top, inset). Emission spectra of Cy3 DNA and Cy5 DNA are shown in Figure 7.34. The emission intensity is two to three times greater between SiFs as compared to between the quartz slides for Cy3 DNA and Cy5 DNA, respectively. The slightly larger increase in emission

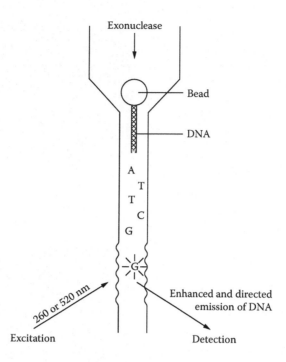

FIGURE 7.33 Notion of DNA sequencing using intrinsic base fluorescence. (Adapted from Lakowicz.[1])

intensity for Cy5 DNA compared to Cy3 DNA is consistent with the results where larger enhancements are observed with low quantum yield fluorophores. Figure 7.34 also shows the photographs of the labeled oligomers on quartz and on SiFs. The emission from the labeled DNA on quartz is almost invisible but is brightly visible on the SiFs. This difference in intensity is due to an increase in the PMD near the fluorophore, which in turn results in an increase in the radiative decay rate and quantum yield of the fluorophores. We note that the photographs are taken through emission filters and the increase in emission intensity is not due to an increased excitation scatter from the silvered plates.[36]

The photostability of the labeled DNA was studied by measuring the emission intensity during continuous illumination at a laser power of 20 mW (Figure 7.35). The intensity initially dropped rapidly but became more constant at longer illumination times. Although not a quantitative result, examination of these plots visually suggests slower photobleaching at longer times in the presence of silver particles compared with quartz slides.

Another widely used fluorophore in DNA detection is 1,19-[1,3-propanediyl-bis[(dimethylimino)-3,1-propanediyl]]bis[4-[(3-methyl-2(3H)-benzoxazolylidene) methyl]]-quinolinum tetraiodide (YOYO-1).[37–39] This fluorophore contains multiple positive charges, binds strongly to DNA, and displays almost no fluorescence in water. As a result, it is useful for the observation of DNA in electrophoretic gels. Figure 7.36 shows the chemical structure of YOYO-1 and the strategy of DNA binding to quartz

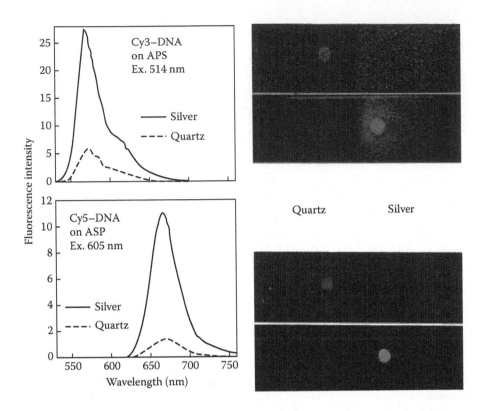

FIGURE 7.34 (See color insert following page 142.) Emission spectra of Cy3 DNA (top left) and Cy5 DNA (bottom left) between quartz plates with and without SiFs. Photographs of the corresponding fluorophores. The emission from Cy3 on quartz (left side of the picture, top right) is very weak (green color) compared to the silver side (right side of the picture, bright green). The emission from Cy5 on quartz (left side of the picture, bottom right) is very weak (red color) compared to the silver side (right side of the picture, bright red). Clearly the presence of silver results in dramatic increases in observed fluorescence intensities. (Adapted from Lakowicz et al.[36])

and silvered quartz. We measured the emission spectra of YOYO-1 DNA when bound to quartz and silver via the protein layers (Figure 7.37). The emission intensity is 15 times greater on the SiFs than on quartz.[40] This dramatic intensity increase can be seen in the upper panels, which are photographs of equal amounts of YOYO-1 DNA spotted on a slide. There are two possible mechanisms responsible for the increase of brightness near the SiFs, an enhanced local field, E_m, and an increase in radiative decay rate, Γ_m.

7.3.3 DNA HYBRIDIZATION USING MEF

Detection of DNA hybridization is the basis of a wide range of biotechnology and diagnostic applications.[41] DNA hybridization is measured on gene chips,[42,43] during polymerase chain reaction (PCR),[44,45] and for fluorescence *in situ* hybridization (FISH),[46] to name a few. In all these applications, increased sensitivity is desirable,

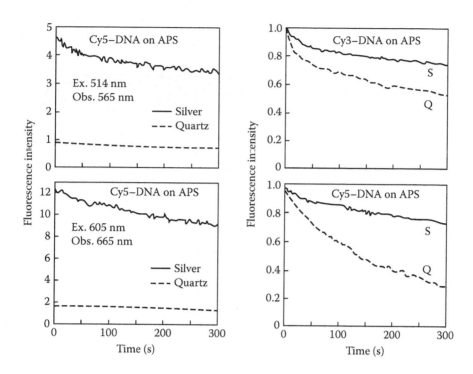

FIGURE 7.35 Photostability of Cy3 DNA and Cy5 DNA between quartz plates with (S) and without (Q) SiFs. The laser power was 20 mW. (Adapted from Lakowicz et al.[36])

particularly for detection of a small number of copies of biohazard agents. Also, it would be valuable to have a general approach to detect the changes in fluorescence intensity upon hybridization. In general, the detectability of a fluorophore is determined by two factors: the extent of background emission from the sample and the photostability of the fluorophore. A highly photostable fluorophore can undergo about 10^6 excitation–relaxation cycles prior to photobleaching, yielding about 10^3 to 10^4 measured photons per fluorophore.[47,48] Background emissions from the samples can easily overwhelm weak emission signals.

In a recent report, we described a simple approach that should provide a readily measurable change in fluorescence intensity in DNA hybridization formats.[49] In addition, our approach increases the intensity relative to the background and increases the number of detected photons per fluorophore molecule by a factor of 10-fold or more. Figure 7.38 (top) shows the sequence and structure of the oligomers used in these experiments. The thiolated oligonucleotide ss DNA–SH was used as the capture sequence which bound spontaneously to the silver particles. The sample containing the silver-bound DNA was positioned in a fluorometer (Figure 7.39, bottom) followed by addition of 18 nM ss Fl–DNA, which is an amount approximately equal to the amount of silver-bound capture DNA.

The fluorescence intensity began to increase immediately upon mixing and leveled off after about 20 min (Figure 7.39, top). We believe this increase in intensity

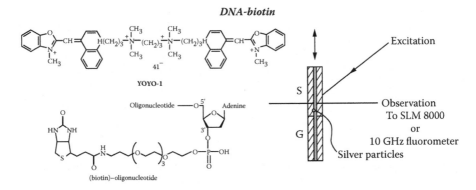

5'–GAA GAT GGC CAG TGG TGT GTG GA-3'–biotin
3'–CTT CTA CCG GTC ACC ACA CAC CT–5'

FIGURE 7.36 Schematic of surface, sequence of the DNA oligomers, structures of YOYO-1 and biotinylated oligonucleotide, and experimental geometry. Note that the sizes in the surface schematic are not to scale. The BSA–avidin protein layer is 80 Å thick and the silver particles are 400 Å high. (Adapted from Lakowicz et al.[40])

is due to localization of ss Fl–DNA near the silver particles by hybridization with the capture oligomer (Figure 7.38, bottom). Because metallic silver particles can increase the emission intensity of many fluorophores, this result suggests that localization of labeled oligomers near silver particles may be used in a wide range of hybridization assays. In control experiments, we hybridized ss DNA–SH with ss Fl–DNA prior to deposition on silver particles. We found a similar 12-fold increase in intensity upon immobilization on silver as compared to an equivalent amount of ds Fl–DNA–SH in solution.

We examined the emission spectra of ss Fl–DNA before and after hybridization to form ds Fl–DNA–SH (Figure 7.40). The fluorescence intensity was found to be 12-fold higher for the bound form. This dramatic increase can be seen visually in Figure 7.40 (bottom). There was no detectable shift in the emission spectra. The intensity increase was reversed by melting the DNA at 80°C (Figure 7.41) and increased once again upon cooling and presumed rehybridization. The intensity did not recover completely upon slow cooling, which may be due to loss of capture DNA from the silver surfaces.

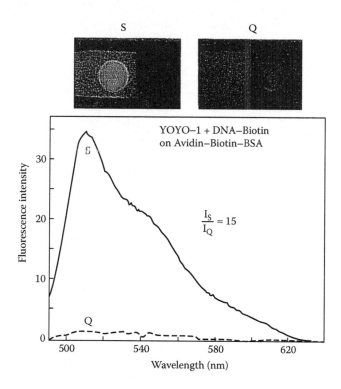

FIGURE 7.37 Emission spectra of YOYO-1-labeled DNA bound to the quartz (Q) and silver (S) surfaces. The upper panels show a real-color photograph of labeled DNA spotted on the silver (left, very bright green) and quartz (right, weak green) surfaces. (Adapted from Lakowicz et al.[40])

Figure 7.42 shows the frequency-domain intensity decays of the single-stranded fluorescein-labeled oligomer in solution and the double-stranded oligomer when bound to silver particles. The lifetime is dramatically shortened for the silver-bound oligomer, which strongly supports our conclusion that the intensity increase is due to localization of the fluorophore near the silver surfaces. In control experiments we found that emissions of fluorescein in the single- and double-stranded oligos were similar to within 10%. The double-stranded form displayed an approximately 10% smaller intensity. Hence, the differences in intensity and lifetime between the solution and silver-bound forms are not due to effects of hybridization on the fluorescein probe. It is interesting to note that there is no detectable 4-ns component for the sample with silver particles, indicating that all the emission is due to the silver-bound DNA.

The differences in the lifetime of fluorescein between the solution and silver-bound form suggested an alternative approach to measuring hybridization. The emission phase angle and modulation measurements could be useful in detecting DNA hybridization; because these values depend on the fluorescence lifetime, they were expected to change upon hybridization. These measurements revealed a rapid

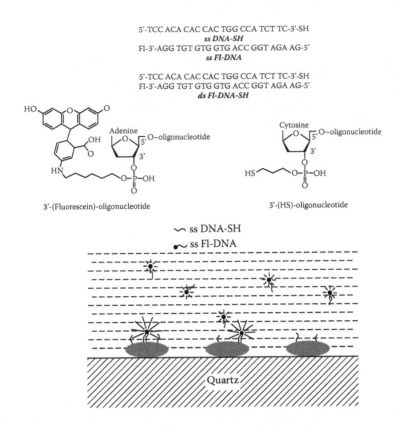

FIGURE 7.38 Structures of DNA oligomers. The lower panel shows a schematic of the oligomers bound to silver particles. (Adapted from Malicka et al.[49])

decrease in phase angle and an increase in modulation following addition of ss Fl–DNA to the silver-bound capture oligomer (Figure 7.43). These changes are due to the decrease in lifetime upon binding to the capture oligomers on the silver particles. The changes in phase and modulation (Figure 7.43) occur somewhat more rapidly than the change in intensity (Figure 7.39). This difference occurs because the phase and modulation are intensity-weighted parameters. It is important to note that phase angle and modulation measurements are mostly independent of total intensity to within the limitations of the instrumentation and the extent of background fluorescence in the sample.[50] This is important because the extent of hybridization may be measured using optical components such as optical fibers, where the intensity may vary, or microwell plates, where the well-to-well intensity may vary due to the plate or adsorbing species in the sample.

We examined the intensity of Fl–DNA–SH on the silver surfaces with continuous illumination (Figure 7.44). As a comparison, we examined a similar biotinylated oligomer bound to a silvered quartz slide using biotinylated bovine serum albumin and avidin. We found that fluorescein in our samples deposited on silver particles photobleaches more slowly than when deposited on a protein monolayer. It should be noted that with the protein monolayer, the Fl–DNA was uniformly deposited on

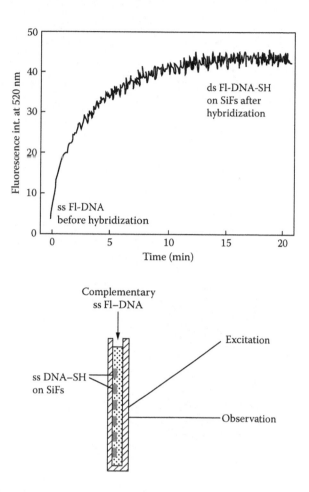

FIGURE 7.39 Time-dependent hybridization of ss Fl–DNA to ss DNA–SH. The lower panel shows the sample configuration. (Adapted from Malicka et al.[49])

the entire silvered slide (i.e., also between the silver particles). The lifetimes measured for the protein monolayer samples contained a long (about 4 ns) component. This component indicates that there is some fluorescein bound to the glass surface between the silver particles. If the increased intensity is mostly due to an increased rate of excitation near the silver particles, then we expect slower photobleaching with the protein monolayers when there are regions without an increased rate of excitation.

7.4 OVERLABELED PROTEINS AS ULTRABRIGHT PROBES

Proteins covalently labeled with fluorophores are widely used as reagents, such as immunoassays or immunostaining of biological specimens with specific antibodies. In these applications, fluorescein is one of the most widely used probes. An unfortunate

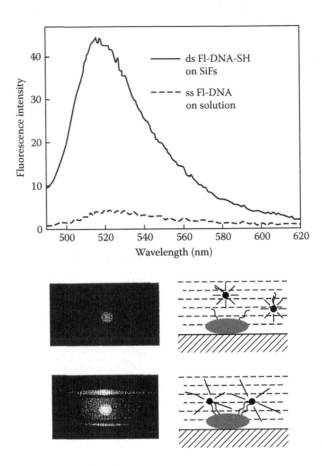

FIGURE 7.40 Emission spectra of ss Fl–DNA in solution (dashed line) and bound (solid line) to silver particles. Roughly the same number of molecules of ss Fl–DNA and ds Fl–DNA–SH were in the illuminated area. The lower panel shows photographs of ss Fl–DNA in solution (top) and ds Fl–DNA–SH on SiFs (bottom). (Adapted from Malicka et al.[49])

property of fluorescein is self-quenching, which is due to Forster resonance energy transfer between nearby fluorescein molecules (homotransfer).[51] As a result, the intensity of labeled protein does not increase with increased labeling but actually decreases. Figure 7.45 shows the spectral properties of FITC–HSA with molar labeling ratios (L) ranging from 1:1 ($L = 1$) to 1:9 ($L = 9$) (the samples had the same optical density at 490 nm). The relative intensity decreased progressively with increased labeling. The insert in Figure 7.45 shows the intensities normalized to the same amount of protein, so the relative fluorescein concentration increases ninefold along the x-axis. It is important to note that the intensity per labeled protein molecule does not increase and, in fact, decreases as the labeling ratio is increased from 1 to 9. However, we found that the self-quenching could be largely eliminated by the proximity to SiFs,[52] as can be seen from the emission spectra for labeling ratios of 1 and 7 (Figure 7.46) and from the dependence of the intensity on the extent of

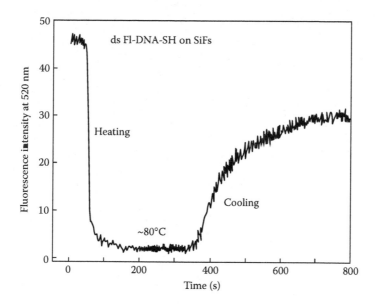

FIGURE 7.41 Thermal melting of ds Fl–DNA–SH on SiFs near 80°C and rehybridization of Fl–DNA–SH on SiFs. (Adapted from Malicka et al.[49])

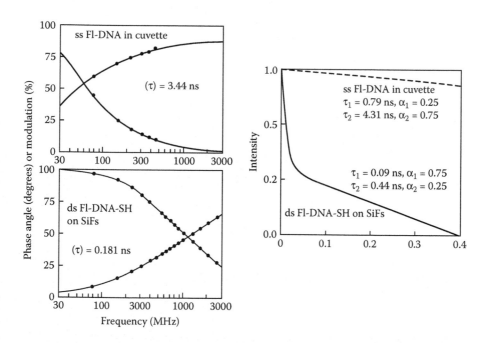

FIGURE 7.42 Intensity decays of ss Fl–DNA in solution and ds Fl–DNA–SH on SiFs measured in the frequency domain (left) and reconstructed in the time domain (right). (Adapted from Malicka et al.[49])

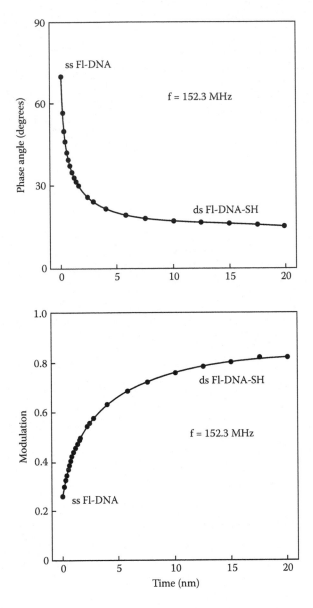

FIGURE 7.43 Phase angle (top) and modulation (bottom) of ss Fl–DNA upon hybridization to silver-bound ss DNA–SH. The light modulation frequency and detection frequency were 152.3 MHz. (Adapted from Malicka et al.[49])

labeling (Figure 7.47). We speculate that the decrease in self-quenching is due to an increase in the rate of radiative decay, Γ_m.

The dramatic difference in the intensity of heavily labeled HSA on glass and on SiFs is shown pictorially in Figure 7.48. The effect is dramatic, as can be seen

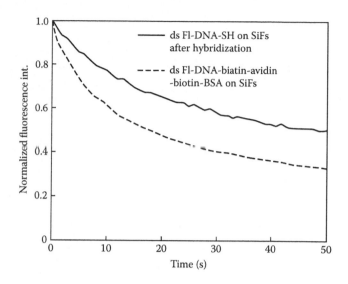

FIGURE 7.44 Photostability of ds Fl–DNA–SH bound to SiFs (solid line). For comparison, we show the photostability of a similar biotinylated oligomer bound to a protein monolayer of BSA–biotin–avidin uniformly deposited on SiFs (dashed line). In this case, some of Fl–DNA–biotin are distant from and not affected by the silver particles. (Adapted from Malicka et al.[49])

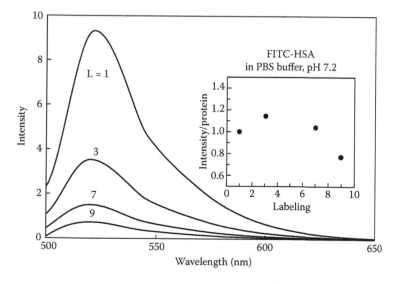

FIGURE 7.45 Dependence of emission intensity of FITC–HSA on the degree of labeling. (Adapted from Lakowicz et al.[52])

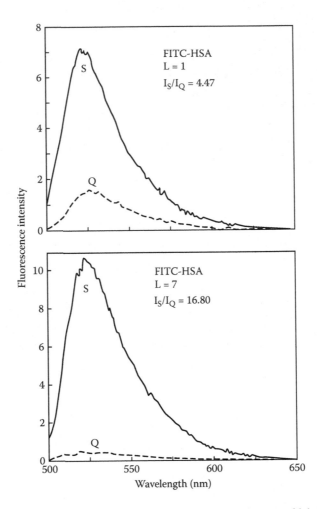

FIGURE 7.46 Emission spectra of FITC–HSA with different degrees of labeling on quartz (Q) and SiFs (S). SiFs release the quenched fluorescence of overlabeled proteins. (Adapted from Lakowicz et al.[52])

from the nearby invisible intensity on quartz (left side) and the bright image on the SiFs (right side) in this unmodified photograph. These results suggest the possibility of ultrabright-labeled proteins based on high labeling ratios and MEF. We conclude that SiFs, and most probably colloidal silver, can be utilized to obtain dramatically increased intensities of fluorescein-labeled macromolecules.[53]

7.4.1 BACKGROUND SUPPRESSION USING OVERLABELED PROTEINS

It is informative to consider how silver-enhanced fluorescence, particularly of a heavily labeled sample, can be used for improved assays and sensing. Figure 7.49 shows emission spectra of a quartz plate coated with FITC–HSA to which we adjusted

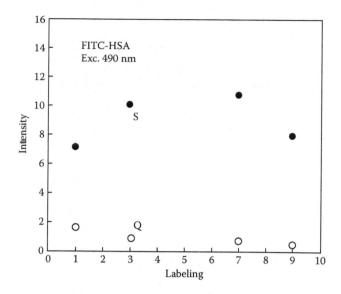

FIGURE 7.47 Emission intensity of FITC–HSA at 520 nm vs. different degrees of labeling on quartz (Q) and on SiFs (S). (Adapted from Lakowicz et al.[52])

the concentration of rhodamine B (0.25 μM) to result in an approximately 1.5-fold-larger rhodamine B intensity. For this example, one may consider the rhodamine B to be simple autofluorescence or any other interference signal. When the same conditions are used for FITC–HSA on silver with $L = 1$, the fluorescein emission is two to three times that of rhodamine B (Figure 7.49, right top). When using the heavily labeled sample ($L = 7$), the fluorescein emission becomes more dominant (Figure 7.49, right bottom).

7.5 SELECTIVE EXCITATION USING MULTIPHOTON EXCITATION AND METALLIC NANOPARTICLES

As we discussed throughout this chapter, proximity to metallic silver nanostructures may alter the radiative decay rate, Γ_m, and/or excitation rate, E_m, of fluorophores. We have shown that the quantum yield of low-quantum-yield fluorophores can be increased, with a maximum predicted increase of $1/Q_0$, where Q_0 is the quantum yield in the absence of metal, whereas significant increases in emission intensity from high-quantum-yield species, in the absence of any nonradiative rate modifications, can only be observed by substantial increases in E_m. Complementary to our interpretations of previous results using one-photon excitation, we also reported that enhanced and localized multiphoton excitation of rhodamine B fluorescence occurs near metallic silver islands.[54]

The increase in fluorescence emission intensity for rhodamine B molecules adjacent to metallic silver islands (Figure 7.50) is accompanied by a reduction in

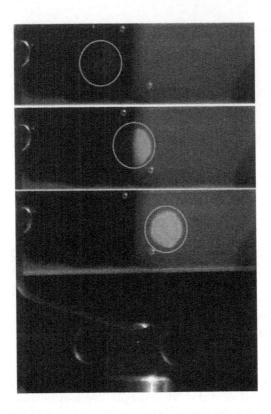

FIGURE 7.48 (See color insert following page 142.) Photograph of fluorescein-labeled HSA (molar ratio of fluorescein/HSA = 7) on quartz (left) and on SiFs (right) as observed with 430-nm excitation and a 480-nm long-pass filter. The excitation was progressively moved from the quartz side to the silver (top to bottom), respectively. The emission from fluorescein is very weak on the quartz side; however, as the excitation is moved toward the silver, emission from fluorescein is brighter. (Adapted from Lakowicz et al.[52])

lifetime compared to that observed using one-photon excitation. Given the high quantum yield of rhodamine B ($Q_0 = 0.48$), these results may be explained by the metallic particles significantly increasing the E_m around the rhodamine B molecules. Moreover, the sample geometry and the absence of any notable increase in emission intensity using one-photon excitation, as well as the fact that the one-photon mean lifetime remained essentially unchanged both in the presence and absence of silver, suggest that enhanced two-photon excitation is localized to regions in proximity to the silver islands. That is, a much more dramatic enhancement is possible for multiphoton excitation. For a two-photon absorption process, the rate of excitation is proportional to the square of the incident intensity. This suggests that two-photon excitation may be enhanced by a factor of 3.8×10^8.[1,2] Such an enhancement in the excitation rate is thought to provide selective excitation of fluorophores near the metal islands or colloids, even if the solution contains a considerable concentration of other fluorophores that could undergo

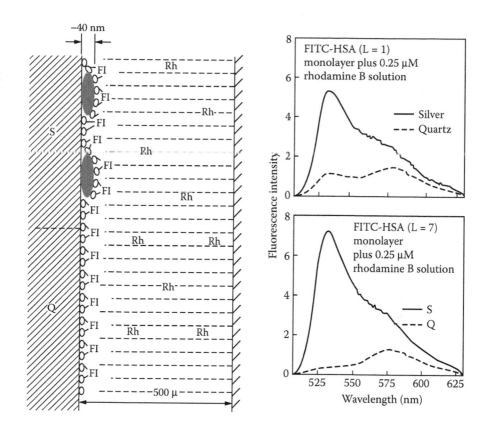

FIGURE 7.49 Emission spectra of a monolayer of FITC–HSA $L = 1$ (right top) and $L = 7$ (right bottom) containing 0.25 mM rhodamine B between the quartz plates (Q) or one SiF (S). Schematic of the sample with bound fluorescein and free rhodamine B (left). (Adapted from Lakowicz et al.[52])

two-photon excitation at the same wavelength but are more distant from the metal surface (Figure 7.51).

This interpretation is borne out by the fact that given the overwhelming excess of high-quantum-yield rhodamine B in this sample geometry (96% of solution is too distant for any fluorophore–metal effect), the fluorescence lifetime is still shorter than that typically observed for bulk solution rhodamine B in the absence of metal. Also, for our samples, we found that for one-photon excitation, the photostability of rhodamine B was not affected by the presence or absence of silver islands (Figure 7.52, bottom). However, for two-photon excitation, an increased photostability was observed for rhodamine B in the presence of silver islands (Figure 7.52, top). These results are consistent with the shorter lifetime observed for rhodamine B between silver islands and with our interpretation that multiphoton excitation occurs preferentially near the silver islands.

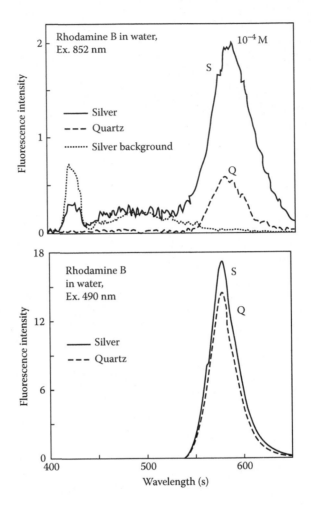

FIGURE 7.50 (Top) Emission spectra of 10^{-4} M rhodamine B between SiFs (S) with two-photon excitation at 852 nm from a Tsunami mode-locked titanium-sapphire laser, 80-MHz repetition rate, 90-fs pulse, and about 0.5 W average power. Also shown are the emission spectra observed from uncoated quartz slides (Q) and silver islands alone without rhodamine B. (Bottom) Emission spectra of rhodamine B between silver islands, S, or quartz plates, Q, with one-photon excitation at 490 nm (bottom). (Adapted from Lakowicz et al.[53])

7.6 MECHANICAL CONTROL OF ENERGY TRANSFER USING MEF

Fluorescence resonance energy transfer (FRET) is widely used in biochemistry[54–57] and in DNA biotechnology.[58–60] FRET is a widely useful phenomenon because the Forster distance (R_0) can be readily calculated from the spectral properties of the donor and acceptor, and the extent of FRET is generally insensitive to the sample

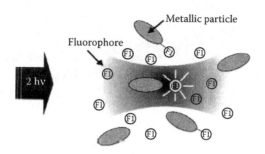

FIGURE 7.51 Pictorial representation of the preferential multiphoton excitation of fluorophores in proximity to metal in the presence of free fluorophore, Fl. (Adapted from Lakowicz et al.[53])

conditions if the sample remains transparent. A disadvantage of FRET is the limited distances over which it occurs. The largest known Forster distances are near 55 Å for organic fluorophores and near 90 Å for lanthanide donors.[61–63] In the case of DNA, a Forster distance of 50 Å corresponds to approximately 16 bp, so that FRET occurs only when the donor and acceptor are closely spaced within the DNA helix. DNA assays based on FRET over greater distances have not been developed because longer-range FRET does not occur.

In a recent report, we described the effects of metallic particles on FRET between donors and acceptors covalently bound to DNA.[64] Theoretical studies have predicted increased rates of energy transfer over distances as large as 700 Å near silver particles of appropriate size and shape.[65,66] To test this prediction, we prepared a double-helical DNA oligomer, 23 base pairs long, with a donor and acceptor placed at opposite sides about 75 Å apart. Because the R_0 value is near 50 Å, little energy transfer is expected under free-space conditions. We used steady-state and time-resolved fluorescence to determine the effects of SiFs on FRET between the widely spaced donor–acceptor pairs. We examined the donor-labeled and acceptor-labeled oligomers separately, in the absence of FRET, to determine the effects of the silver particles.

Figure 7.53 (top) shows the emission spectra of 7-amino-3-((((propyl) amino)carbonyl)methyl)-4-methyl coumarin-6-sulfonic acid (AMCA)–DNA between uncoated quartz plates and between SiFs. Figure 7.53 (bottom) also shows the emission spectra of the donor–acceptor pair of AMCA–DNA–Cy3 with no excitation of the donor and direct excitation of the acceptor at 514 nm. The intensity of the donor-alone AMCA–DNA was essentially unchanged by the silver particles. In contrast, the intensity of DNA-Cy3 was increased several-fold. The different effects are consistent with the effects expected for high- and low-quantum-yield fluorophores. It is not possible to increase a quantum yield above unity, and the larger increases in quantum yield (observed fluorescence intensity) are obtained for lower quantum yield fluorophores.

The oligomer length of 23 bp was chosen so there would be little energy transfer between AMCA and Cy3 under free-space conditions. The emission from the donor is nearly unaffected by the presence of acceptor and is decreased about 2% (not

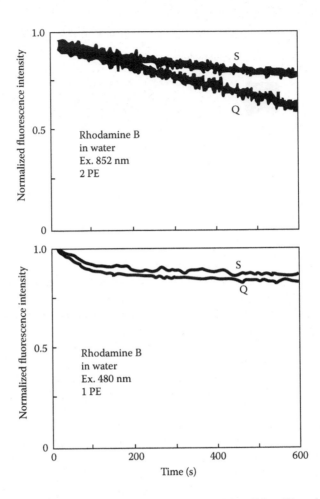

FIGURE 7.52 Photostability of rhodamine B between quartz slides (Q) and SiFs (S) with two-photon excitation at 852 nm (top) and one-photon excitation at 490 nm (bottom). The 490-nm excitation was from an argon ion laser attenuated to about 10 mW. Fluorescence was observed at 580 nm. (Adapted from Lakowicz et al.[53])

shown). We also examined the intensity decay of the donor alone and the donor–acceptor pair. The frequency response of the donor–acceptor pair is just slightly shifted to higher frequencies, which reflects a slightly shortened donor lifetime due to the acceptor. In the case of FRET, the efficiency of energy transfer may be calculated simply from the amplitude-weighted lifetimes:

$$E = 1 - \frac{(\tau_{DA})}{(\tau_D)}$$

These values are given in Table 7.1. Hence, the time-resolved donor decays also indicate low transfer efficiency near 5.4%.

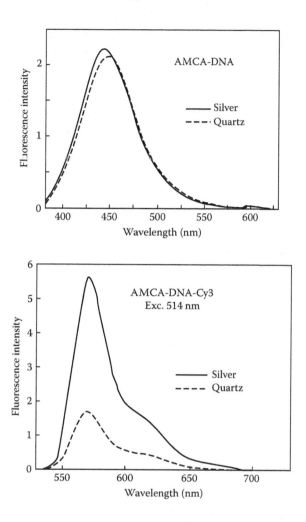

FIGURE 7.53 Emission spectra of the donor AMCA–DNA between quartz plates (– –) and between SiFs (—) (top). Emission spectra of the directly excited acceptor in AMCA–DNA–Cy3 between quartz plates (– –) and between SiFs (—) (bottom). (Adapted from Malicka et al.[64])

We subsequently examined the emission spectra of the donor- and acceptor-labeled DNA between quartz plates and SiFs (Figure 7.54). An increase in the acceptor emission is seen in the donor-normalized spectra (Figure 7.54, bottom). It is difficult to judge the extent to which the increased acceptor emission is due to increased energy transfer or due to the increased intensity of Cy3, shown in Figure 7.53. Nevertheless, it is clear that much of the Cy3 emission from AMCA–DNA–Cy3 between the silver islands is due to FRET.

Our analysis of the frequency-domain intensity decays of AMCA–DNA–Cy3 suggested that R_0 contributed to the 30% of donor emission (not shown). The 30% fraction of the strongly affected donor–acceptor pairs is in disagreement with the

TABLE 7.1

Multiexponential Analysis of Fluorescence Intensity Decay of AMCA–DNA (Donor), Cy3 DNA (Acceptor), and AMCA–DNA–Cy3 (Donor–Acceptor)

Compound/condition	Excitation (nm)	Emission (nm)	$\bar{\tau}$(ns)	$\langle\tau\rangle$(ns)	$\alpha_1(f_1)$	τ_1 (ns)	$\alpha_2(f_2)$	τ_2 (ns)	$\alpha_3(f_3)$	τ_3(ns)	χ^2?
AMCA–DNA, cuvette	345	450	4.78[a]	4.78[b]	1.0	4.78	—	—	—	—	1.7
AMCA–DNA–Cy3, cuvette	345	450	4.52	4.52	1.0	4.52	—	—	—	—	1.9
Cy3–DNA, cuvette	345	>570	1.03	0.74	0.623 (0.318)	0.91	0.377 (0.682)	1.33	—	—	1.9
AMCA–DNA, quartz	345	450	3.24	2.63	0.369 (0.136)	0.97	0.631 (0.864)	3.61	—	—	0.9
AMCA–DNA, two silver slides	345	450	3.20	2.34	0.445 (0.137)	0.72	0.555 (0.863)	3.64	—	—	1.0
AMCA–DNA, one silver slide	345	450	3.27	2.38	0.501 (0.192)	0.91	0.499 (0.808)	3.84	—	—	1.4
Cy3–DNA, quartz	514	>570	1.22	0.90	0.432 (0.139)	0.29	0.568 (0.861)	1.36	—	—	1.7
Cy3–DNA, two silver slides	514	>570	0.72	0.11	0.888 (0.310)	0.04	0.078 (0.263)	0.38	0.034 (0.427)	1.41	0.9
AMCA–DNA–Cy3, quartz	345	450	2.93	2.37	0.371 (0.136)	0.87	0.629 (0.864)	3.26	—	—	1.3
AMCA–DNA–Cy3, two silver slides	345	450	2.12	0.66	0.580 (0.148)	0.17	0.309 (0.290)	0.63	0.111 (0.562)	3.40	1.4
AMCA–DNA–Cy3, one silver slide	345	450	3.01	2.29	0.504 (0.200)	0.87	0.496 (0.800)	3.54	—	—	0.9
AMCA–DNA–Cy3, quartz	514	>570	1.25	1.03	0.272 (0.069)	0.26	0.728 (0.931)	1.32	—	—	1.8
AMCA–DNA–Cy3, two silver slides	514	>570	0.93	0.07	0.948 (0.301)	0.02	0.030 (0.159)	0.329	0.022 (0.540)	1.56	1.1
AMCA–DNA–Cy3, two silver slides	345	>570	1.50	0.55	0.578 (0.072)	0.07	0.296 (0.407)	0.76	0.125 (0.521)	2.30	2.2

$$^a \bar{\tau} = \sum_i f_i\tau_i, \quad f_i = \frac{\alpha_i\tau_i}{\sum \alpha_i\tau_i}$$

$$^b \langle\tau\rangle = \sum_i \alpha_i\tau_i$$

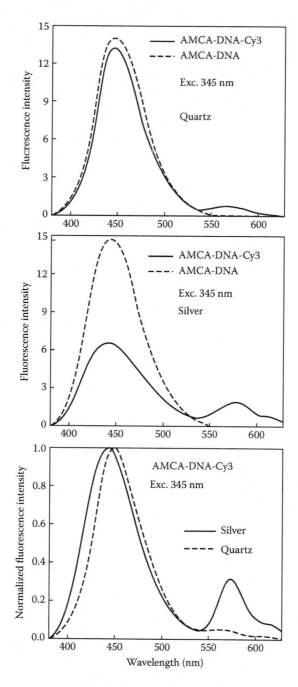

FIGURE 7.54 Emission spectra of AMCA–DNA donor and AMCA–DNA–Cy3 donor–acceptor pair on quartz (top) and silver (middle). The bottom panel shows normalized emission spectra of AMCA–DNA–Cy3 donor–acceptor between quartz (– –) and between SiFs (—). (Adapted from Malicka et al.[64])

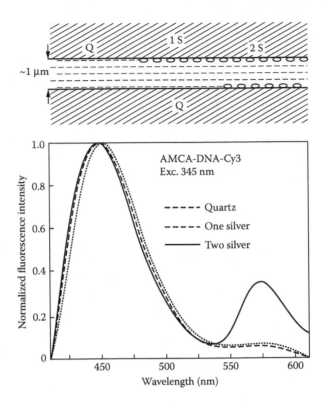

FIGURE 7.55 Normalized emission spectra of AMCA–DNA–Cy3 recorded on quartz (Q), one silvered slide (1S), and two silvered slides (2S). (Adapted from Malicka et al.[64])

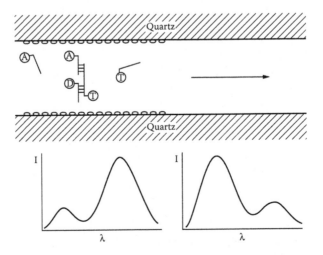

FIGURE 7.56 Detection of DNA sequences using long-range FRET on a silver particle surface. D, A, and T are the donor, acceptor, and target oligonucleotides, respectively.

estimated 4% volume of the sample that is close to the silver particles. To clarify this discrepancy, we repeated our experiments with different sample geometries, using one SiF instead of two SiFs. That is, one side of the sandwich was an SiF and the other was an unsilvered quartz plate. This was accomplished by coating different slides with silver over one third or two thirds of the area, so that there were regions of the sample between two quartz plates, between one quartz and one SiF, or between two SiFs (Figure 7.55, top). When there was only one SiF surface, we found no increase in energy transfer (Figure 7.55) and no decrease in the donor lifetime (Table 7.1).

7.6.1 Opportunities for Long-Range Energy Transfer

It is difficult to anticipate the future uses of long-range FRET because all present FRET assays have been designed to position the donors and acceptors within the upper range of Forster distances near 50 Å. One possibility for metal-enhanced FRET is detection of target sequences with larger numbers of base pairs (Figure 7.56). Shorter donor–acceptor distances may be detected between quartz and larger distances between two silvered plates. One may also imagine the use of induced long-range FRET for analysis of chromosomes with FISH. As currently performed, the emission spectra of the FISH samples reflect the location of specific sequences, and FRET does not normally occur between the labeled oligonucleotides used in these hybridizations. This situation, however, may change for labeled chromosomes, especially if placing a solution in a microcavity-type system results in FRET over hundreds of angstroms. Further studies by our laboratories are under way.

7.7 THE ROLE OF MEF IN DRUG DISCOVERY

Metal-enhanced fluorescence appears to be most suitable to the fluorescence assays used in drug discovery and DNA analysis. As we have shown, silver may be readily deposited on glass or polymer substrates by a variety of methods. Silver colloids are easily prepared and can be attached to surfaces functionalized with amine or sulfhydryl groups. We may imagine the bottom of multiwell plates or DNA arrays being coated with silver particles. A variety of new assay formats is possible. Assays could be based on E_m modifications. Biochemical affinity interactions could bring the fluorophore close to the metal surface, for localized excitation, eliminating the washing steps. Another approach could be to use low-quantum-yield fluorophores brought near the metal (Figure 7.57, top). These effects might be coupled with another remarkable property of metal–fluorophore interactions. If the metal is close to a semitransparent metallic surface, the emission can couple into the metal and become directional rather than isotropic (Figure 7.57, bottom).

Metal-enhanced fluorescence is not limited to single metallic particles. Theory predicted that the lightening rod effect is much stronger between two metallic spheres than for isolated spheres. If the biochemical affinity reaction brings particles closer together, then excitation fields may be substantially increased in the spaces between particles (Figure 7.58, top). Multiphoton excitation is also known to be increased near

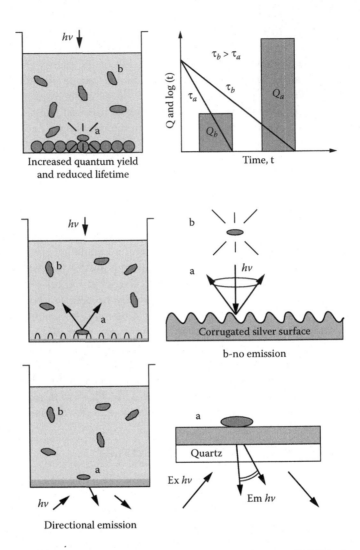

FIGURE 7.57 (See color insert following page 142.) Potential uses of MEF in drug discovery based on local increases in quantum yield (top) and directional emission (bottom). (Adapted from Geddes et al.[67])

metallic surfaces and may be even more efficient between metal particles. Also, proximity to the particles may result in long-range energy transfer (Figure 7.58, bottom), which would allow selective detection of macromolecule complexes.

7.8 CONCLUSION

In this chapter we have shown the favorable effects of fluorophores near metallic nanostructures. These favorable effects include increased fluorescence intensities (increased quantum yields), increased probe photostabilities

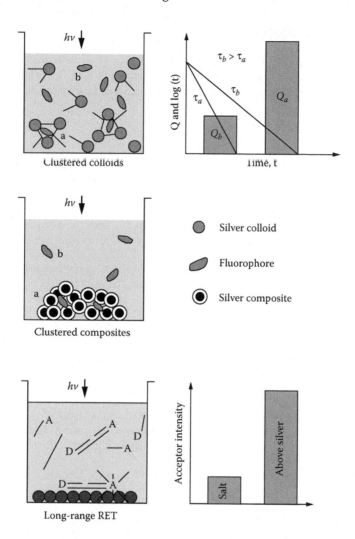

FIGURE 7.58 (See color insert following page 142.) Potential uses of MEF based on colloid clustering multiphoton excitation (top) and long-range FRET (bottom). (Adapted from Geddes et al.[67])

(reduced fluorescence lifetimes), and increased rates of excitation and energy transfer.

Metallic silver may be deposited by a variety of methods, depending on the sensing application, on demand if required, and using fairly simple apparatuses and biologically inert chemicals. The metal–fluorophore effects we have observed to date offer unique perspectives in fluorescence sensing, providing for improved background suppression, increased detection limits, and localized excitation near the silver nanostructures. Although our results to date are for the most part preliminary, we believe MEF will find many applications in sensing, including medical diagnostics and biochemistry.

ACKNOWLEDGMENTS

This work was supported by the National Institutes of Health, National Center for Research Resources, RR-01889. Financial support to J.R. Lakowicz and I. Gryczynski from the University of Maryland Biotechnology Institute (UMBI) is also acknowledged.

REFERENCES

1. Lakowicz, J.R., Radiative decay engineering: biophysical and biomedical applications. *Anal Biochem,* 298, 1–24, 2001.
2. Lakowicz, J.R., Shen, Y., D'Auria, S., Malicka, J., Fang, J., Gryczynski, Z., and Gryczynski, I., Radiative decay engineering. 2. Effects of silver island films on fluorescence intensity, lifetimes, and resonance energy transfer. *Anal Biochem,* 301, 261–277, 2002.
3. Lakowicz, J.R., Shen, Y., Gryczynski, Z., D'Auria, S., and Gryczynski, I., Intrinsic fluorescence from DNA can be enhanced by metallic particles. *Biochem Biophys Res Commun,* 286, 875–879, 2001.
4. Gryczynski, I., Malicka, J., Shen, Y., Gryczynski, Z., and Lakowicz, J.R., Multiphoton excitation of fluorescence near metallic particles: enhanced and localized excitation. *J Phys Chem B,* 106, 2191–2195, 2002.
5. Gersten, J. and Nitzan, A., Spectroscopic properties of molecules interacting with small dielectric particles. *J Chem Phys,* 75, 1139–1152, 1981.
6. Weitz, D.A., Garoff, S., Gersten, J.I., and Nitzan, A., The enhancement of Raman scattering, resonance Raman scattering and fluorescence from molecules absorbed on a rough silver surface. *J Chem Phys,* 78, 5324–5338, 1983.
7. Kummerlen, J., Leitner, A., Brunner, H., Aussenegg, F.R., and Wokaun, A., Enhanced dye fluorescence over silver island films—analysis of the distance dependence. *Mol Phys,* 80, 1031–1046, 1993.
8. Malicka, J., Gryczynski, I., Geddes, C.D., and Lakowicz, J.R., Metal-enhanced emission from indocyanine green: a new approach to *in vivo* imaging. *J Biomed Opt,* 8, 472–478, 2003.
9. Geddes, C.D., Cao, H., Gryczynski, I., Gryczynski, Z., Fang, J., and Lakowicz, J.R., Metal-enhanced fluorescence due to silver colloids on a planar surface: potential applications of indocyanine green to *in vivo* imaging. *J Phys Chem A,* 107, 3443–3449, 2003.
10. Link, S. and El-Sayed, M.A., Shape and size dependence of radiative, nonradiative and photothermal properties of gold nanocrystals. *Int Rev Phys Chem,* 19, 409–453, 2000.
11. Kreibig, U., Vollmer, M., and Toennies, J.P., *Optical Properties of Metal Clusters.* Berlin: Springer-Verlag, 1995.
12. Chew, H., Transition rates of atoms near spherical surfaces. *J Chem Phys,* 87, 1355–1360, 1987.
13. Philpott, M.R., Effect of surface plasmons on transitions in molecules. *J Chem Phys,* 62, 1812–1817, 1975.
14. Chance, R.R., Prock, A., and Silbey, R., Molecular fluorescence and energy transfer near interfaces. *Adv Chem Phys,* 37, 1–65, 1978.
15. Geddes, C.D., Parfenov, A., and Lakowicz, J.R., Photodeposition of silver can result in metal-enhanced fluorescence. *Appl Spectrosc,* 57, 526–531, 2003.

16. Geddes, C.D., Parfenov, A., Roll, D., Fang, J., and Lakowicz, J.R., Electrochemical and laser deposition of silver for use in metal-enhanced fluorescence. *Langmuir,* 19, 6236–6241, 2003.
17. Geddes, C.D., Parfenov, A., Roll, D., Gryczynski, I., Malicka, J., and Lakowicz, J.R., Silver fractal-like structures for metal-enhanced fluorescence: enhanced fluorescence intensities and increased probe photostabilities. *J Fluoresc,* 13, 267–276, 2003.
18. Parfenov, A., Gryczynski, I., Malicka, J., Geddes, C.D., and Lakowicz, J.R., Enhanced fluorescence from fluorophores on fractal silver surfaces. *J Phys Chem B,* 107, 8829–8833, 2003.
19. Michaels, A.M., Jiang, J., and Brus, L., Ag nanocrystal junctions as the site for surface-enhanced Raman scattering of single rhodamine 6G molecules. *J Phys Chem B,* 104, 11965–11971, 2000.
20. Michaels, A.M., Nirmal, M., and Brus, L.E., Surface enhanced Raman spectroscopy of individual rhodamine 6G molecules on large Ag nanocrystals. *J Am Chem Soc,* 121, 9932–9939, 1999.
21. Georghiou, S., Nordlund, T.M., and Saim, A.M., Picosecond fluorescence decay time measurements of nucleic acids at room temperature in aqueous solution. *Photochem Photobiol,* 41, 209–212, 1985.
22. Georghiou, S., Braddick, T.D., Philippetis, A., and Beechem, J.M., Large-amplitude picosecond anisotropy decay of the intrinsic fluorescence of double-stranded DNA. *Biophys J,* 70, 1909–1922, 1996.
23. Steiner, R.F. and Kubota, Y., Fluorescent dye–nucleic acid complexes excited states of biopolymers. In: *Excited States of Biopolymers,* Steiner, R.F., ed. New York: Plenum Press, 1983:203–254.
24. Georghiou, S., Interaction of acridine drugs with DNA and nucleotides. *Photochem Photobiol,* 26, 59–68, 1977.
25. Murphy, C.J., Photophysical probes of DNA sequence directed structure and dynamics. In: *Advances in Photochemistry,* Neckers, D.C., Von Bunau, G., and Jenks, W.S., eds. New York: John Wiley & Sons, 2001:145–217.
26. Timtcheva, I., Maximova, V., Deligeorgiev, T., Gadjev, N., Drexhage, K.H., and Petkova, I., Homodimeric monomethine cyanine dyes as fluorescent probes of biopolymers. *J Photochem Photobiol B,* 58, 130–135, 2000.
27. The human genome. *Nature,* 409, 813–958, 2001.
28. The human genome. *Science,* 291, 1177–1351, 2001.
29. Enderlein, J., Robbins, D.L., Ambrose, W.P., and Keller, R.A., Molecular shot noise, burst size distribution, and single molecule detection in fluid flow: effects of multiple occupancy. *J Phys Chem A,* 102, 6089–6094, 1998.
30. VanOrden, A., Machara, N.P., Goodwin, P.M., and Keller, R.A., Single-molecule identification in flowing sample streams by fluorescence burst size and intraburst fluorescence decay rate. *Anal Chem,* 70, 1444–1451, 1998.
31. Garcia-Ramos, J.V. and Sanches-Cortes, S., Metal colloids employed in the SERS of biomolecules: activation when exciting in the visible and near-infrared regions. *J Mol Struct,* 405, 13–28, 1997.
32. Jensen, T.R., Malinsky, M.D., Haynes, C.L., and Van Duyne, R.P., Nanosphere lithography: tunable localized surface plasmon resonance spectra of nanoparticles. *J Phys Chem, B,* 104, 10549–10556, 2000.
33. Malinsky, M.D., Kelly, K.L., Schatz, G.C., and Van Duyne, R.P., Chain length dependence and sensing capabilities of the localized surface plasmon resonance of silver nanoparticles chemically modified with alkanethiol self-assembled monolayers. *J Am Chem Soc,* 123, 1471–1482, 2001.

34. Eck, D., Helm, C.A., Wagner, N.J., and Vaynberg, K.A., Plasmon resonance measurements of the adsorption and adsorption kinetics of a biopolymer onto gold nanocolloids. *Langmuir,* 17, 957–960, 2001.

35. Valina-Saba, M., Bauer, G., Stich, N., Pittner, F., and Schalkhammer, T., A self-assembled shell of 11-mercaptoundecanoic aminophenylboronic acids on gold nanoclusters. *Mater Sci Eng C,* 8–9, 205–209, 1999.

36. Lakowicz, J.R., Malicka, J., and Gryczynski, I., Silver particles enhance emission of fluorescent DNA oligomers. *BioTechniques,* 34, 62–68, 2001.

37. Rye, H., Yue, S., Wemmer, D., Quesada, M.A., Haugland, R.P., Mathies, R.A., and Glazer, A.N., Stable fluorescent complexes of double-stranded DNA with bis-intercalating asymmetric cyanine dyes: properties of applications. *Nucleic Acid Res,* 20, 2803–2812, 1992.

38. Benson, S.C., Mathies, R.A., and Glazer, A.N., Heterodimeric DNA binding dyes designed for energy transfer: stability and applications of the DNA complexes. *Nucleic Acids Res,* 21, 5720–5726, 1993.

39. Benson, S.C., Zeng, Z., and Glazer, A.N., Fluorescence energy transfer cyanine heterodimers with high affinity for double-stranded DNA. *Anal Biochem,* 231, 247–255, 1995.

40. Lakowicz, J.R., Malicka, J., and Gryczynski, I., Increased intensities of YOYO-1-labeled DNA oligomers near silver particles. *Photochem Photobiol,* 77, 604–608, 2003.

41. Morrison, L.E., Fluorescence in nucleic acid hybridization assays. In: *Topics in Fluorescence Spectroscopy,* vol. 7, *DNA Technology,* Lakowicz, J.R., ed. New York: Kluwer Academic, 2003:69–103.

42. Brown, P.O. and Botstein, D., Exploring the new world of the genome with DNA microarrays. *Nat Genet,* 21(1 suppl), 33–37, 1999.

43. Schena, M., Heller, R.A., Theriault, T.P., Konrad, K., Lachenmeier, E., and Davis, R.W., Microarrays: biotechnology's discovery platform for functional genomics. *Trends Biotechnol,* 16, 301–306, 1998.

44. Komurian-Pradel, F., Paranhos-Bacala, G., Sodoyer, M., Chevallier, P., Mandrand, B., Lotteau, V., and Andre, P., Quantitation of HCV RNA using real-time PCR and fluorimetry. *J Virol Methods,* 95, 111–119, 2001.

45. Walker, N.J., A technique whose time has come. *Science,* 296, 557–559, 2002.

46. Difilippantonio, M.J. and Ried, T., Technicolor genome analysis. In: *Topics in Fluorescence Spectroscopy,* vol. 7, *DNA Technology,* Lakowicz, J.R., ed. New York: Kluwer Academic, 2003:291–316.

47. Soper, S.A., Nutter, H.L., Keller, R.A., Davis, L.M., and Shera, E.B., The photophysical constants of several fluorescent dyes pertaining to ultrasensitive fluorescence spectroscopy. *Photochem Photobiol,* 57, 972–977, 1993.

48. Amrbose, W.P., Goodwin, P.M., Jett, J.H., VanOrden, A., Werner, J.H., and Keller, R.A., Single molecule fluorescence spectroscopy at ambient temperature. *Chem Rev,* 99, 2929–2956, 1999.

49. Malicka, J., Gryczynski, I., and Lakowicz, J.R., DNA hybridization assays using metal-enhanced fluorescence. *Biochem Biophys Res Commun,* 306, 213–218, 2003.

50. Szmacinski, H. and Lakowicz, J.R., Lifetime-based sensing. In: *Topics in Fluorescence Spectroscopy,* Lakowicz, J.R., ed. New York: Plenum Press, 1994:295–334.

51. Forster, T., Intermolecular energy migration and fluorescence. *Ann Phys,* 2, 55–75, 1948.

52. Lakowicz, J.R., Malicka, J., D'Auria, S., and Gryczynski, I., Release of the self-quenching of fluorescence near silver metallic surfaces. *Anal Biochem,* 320, 13–20, 2003.

53. Lakowicz, J.R., Gryczynski, I., Malicka, J., Gryczynski, Z., and Geddes, C.D., Enhanced and localized multiphoton excited fluorescence near metallic silver islands: metallic islands can increase probe photostability. *J Fluoresc,* 12, 299–302, 2002.

54. Wu, P. and Brand, L., Resonance energy transfer: methods and applications. *Anal Biochem,* 218, 1–13, 1994.

55. Dos Remedios, C.G. and Moens, P.D.J., Fluorescence resonance energy transfer spectroscopy is a reliable "ruler" for measuring structural changes in proteins. *J Struct Biol,* 115, 175–185, 1995.

56. Lilley, D.M.J. and Wilson, T.J., Fluorescence resonance energy transfer as a structural tool for nucleic acids. *Curr Opin Chem Biol,* 4, 507–517, 2000.

57. Walter, F., Murchie, A.I.H., Duckett, D., and Lilley, D.M.J., Global structure of four-way RNA junctions studied using fluorescence resonance energy transfer. *RNA,* 4, 719–728, 1998.

58. Mitsui, T., Nakano, H., and Yamana, K., Coumarin–fluorescein pair as a new donor–acceptor set for fluorescence energy transfer study of DNA. *Tetrahedron Lett,* 41, 2605–2608, 2000.

59. Norman, D.G., Grainger, R.J., Uhrin, D., and Lilley, D.M.J., Location of cyanine-3 on double-stranded DNA: importance for fluorescence resonance energy transfer studies. *Biochemistry,* 39, 6317–6324, 2000.

60. Sueda, S., Yuan, J., and Matsumoto, K., Homogenous DNA hybridization assay by using europium luminescence energy transfer. *Bioconjugate Chem,* 11, 827–831, 2000.

61. Selvin, P.R., Rana, T.M., and Hearst, J.E., Luminescence resonance energy transfer. *J Am Chem Soc,* 116, 6029–6030, 1994.

62. Selvin, P.R., Lanthanide-based resonance energy transfer. *IEEE J Selected Topics Quantum Electron,* 2, 1077–1087, 1996.

63. Mathis, G., Rare earth cryptates and homogenous fluoroimmunoassays with human sera. *Clin Chem,* 39, 1953–1959, 1993.

64. Malicka, J., Gryczynski, I., Fang, J., Kusba, J., and Lakowicz, J.R., Increased resonance energy transfer between fluorophores bound to DNA in proximity to metallic silver particles. *Anal Biochem,* 315, 160–169, 2003.

65. Hua, X.M., Gersten, J.I., and Nitzan, A., Theory of energy transfer between molecules near solid state particles. *J Chem Phys,* 83, 3650–3659, 1985.

66. Gersten, J.I. and Nitzan, A., Accelerated energy transfer between molecules near a solid particle. *Chem Phys Lett,* 104, 31–37, 1984.

67. Geddes, C.D., Gryczynski, I., Malicka, J., Gryczynski, Z., and Lakowicz, J.R., Metal-enhanced fluorescence: potential applications in HTS. *Comb Chem High Throughput Screen,* 6, 109–117, 2003.

8 Subpicomolar Assays of Antibodies and DNA Using Phosphorescence Labels

A.P. Savitsky, Ph.D., D.Sc.

CONTENTS

8.1 INTRODUCTION

Over the last few decades, analytical chemistry has undergone a series of substantial methodological changes. This can be explained by the fact that, at the present time, methods of analytical biochemistry, mainly immunochemical analysis and DNA-based diagnostics, are widely used in biotechnology, clinical diagnostics, environmental protection, and food quality control. These methods are based on the use of labeled compounds, and in this case very high sensitivity and selectivity are achieved. However, because of their high sensitivity, all these methods face one very important problem, that is, the necessity to minimize nonspecific signals from a sample.

Different compounds — radioactive atoms, enzymes, and fluorescent compounds — are used as labels. In recent years, sufficiently general and, at the same time, extremely effective instrumental ways to minimize nonspecific signals have been developed for fluorescence methods. Time-resolved fluorescence (TRF) analysis is one of the most promising approaches for biological and chemical applications because this method is one of a few methods that successfully compete with radio-isotopic analysis in measurement sensitivity and accuracy. TRF allows one to minimize background signals and measure weak photoluminescence signals from labeled compounds in complex biological samples. TRF differs fundamentally from other fluorescence assay methods at the photoluminescence measurement step. After the sample is excited with a short light pulse (usually 1 to 2 μs), the collection of the emission is delayed by several dozens or hundreds of microseconds. Background signals decay almost completely during this delay and the signal of the long-lived label is measured with almost no background. In order to increase the integral signal, the recording cycle may be repeated many times (1000 times or more) (Figure 8.1).

Scattered light and self-luminescence of biological samples are the main contributors to the background signal. Strong light scattering occurs due to a high concentration of macromolecules in biological samples. Scattered light generally concurs with the flash profile. Background luminescence of biological samples is caused by chromophore-containing proteins and other molecules. For example, the time for the intrinsic emission of blood serum to decay is 50 to 100 ns.[1] However, in addition to light scattering and self-luminescence of the sample, the background emission of cuvettes (plastic), solutions (different admixtures in salts, etc.), and construction materials of the cuvette compartment itself may contribute to the background signal. The decay time of these background signals does not usually exceed 100 to 200 μs.[1-3]

Thus, in spite of some special additional requirements for photoluminescent labels, the introduction of a microsecond delay before recording of the label luminescence reduces the background signals to a minimum. This approach allows one to increase 100- to 1000-fold the sensitivity of detection compared with that of steady-state fluorescence. For the TRF method, two main classes of photoluminescence labels have been proposed. These are lanthanide labels[1,2] and phosphorescence labels.[3,4]

The phosphorescence immunoassay method based on metalloporphyrins as labels was proposed several years ago[5] and has several advantages over lanthanide-based assays. Metalloporphyrin labels are individual chemical compounds that are thermodynamically (metal ion dissociation constants less than 10^{-34}) and kinetically stable as well as resistant to acids, alkalies, and heavy metal ions.

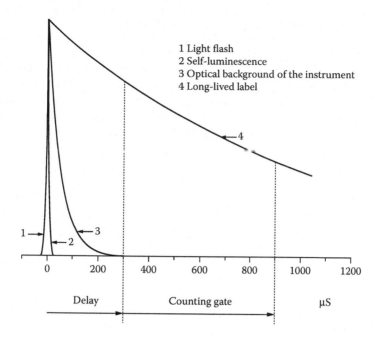

1 Light flash
2 Self-luminescence
3 Optical background of the instrument
4 Long-lived label

FIGURE 8.1 The principle of detection of phosphorescence with a pulsed light source. (1) Flash with 10–1000-Hz frequency, duration 1–2 μs. (2) Fluorescence of biological samples (not longer than 100 ns). (3) Phosphorescence of the optical construction elements and cells (not longer than 200 μs). (4) Luminescence of a long-lived label.

Methods for the measurement of metalloporphyrin phosphorescence at room temperature, both in aqueous solutions and on dry surfaces, are well developed and are described in detail below. These techniques are simple and in good agreement with practically all methods and techniques used in modern immunochemistry and DNA diagnostics. This review provides a detailed description of the application of phosphorescence labels for such assays.

8.2 PHOSPHORESCENCE LABELS: STRUCTURE AND SPECTRA

At present, the phenomenon of high-temperature phosphorescence (room-temperature phosphorescence [RTP]) is attracting the attention of specialists working in the field of analytical chemistry and biochemistry.[6] Biological samples (blood sera) do not give noticeable background signals under the conditions of high-temperature phosphorescence measurement and, therefore, phosphorescence labels may be efficiently used in immunoassay. Some derivatives of xanthene dyes containing heavy atoms in their structures (e.g., erythrosin B) are able to phosphoresce at room temperature in aqueous solutions with luminescence times of about 100 μs.[3,7] This property is conserved in their conjugates with proteins,[7] but low absolute sensitivity limits their use in immunoassays.

Metalloporphyrins that phosphoresce at room temperature in aqueous solutions were first proposed as luminescence markers for antigens and antibodies by Savitsky et al.[5] It is well known that many metalloporphyrins phosphoresce at liquid nitrogen temperature.[8] Palladium, platinum, and lutetium metalloporphyrins phosphoresce with high quantum yields at room temperature in deoxygenated aqueous and organic solutions.[8,9] Zinc, aluminum, yttrium, and tin metalloporphyrins also manifest delayed fluorescence[9] in addition to high-temperature phosphorescence.

Metalloporphyrins suitable for immunoassays should be easily soluble in water and photochemically stable; there should also be relatively simple methods to synthesize their conjugates with proteins and other physiologically active compounds. The structural formulas of different derivatives of the three main classes of metalloporphyrins (derivatives of porphyrin, tetraphenylporphyrin, and tetraphenyl-(tetrabenzo)-porphyrin)[10,11] are shown in Figure 8.2, and the spectral and luminescence properties of these porphyrins are summarized in Table 8.1.

From the spectral point of view, the compounds in Table 8.1 may be divided into four groups:

- Platinum porphyrins (compound 1)
- Palladium porphyrins (compound 2)
- Palladium tetraphenylporphyrins (compounds 5 and 6)
- Palladium tetrabenzo-(tetratolyl) porphyrins (compound 7)

Typical absorption and fluorescence spectra are shown in Figure 8.3.

The most efficient phosphor from the spectral point of view is the platinum coproporphyrin (PtCP; compound 1). However, its lifetime in the excited state is approximately 100 μs, and the efficiency of its use in specific types of immunoassays strongly depends on the quality of optical filters and the background phosphorescence of the samples and spectral cells. Palladium coproporphyrin (PdCP; compound 3) is less efficient than PtCP, but the lifetime of PdCPs is more than 900 μs. Therefore, using the same filter set but a larger delay time, it is possible to reach more efficient discrimination of background phosphorescence, and the detection limit for these two compounds obtained on a specially modified DELFIA Plate Fluorometer 1232 (Wallac, Inc., Turku, Finland) is practically the same.[10] Platinum coproporphyrin has an excitation wavelength peak at 532 nm that can be efficiently used for excitation by frequency-doubled neodymium:yttrium-aluminium-garnet (Nd:YAG) pulsed laser.[12]

The substituents in the mesoposition of the porphyrin macrocycle have a strong negative effect on the metalloporphyrin quantum yield. Even the presence of one substituent in a mesoposition results in a three- to fivefold decrease in the quantum yield and a shorter lifetime (compound 3). Tetraphenylporphyrins, which are very easy to synthesize (compounds 5 and 6), are much less efficient than coproporphyrins.

Emission spectra of palladium tetrabenzo-(tetratolyl) porphyrin tetrasulfonic acid (compound 7) are rather unusual for palladium complexes with porphyrins (Figure 8.3b). The first emission band of these compounds has a small shift (approximately 10 nm) from the last absorbance band that is typical of emission from a singlet state. At the same time, both emission peaks at 642–643 nm and 790–795 nm

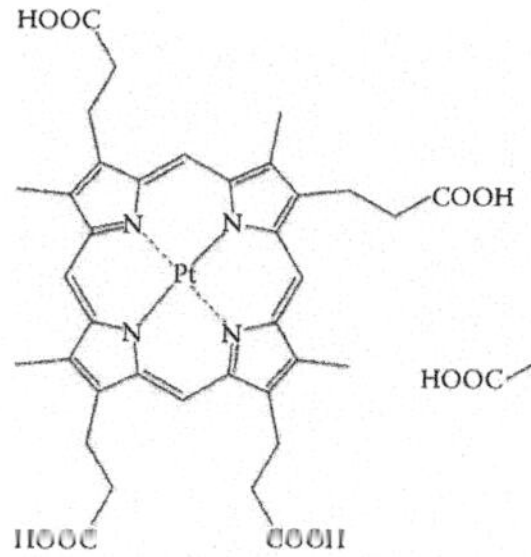

1. PtCP (III) 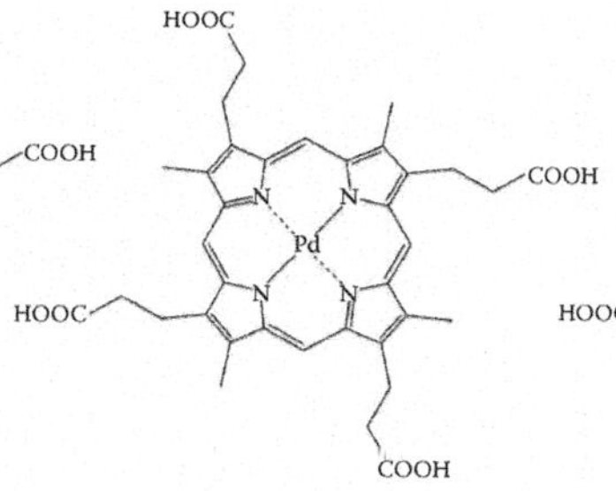 2. PdCP (I) 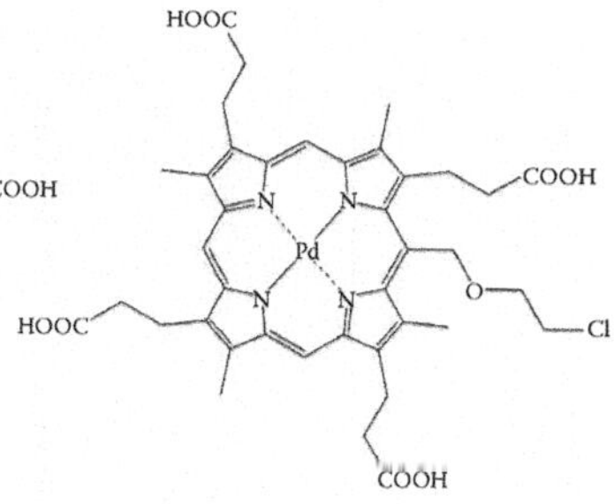 3. PdCP (R)Cl

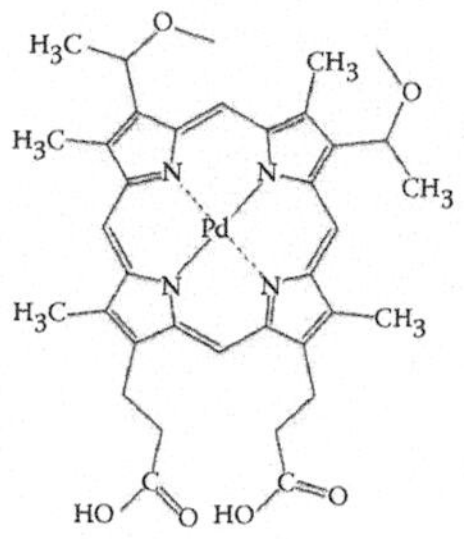

4. PdDMHP 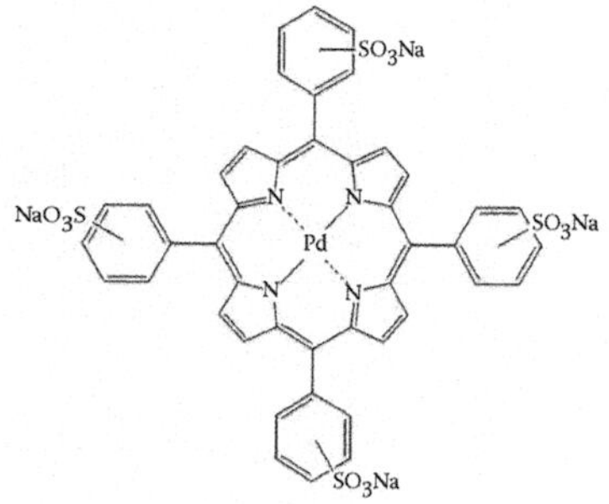 5. PdTSPP

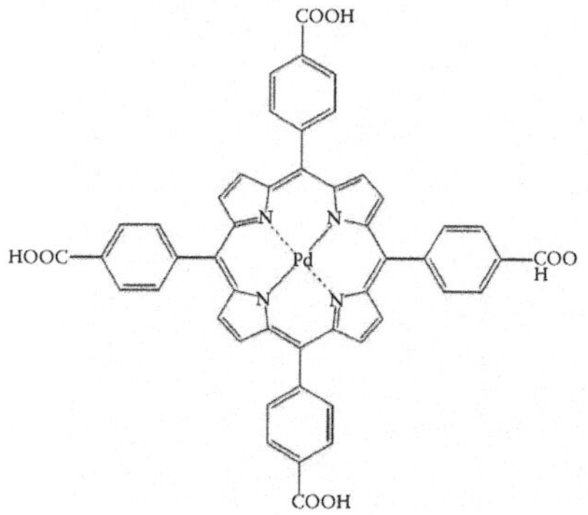

6. PdTCTP 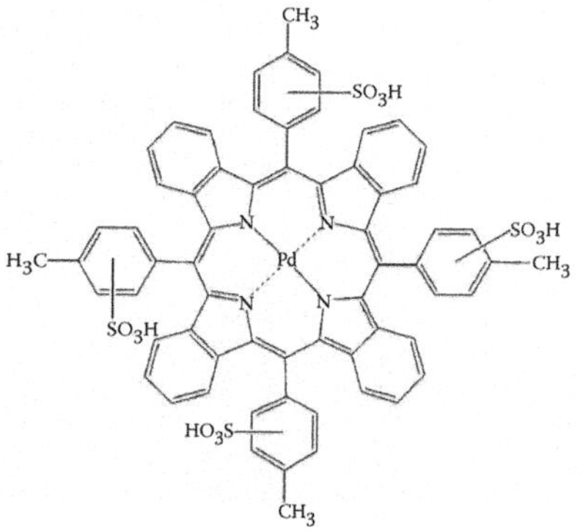 7. PdTTTPTS

FIGURE 8.2 The structures of phosphorescent porphyrin derivatives.

decay simultaneously with a lifetime of 325–330 μs, which is typical of phosphorescence. This means that the first emission band at 642–643 nm may be attributed to delayed fluorescence, most probably E-type. Both absorbance bands of this compound are suitable for laser excitation. In the near-infrared range, fluorescent background signals of biological samples are very small and it is possible to achieve high sensitivity in the detection of labeled compounds. With palladium tetrabenzo-(tetratolyl) porphyrin tetrasulfonic acid as a label, it is possible to use microsecond time resolution to cut off background signals for near-infrared labels and thus improve sensitivity.[11]

TABLE 8.1
Spectral and Luminescence Properties of Phosphorescing Metalloporphyrins at Room Temperature in Deoxygenated Micellar Solutions

	Sample	Detergent	Molar Extinction (per M/cm × 10^5)	Excitation (nm)	Emission (nm)	Quantum Yield	Lifetime (μs)	Reference
1	PtCP(III)	TX-100	1.76	381	646	0.76	108	37
		CTAB	1.71	381	647	0.55	82	37
2	PdCP(I)	TX-100	1.81	391.5	666	0.17	984	37
		CTAB	1.79	393	667	0.064	490	37
3	PdCP-(R-)Cl	TX-100	2.04	397.5	688.5	0.056	592	37
		CTAB	1.75	398.5	684.5	0.089	457	37
4	PdDMHP	TX-100	1.64	396	668	0.214	1100	37
5	PdTSPP	TX-100	1.94	416	694	0.069	675	37
6	PdTCPP	TX-100	2.26	418	699	0.0084	586	37
7	PdTTTPTS	Cyclodextrin		449	790	0.13	330	11
				630				

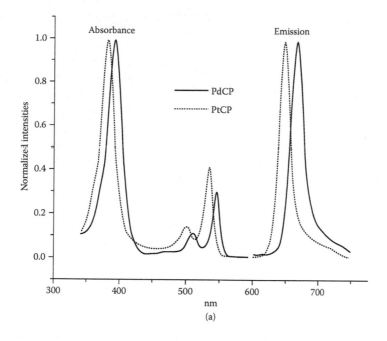

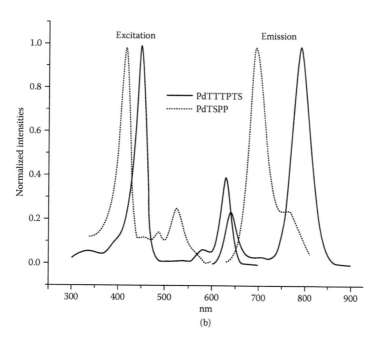

FIGURE 8.3 Absorbance and emission spectra of porphyrin derivatives in 2% TX-100 solution. (a) Solid lines: PdCP, Compound 2; dashed lines: PtCP, Compound 1. (b) Solid lines: PdTTTPTS, Compound 7; dashed lines: PdTSPP, Compound 5.

8.3 METHODS FOR MEASURING PHOSPHORESCENCE AT ROOM TEMPERATURE

A general problem for all types of phosphorescence measurements at room temperature is the sensitivity of the phosphorescence to the solvent type and quenching with oxygen. However, only a combination of these two factors (liquid phase + oxygen) leads to phosphorescence quenching at room temperature. Removal of oxygen makes it possible to measure phosphorescence of metalloporphyrins in solutions at room temperature. Removal of the solvent allows one to measure phosphorescence in the presence of oxygen. Both general approaches can be used in special schemes for immunochemical assay.

8.3.1 DEOXYGENATION OF SOLUTIONS

There are a large number of chemical methods for the removal of oxygen from aqueous solutions, for example, using strong reducing agents such as sulfite and dithionite. Different sulfite salts (sulfite, bisulfite, metabisulfite) are the most convenient for phosphorescence measurements. Oxygen is removed as a result of the following chemical reaction:

$$2SO_3^{2-} + O_2 \Rightarrow 2SO_4^{2-}$$

Depending on the sulfite concentration, deoxygenation occurs in several seconds (15–30 s) and the reverse diffusion of oxygen to the solution is very slow. The concentration of oxygen in water is usually about 2×10^{-4} M and sulfite concentrations used are several milligrams per milliliter or 0.1 to 0.01 M. The excess sulfite quickly enters into the reaction with newly dissolved oxygen at the surface; the concentration of oxygen is negligibly low already at a small distance from the surface. Anaerobic conditions may be retained for several hours. Figure 8.4 shows the dependence of phosphorescence vs. time for different concentrations of PdCP in the wells of a standard microplate for immunological tests. The solution volume was 200 µl. As can be seen, the phosphorescence intensity remained practically unchanged over 3 h.[13] Some other biochemical[14] and physical (flushed with nitrogen)[15] methods for deoxygenation of solution may be used for phosphorescent assays.

8.3.2 MICELLAR SOLUTIONS

Water is known to be an effective phosphorescence quencher for metalloporphyrins at room temperature. Direct confirmation of this fact is the increase in the quantum yield of PdCP in heavy water compared with plain water and the lowering of this effect by detergent (Figure 8.5).

As can be seen from Figure 8.5, the intensity of phosphorescence of PdCP in deuterium water is five times that of plain water, whereas the addition of detergent to both water types results in almost identical values of phosphorescence intensity of free PdCP.[16]

For a series of phosphorescing dyes, it has been shown that micellar solutions minimize the quenching effect of water.[6] The tetrapyrrole macrocycle in water-soluble

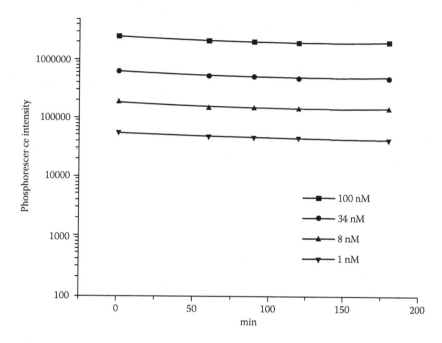

FIGURE 8.4 Time dependence of phosphorescence intensity of 100 nM (squares), 34 nM (circles), 8 nM (triangles), and 1 nM (inverted triangles) PdCP in 20 mg/ml (0.16 M) Na$_2$SO$_3$, 40 mg/ml (0.11 M) NaH$_2$PO$_4$·12H$_2$O buffer, 2% TX-100, pH 7.2.

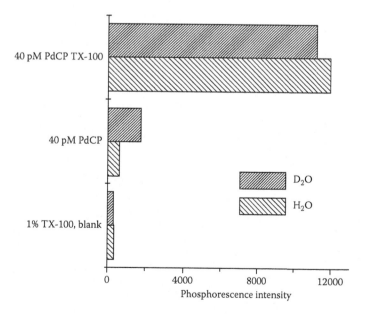

FIGURE 8.5 Phosphorescence intensity in D$_2$O (fine grid) and H$_2$O (coarse grid) of 40 pM PdCP in TX-100, in solution, and of TX-100 alone.

TABLE 8.2
Enhancement Factors for PdCP Phosphorescence in the Presence of Different Detergents at pH 7.2

	Detergent	Abbreviation	Molecular Weight	Concentration (% or M)	Charge	Enhancement Factors
1	Cetyltrimethylammonium bromide	CTAB	364.46	1.000 mM	+	6.4
2	Benzyldimethylammonium chloride	BDHD	396.1	0.011 mM	+	12.1
3	Phemerol chloride	BTC	448.1	0.039 mM	+	14.7
4	Methylphemerol chloride	MBTC	462.1	0.020 mM	+	14.2
5	TX-100	TX-100	646.86	2.31%	0	12.8
6	TX-405	TX-405	1966.86	6.94%	0	14.1
7	Tween-20	TW-20	1227.54	0.26%	0	7.5
8	Polyoxyethylene-10 dodecyl ether	POELE	626.87	2.31%	0	13.0
9	Dodecyldimethylamino-propane sulfonate	DDAPS	335.5	0.07%	+	10.4
10	Sodium dioctylsulphosuccinate	AOT	444.57	6.94%	−	2.9

Data from Savitsky.[37]

porphyrins is shown to be buried deep inside the hydrophobic nucleus of the micelle and the side substituents are presumably localized in the Stern layer. Therefore, the porphyrin macrocycle, which is responsible for light absorption and luminescence, is shielded from the interaction with water and water-dissolved quenchers.[17]

Almost all studied surface-active compounds (detergents) have a pronounced effect on the quantum yield of metalloporphyrin phosphorescence at room temperature (Table 8.2). However, the phosphorescence enhancement factors are different for different classes of detergents. For PdCP, negatively charged detergents give the weakest enhancement, which can be explained by the electrostatic factor, because this porphyrin is negatively charged (detergent 10). The rest of the detergents — positively charged, neural, or zwitter-ionic — have more or less identical effects. The hydrophobic portion of the detergent has a pronounced effect on the enhancement of phosphorescence; however, it is difficult to single out any preferable structure.

The values of the enhancement factors in Table 8.2 are given for the concentrations of the detergents when the maximum phosphorescence intensities are observed. Figure 8.6 shows a plot of phosphorescence intensity vs. the concentration of PdCP in a 5% solution of Triton X-405 (TX-405). The phosphorescence intensity is linear over six orders of magnitude and the detection limit is 10^{-13} M, or 10 amol in the sample.[18] As can be seen in Figure 8.6, the effect of self-quenching typical of aqueous solutions of PdCP is not observed in the solution of TX-405, even at high concentrations of this porphyrin.[9]

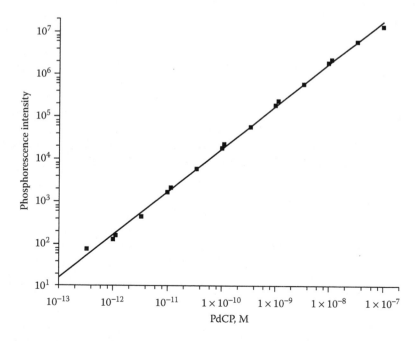

FIGURE 8.6 Phosphorescence intensity as a function of PdCP concentration in 5% solution of TX-405 at pH 7.2.

One of the most common detergents, Triton X-100 (TX-100), is quite suitable for carrying out phosphorescence measurements in solution. However, noticeable enhancement of phosphorescence is observed at concentrations of the detergent greater than 1% (i.e., at concentrations that destroy specific antibody–antigen interactions). Such high concentrations are also able to destroy complex biological structures *in vivo*, therefore, detergents may be used only for *in vitro* phosphorescence analysis.

8.3.3 CYCLODEXTRINS

Cyclodextrins are cyclic oligomers of glucose. Small molecules may be placed inside the cavity that is formed inside the ring to form inclusion complexes. The binding in these complexes is predominantly hydrophobic in nature.[19]

There are three main types of cyclodextrins: α-cyclodextrin, a six-membered ring; β-cyclodextrin, a seven-membered ring; and γ-cyclodextrin, an eight-membered ring. Only α-cyclodextrin (unless derivatized) is soluble in concentrations sufficient to form inclusion complexes. Its solubility in water is up to 10%. The solubility of β- and γ-cyclodextrins does not exceed 1%. To increase solubility, these cyclodextrins are subjected to methylation or hydroxypropylation, and the solubility of such modified derivatives is usually about 20%. The exact locations of modified sites are usually unknown. All three types of cyclodextrins significantly enhance the phosphorescence of metalloporphyrins at room temperature.

Enhancement factors of PdCP phosphorescence with different cyclodextrins vary depending on the cycle size (Table 8.3). The most efficient is met-β-cyclodextrin-1.8; its enhancement factor is greater than those for the most efficient detergents, TX-405 and phemerol chloride (BTC) (compare Table 8.2 and Table 8.3). However, the dissociation constant of the complex of PdCP with met-β-cyclodextrin-1.8 is rather low. Therefore to achieve a noticeable effect of phosphorescence enhancement, it is necessary to use decimolar concentrations of this cyclodextrin. It should be noted that cyclodextrins at such concentrations do not destroy specific protein–protein interactions, and they are suitable for nondestructive methods of phosphorescence analysis of biological objects involving phosphorescence microscopy, as in studies of live cells in the so-called vital methods.

TABLE 8.3
Enhancement Factors and Dissociation Constants for PdCP at pH 7.2

	Cyclodextrin	Concentration (M)	PdCP	
			Enhancement	K_{diss} (per M)
1	α-cyclodextrin	0.084	10.3	
2	met-β-cyclodextrin-1.8	0.172	15.5	$2.26 \cdot 10^2$
3	2-hydroxypropyl-γ-cyclodextrin-0.6	0.097	14.4	

8.3.4 pH DEPENDENCE OF PHOSPHORESCENCE

Solution pH has a great impact on the phosphorescence of metalloporphyrins.[18] At acidic pH, the intensity of PdCP phosphorescence is decreased to almost zero (Figure 8.7). Detergents increase the phosphorescence quantum yield, but at the same time they have practically no influence on the form of the curve in the acidic region. The nature of the detergent is immaterial. Such a decrease in fluorescence intensity at acidic pH was observed for a series of nonmetal porphyrins and at that time it was explained by the protonation of the porphyrin carboxy groups at acidic pH values.[17]

The pH dependence of the phosphorescence of PdCP at alkaline pH values is rather unusual (Figure 8.8).[18] The phosphorescence intensity under these conditions depends strongly on the mode of sample irradiation, but it does not matter what type of irradiation—impulse or continuous—is used. The curves for lasers in Figure 8.8 correspond to excitation with a pulsed nitrogen laser and continuous helium–cadmium laser with mechanical chopper (the light dose per second differs insignificantly) (Table 8.4). The pH dependencies for these two lasers are practically the same. In the case of excitation with the pulsed xenon lamp, which gives significantly

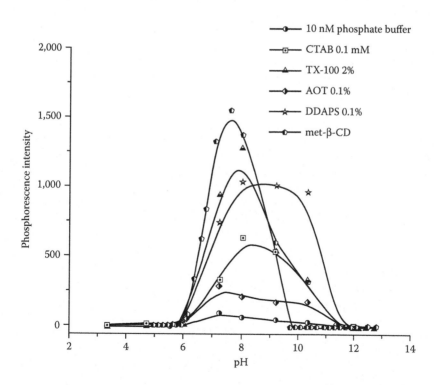

FIGURE 8.7 The pH dependence of phosphorescence intensity of PdCP in 10 mM phosphate buffer (circles), 0.1 mM CTAB (squares), 2% TX-100 (triangles), 0.1% sodium dioctylsulphosuccinate (AOT) (diamonds), 0.1% dodecyldimethylaminopropane sulfonate (DDAPS) (stars), and met-β-cyclodextrin (pentagons).

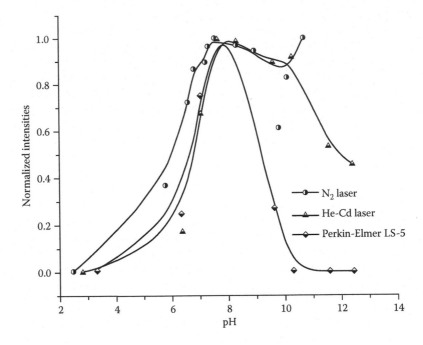

FIGURE 8.8 The pH dependence of PdCP phosphorescence intensity excited with a pulsed nitrogen laser (circles), continuous-wave helium–cadmium laser (triangles), and pulsed xenon lamp (diamonds).

less light, the dependence is noticeably different (Figure 8.8). Dependence on the light dose rather than on the energy of single impulses or their frequencies allows one to assume the photochemical nature of this phenomenon. The most probable explanation is photodissociation of porphyrin dimers, which can exist both in micellar and cyclodextrin solutions at alkaline pH.

The most important feature of the pH dependence of metalloporphyrin phosphorescence is the existence of the optimum at neutral, or so-called physiological,

TABLE 8.4
Comparison Characteristics of Lasers Used in the Studies of pH Dependence of PdCP (See Figure 8.10)

Laser	Helium-Cadmium	Nitrogen
Wavelength (nm)	325	337
Operating mode	Continuous with mechanical chopper	Impulse
Quanta per impulse	1.6×10^{12}	5×10^{13}
Impulse breadth (μs)	100	0.01
Impulse frequency	1000	100
Light dose per second	1.6×10^{15}	5×10^{15}

pH values. It is essential that the optimum for cyclodextrins is also observed in this pH region, which makes cyclodextrins ideal compounds for enhancement of metalloporphyrin phosphorescence and avoids the quenching effect of water.

8.3.5 Measurements on Dry Surfaces

It is well known that phosphorescence of some aromatic compounds at room temperature may be easily detected from the surface of different solid supports such as filter paper, silica gel, aluminum oxide, silicon rubber, and crystals of sodium acetate or potassium bromide.[20] However, even if a fluorophore is bound to an appropriate solid support, its fluorescence is still sensitive to both oxygen and residual moisture on the surface. The phosphorescence signal is strongly affected by such factors as the chemical composition of the solution, procedures of solvent vaporization and drying of the solid support, and the interaction of the fluorophore with the solid support. To minimize the influence of the above-mentioned factors, the cuvette compartment is usually purged with dry nitrogen or air.[20]

   In contrast to the conditions described above, the intensity of phosphorescence of PdCP conjugated with streptavidin measured from dry surfaces in the presence of oxygen is practically the same as the signal intensity of the same conjugate in deoxygenated micellar solutions (Figure 8.9). To obtain such a signal,

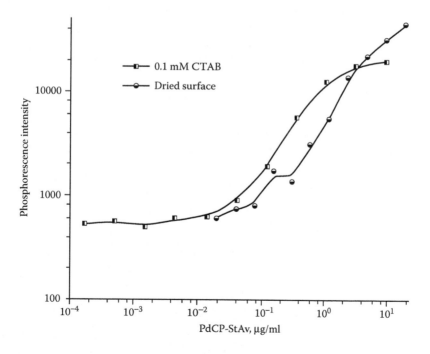

FIGURE 8.9 Concentration dependence of PdCP–streptavidin conjugate bound to biotinylated mouse antibodies adsorbed to a dry polystyrene microwell plate surface (circles) and in 0.1 mM CTAB solution with 20% polyethylene glycol 6000 (squares).

the sample was simply dried in air at normal humidity for 15 min without any special precautions.[16]

The chemical composition of a solid support (polystyrene, glass, silicon, cell surface) does not play any role and has practically no effect on the intensity of the phosphorescence signal. The method of coupling the PdCP to the protein (covalent or specific interaction with antibodies) as well as addition of the protein to the support (adsorption on polystyrene or chemical fixation on glass in a Langmuir-Blodgett film) also does not exert a noticeable influence on the phosphorescence signal.[10,16,21]

Different polymer compounds similar to the polymer-fixing media for microscopy may be efficiently used for the enhancement of phosphorescence signals obtained from a surface. After solidification of a polymer-fixing medium, the phosphorescence signal of PdCP is even higher than that obtained from a dry surface. However, composition of the polymer is very important for phosphorescence enhancement. One of the most efficient polymer-fixing media for phosphorescent compounds is Merckoglas (Merck, Darmstadt, Germany) (Figure 8.10).[16,22]

However, in this case the concentration dependence of the phosphorescence is more complex than for micellar solutions. For example, the detection limit for biotinylated rabbit anti-mouse antibodies in deoxygenated micellar solutions is almost five times higher than for dry surfaces (Figure 8.11, panel a), but the signal intensity at high concentrations of the streptavidin conjugate of PdCP is significantly higher for dry surfaces compared with the detergent solution (Figure 8.9 and Figure 8.11, panel b). A similar result was obtained for noncovalent binding of PdCP to proteins. An immune complex of PdCP with specific antibodies is able to form a secondary immune complex with rabbit anti-mouse antibodies adsorbed on a polystyrene surface. In this case, the

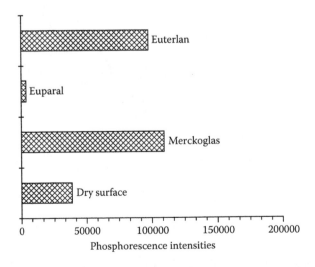

FIGURE 8.10 Phosphorescence intensities of PdCP–streptavidin conjugate bound to biotinylated antibodies on the surface of a microwell plate in embedding media.

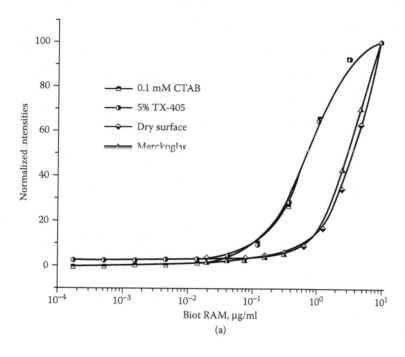

(a)

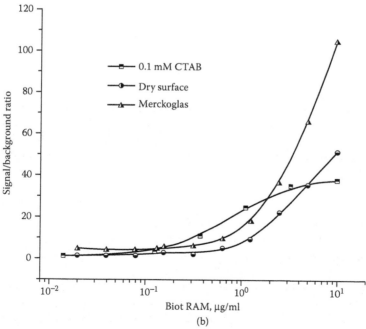

(b)

FIGURE 8.11 Phosphorescence intensities of biotinylated rabbit anti-mouse antibody determined by PdCP (panel a), streptavidin-conjugated PdCP (panel b), and rabbit anti-mouse antibodies adsorbed to a surface (panel c).

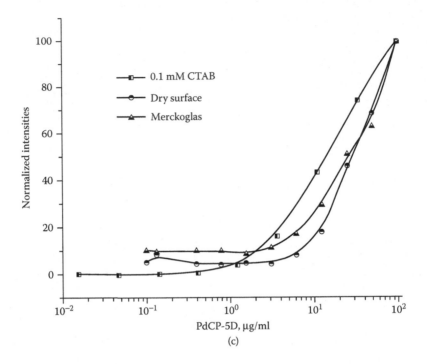

FIGURE 8.11 (Continued).

detection limits of PdCP in micellar solutions and on dry surfaces differ twofold (Figure 8.11, panel c).

Therefore, measuring phosphorescence from dry surfaces using polymer-fixing media and without them at high surface concentrations of PdCP allows one to obtain higher signals from the surface when compared with detergent solutions. At low surface concentrations of PdCP, the signals from dry surfaces are lower than those in detergent solutions.

8.4 METHODS FOR THE PREPARATION OF CONJUGATES AND FLUORESCENCE PROPERTIES OF CONJUGATES

Metalloporphyrins with sulfo or carboxy groups as side substituents are quite soluble in water. Different activated derivatives of sulfo and carboxy groups may be used to synthesize conjugates of metalloporphyrins with proteins. However, because phosphorescence quantum yields are highest for the derivatives containing carboxy groups,[10] the main attention in the preparation of conjugates has been drawn to this group of porphyrins.

8.4.1 SELECTION OF METALLOPORPHYRINS

The simplest and best-developed methods include the preparation of conjugates through activated esters using N-hydroxysuccinimide.[23] The number of carboxy groups in a porphyrin is critically important for hydrophobic–hydrophilic balance and, hence, for nonspecific adsorption of porphyrin conjugates. This is seen most clearly in the conjugates of streptavidin with PdCP and palladium-dimethoxyhematoporphyrin (PdDMHP) (Figure 8.12 and Figure 8.13).[10] The maximal signal for both adsorbed conjugates is almost the same and equals 700,000 to 800,000 impulses, but the signal:background ratio for the streptavidin–PdCP conjugate is almost 500 (Figure 8.12), whereas this ratio is less than 10 for the streptavidin–PdDMHP conjugate (Figure 8.13). Thus, the optimal compound from the viewpoint of nonspecific sorption and phosphorescence quantum yield is PdCP. *p*-isothiocyanatophenyl derivatives of Pt(II)- and Pd(II)-coproporphyrin I are also described as stable monofunctional derivatives for conjugate preparation.[24]

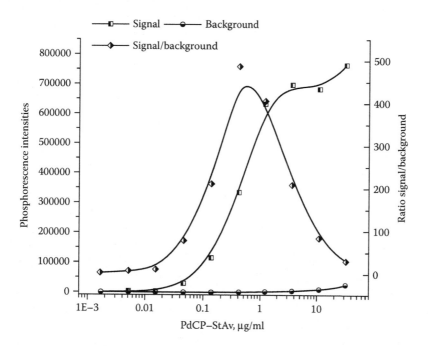

FIGURE 8.12 Micelle-stabilized phosphorescence immunoassay of biotinylated antibodies. Indirect assay using PdCP-labeled streptavidin. Mouse monoclonal antibodies were adsorbed in the microplate wells and biotinylated rabbit anti-mouse antibodies (40 ng per well) and titrated streptavidin-PdCP were added. For the desorption, 100 µl of 0.1 M potassium hydroxide and 5% TX-405 were added for 15 min. Before the measurements, 100 µl of Na_2SO_3 (20 mg/ml) and NaH_2PO_4 (20 mg/ml) were added. The concentration dependence of PdCP–streptavidin conjugate phosphorescence intensity (squares), background levels (circles), and the signal:background ratio (diamonds).

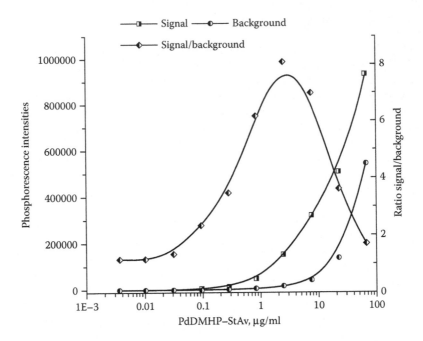

FIGURE 8.13 Micelle-stabilized phosphorescence immunoassay of biotinylated antibodies. Concentration dependence of PdDMHP–streptavidin conjugate phosphorescence intensity (squares), background levels (circles), and the signal:background ratio (diamonds).

8.4.2 QUANTUM YIELD OF CONJUGATES

The quantum yield of fluorescent dyes is often changed upon binding with proteins. By comparison, PdCP covalently conjugated to proteins by the carbodiimide method does not exhibit a change in its quantum yield, but the degree of its phosphorescence enhancement by different detergents varies substantially.[25] The relative quantum yields of different PdCP conjugates are shown in Figure 8.14. The degree of phosphorescence enhancement with TX-405 for the conjugates is significantly lower as compared with free PdCP, resulting in a decrease in the phosphorescence quantum yield of the conjugates. Decreases in the quantum yields of platinum and PdCP conjugates with proteins are described elsewhere.[26]

Conjugation of PdCP with proteins leads to the distortion of the micelle structure around the porphyrin and to violation of the localization site of the porphyrin macrocycle in the micelle: a structure similar to the Stern layer may appear instead of the micelle nucleus. The distortion of the micelle structure and localization of porphyrin in the micelle are clearly shown in experiments with biotinylated PdCP. Four amine derivatives of biotin (biotin cadaverine) may be added to PdCP at its four carboxy groups by the carbodiimide method. As shown in Table 8.5, chemical attachment of PdCP to streptavidin results not in a decrease but in an increase in the phosphorescence quantum yield in the absence of TX-100 (compare positions 2 and 4 in Table 8.5). However, as was noted above, the enhancement factor of the PdCP phosphorescence

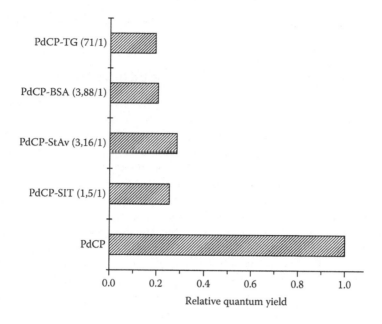

FIGURE 8.14 Relative quantum yield of PdCP–SIT, PdCP–streptavidin, PdCP–BSA, and PdCP–TG. The number of PdCP molecules per protein molecule is indicated in parentheses. Buffer: 30 mg/ml Na_2SO_3, 12 mg/ml NaH_2PO_4, 5% TX-405, 7.27.

with TX-100 in the conjugate is significantly reduced (compare positions 1 and 3). Chemical modification of all four carboxyl groups of PdCP with biotin cadaverine has a weak effect on the quantum yield of PdCP phosphorescence (compare positions 2 and 6). The same weak effect of the modification by this low-molecular-weight compound (biotin cadaverine) is observed for the enhancement of PdCP phosphorescence with TX-100 (compare positions 1 and 5). Thus, biotin cadaverine (unlike

TABLE 8.5
Relative Quantum Yields of PdCP and Its Conjugates with Biotin Cadaverin and Streptavidin

	PdCP or Its Conjugate	Relative Quantum Yield
1	PdCP + 2% TX-100	1.00
2	PdCP	0.07
3	PdCP–streptavidin + 2% TX-100	0.43
4	PdCP–streptavidin	0.11
5	(PdCP/biotin cadaverin = 1/4) + 2% TX-100	0.75
6	(PdCP/biotin cadaverin = 1/4)	0.04
7	(PdCP/biotin cadaverin = 1/4) + streptavidin	0.08
8	(PdCP/biotin cadaverin = 1/4) + streptavidin + 2% TX-100	0.74

high-molecular-weight compounds such as proteins) has only a weak influence on the structure of the micelles around PdCP.

8.4.3 LIFETIMES OF PdCP CONJUGATES IN THE EXCITED STATE

Covalent attachment of palladium and platinum porphyrins to proteins noticeably influences their lifetimes in the excited state.[22,25] For most conjugates, the phosphorescence decay becomes multiexponential (Table 8.6). The existence of two or three exponents is typical of PtCP conjugates, and their multiexponential character becomes more pronounced with an increase in the degree of protein labeling with the porphyrin (see the PtCP–streptavidin entry in Table 8.6). The same regularity is typical of PdCP conjugates; that is, the increase in protein molecular mass and the degree of protein labeling result in the pronounced multiexponential character of PdCP phosphorescence decay, even in TX-100 micellar solution. The distribution between the exponents for the conjugate may depend on the nature of the detergent. The lifetimes in cetyl trimethyl ammonium bromide (CTAB) are, as a rule, shorter and the distribution between fast, medium, and slow exponents is different than in TX-100. Generally, a multiexponential character is a negative feature because it complicates the calibration curves and may cause differences in the calibration curves for different batches of the conjugate.

PdCP–streptavidin is an exception to the rule described above. Either one-exponential decay or pronounced domination of one of the exponents (with the ratio of the dominating exponent exceeding 70%) is typical of this conjugate with three to four labels. This may be due to the fact that streptavidin is composed of four subunits, and one amino group of each subunit preferably enters into the reaction. Thus, the microenvironment of PdCP in the conjugate becomes homogeneous and this is reflected in the kinetics of phosphorescence decay.

Taking into account all the properties mentioned above, it may be concluded that PdCP–streptavidin conjugates meet in the best way the requirements of conjugates for immunochemical experiments. Conjugates of isothiocyanatophenyl derivatives of Pt(II)- and Pd(II)-coproporphyrin I with streptavidin and neutroavidin also may be efficiently used for binding assay of different types.[26]

8.5 SOLID-PHASE ANALYSIS WITH THE DESORPTION OF CONJUGATE FROM THE SURFACE

At the present time, the most widespread type of immunochemical assay is solid-phase analysis on polystyrene plates. All these procedures are similar to the enzyme-linked immunosorbent assay (ELISA). A wide spectrum of universal measurement devices, so-called plate readers, has been developed for these methods. The measurements described below were carried out on Arcus 1230 and DELFIA Plate Fluorometer 1232 or 1234 (Wallac, Inc., Turku, Finland) commercial instruments with specially installed excitation (400 nm) and emission (660 nm) filters.

TABLE 8.6
Photoluminescence Properties of Metalloporphyrin Conjugates

Conjugate	Number of Porphyrin Molecules per Protein Molecule	Measuring Conditions	Exponents						Relative Quantum Yield
			Fast		Medium		Slow		
			Lifetime (µs)	Pre-exponent (%)	Lifetime (µs)	Pre-exponent (%)	Lifetime (µs)	Pre-exponent (%)	
1 PtCP–glucose oxidase (MW = 55,000)	1.3	Buffer			30	47	70	53	—
		TX-100			80	51.0	120	49.0	0.91
2 PtCP–streptavidin (MW = 60,000)	11	TX-100	40.0	7.0	80	54.0	120	39.0	0.12
		Dry surface	40.0	14.9	80	75.5	120	9.6	
		Merckoglas			80	79.9	120	20.1	
3 PdCP–streptavidin	3.16	TX-100					920	100	0.31
		Dry surface	80.0	5.2	800	59.3	1080	35.5	
		Merckoglas	120.0	9.3	960	9.2	1000	81.4	
4 PdDMHP–streptavidin	2	TX-100	200.0	16.0	600	23.0	800	57.0	0.49
5 PdCP–SIT (MW = 25,000)	1.5	Buffer	40.0	18.0	159	40.0	440	38.0	—
		TX-100			279	21.0	783	75.0	0.25
		CTAB	40.0	24.0	147	39.0	455	37.0	—
6 PdCP–BSA (MW = 65,000)	3.88	Buffer	64.0	29.0	240	33.0	640	31.0	—
		TX-100	138.0	20.0	680	31.0	760	47.0	0.20
		CTAB	40.0	39.0	158	48.0	437	14.0	—
7 PdCP–TG (MW = 669,000)	77	Buffer			40	45.0	246	55.0	—
		TX-100	77.0	15.0	520	37.0	680	38.0	0.19
		CTAB	40.0	39.0	144	55.0	387	11.0	—

8.5.1 Micelle-Stabilized Phosphorescence Immunoassay

To desorb a label from the surface of a polystyrene microplate, it is necessary to break the antibody–antigen interaction. It is known that antibody–antigen complexes are destroyed at very low (acidic) or very high (alkaline) pH values. However, the phosphorescence of PdCP at these pH levels is completely quenched[18]; therefore, it is necessary to separate the steps of desorption and phosphorescence signal measuring. Alkaline desorption is more universal because it results in the destruction of not only antibody–antigen complexes but also the interaction of protein with plastic. As may be seen in Figure 8.15, the pH optimum for desorption is pH 12, which corresponds to a 0.01-M alkali concentration. The presence of detergent in alkaline solution has a noticeable effect on desorption efficiency (Figure 8.16). Thus, the standard two-step micelle-stabilized phosphorescence immunoassay (MS-PhIA) procedure is optimal:

1. Desorption of the PdCP conjugate with 100 μl of 0.1 M alkali solution containing 1% TX-100
2. Neutralization to pH 7.2 to 7.4 and deoxygenation of the alkali solution by the addition of an equal volume (100 μl) of buffer containing 20 mg/ml sodium metabisulfite ($Na_2S_2O_5$) and 20 mg/ml sodium dihydrogen orthophosphate (NaH_2PO_4)

This procedure is very reproducible and, as was mentioned above, deoxygenation conditions are retained for several hours in the 200-μl volume of microplate wells.[13,18]

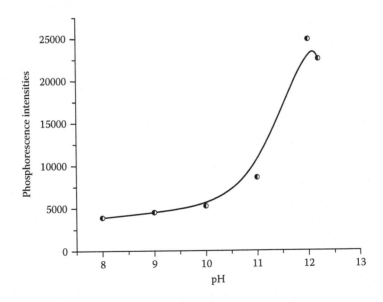

FIGURE 8.15 pH dependence of the desorption efficiency of PdCP conjugate with monoclonal antibodies against human chorionic gonadotropin (anti-hCG–PdCP) in the presence of 1% TX-100. Human chorionic gonadotropin (10 μg/ml) was adsorbed in microplate wells, and anti-hCG–PdCP conjugate was added (50 ng per well); incubation 1 h, measurements at pH 7.4.

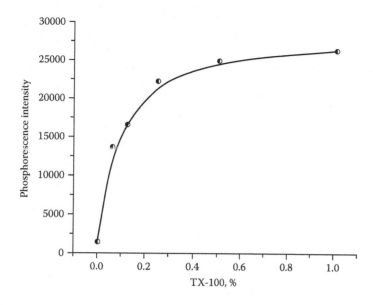

FIGURE 8.16 Dependence of the desorption efficiency of PdCP conjugate with monoclonal antibodies against human chorionic gonadotropin (anti-hCG–PdCP) on the concentration of TX-100 at pH 12. Human chorionic gonadotropin (10 µg/ml) was adsorbed in microplate wells and anti-hCG–PdCP conjugate was added (50 ng per well); incubation 1 h, measurements at pH 7.4.

8.5.2 PHOSPHORESCENCE IMMUNOASSAY OF HUMAN THYROTROPIN

Human thyrotropin (thyroid-stimulating hormone, TSH) is a glycoprotein that stimulates biosynthesis of thyroid hormones by the thyroid gland. Analysis of this hormone is important for early diagnosis of a series of congenital diseases, particularly hypothyroidism. A pair of monoclonal antibodies from a DELFIA neonatal TSH kit (Wallac, Inc., Turku, Finland) was used for the sandwich assay.[10] The only difference was that the second antibodies were biotinylated and a PdCP–streptavidin conjugate was used for the development. To the microplate wells with the adsorbed first antibodies were simultaneously added the TSH standards and the second biotinylated antibodies, and the plate was incubated for 2 h. After washing, PdCP–streptavidin conjugate was added (20 ng per well) and the plate was incubated for 15 min. After that, desorption and measurement were carried out according to the procedure described above. The total assay time was 2.5 h. A calibration curve obtained on an Arcus 1230 is shown in Figure 8.17. It has a good dynamic range that corresponds to the clinically important range of TSH concentrations. The coefficient of variation in this range does not exceed 7.5%. A similar approach was used for the detection of α-fetoprotein with *p*-isothiocyanatophenyl derivatives of Pt(II)- and Pd(II)-copro-porphyrin I as labels.[12]

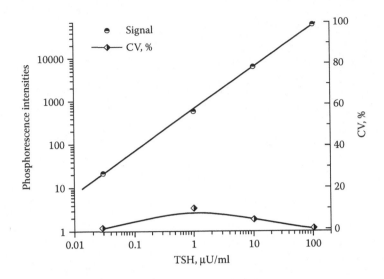

FIGURE 8.17 Thyroid-stimulating hormone concentration dependence of phosphorescence intensity (circles) and coefficients of variation (diamonds). The calibration curve for TSH assay was obtained by MS–PhIA on an Arcus 1230 (Wallac, Inc., Turku, Finland) with 400-nm excitation and 660-nm emission filters. The delay time was 200 μs, measuring time 800 μs, and signal accumulation time 1 s.

8.5.3 Cyclodextrin Phosphorescence Immunoassay

The cyclodextrin system is an alternative to detergents. One of the main advantages of cyclodextrins is higher phosphorescence enhancement factors as compared with detergents. The efficiency of the application of met-β-cyclodextrin-1.8 was demonstrated by thyrotropin assay using a procedure that is essentially identical to that described in the previous section. In the last step of the assay, 100 μl of 0.1 M potassium hydroxide was added, and in 15 min, directly before the measurement, an equal volume (100 μl) of buffer containing 20 mg/ml $Na_2S_2O_5$, 20 mg/ml NaH_2PO_4, 2% met-β-cyclodextrin-1.8 was added (Figure 8.18). Cyclodextrins, like detergents, are sensitive to the presence of the covalent bond in PdCP for protein attachment, and the efficiency of their application for phosphorescence enhancement with PdCP conjugates is less than with free PdCP.

8.5.4 Universal Phosphorescence Immunoassay

Monoclonal antibodies with high affinity constants can be obtained against porphyrins.[27] With these antibodies it is possible to develop immunoassay schemes without using covalent links between PdCP and a protein.[13,28] The advantage of this approach is that the immune complex of PdCP with antibody can be destroyed in the last step of the assay. After that, there are no obstacles to the localization of PdCP in the hydrophobic nucleus of the micelle and, thus, maximal phosphorescence enhancement factors can be reached.

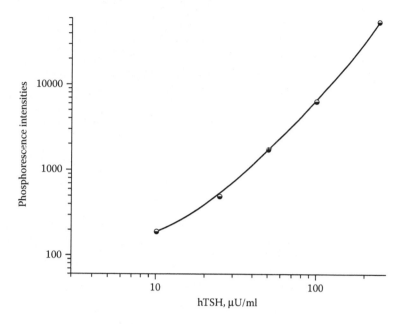

FIGURE 8.18 Thyroid-stimulating hormone concentration dependence of phosphorescence intensity (circles) in the presence of 2% met-β-cyclodextrin. Calibration curve for TSH assay obtained by cyclodextrin phosphorescence immunoassay on an Arcus 1232 (Wallac, Inc., Turku, Finland) with a standard filter for europium for excitation and 660-nm emission filter. The delay time was 300 μs, measuring time 700 μs, and signal accumulation time 1 s.

Hybridomas producing antibodies against porphyrins were obtained by a standard procedure. A PdCP–bovine serum albumin (BSA) conjugate with a label:protein ratio of 10:1 was used for the immunization of BALB/C mice. As can be seen from Table 8.7, the 3D antibodies with the highest affinity are bound equally well by both PdCP and its conjugates. Endogenous (natural) porphyrins bind 3D and 5D antibodies several orders of magnitude less well than PdCP; thus, the probability of cross-reactions is rather low.[27]

TABLE 8.7
Main Properties of Monoclonal Antibodies against PdCP

	Clone	Class, Subclass	Association Constants (M)	
			Free PdCP	PdCP–SiT Conjugate (3:1)
1	D3.F5	IgG1	$4.76 \cdot 10^{-10}$	3.15×10^{-10}
2	D5.E3	IgG1	$4.99 \cdot 10^{-9}$	5.48×10^{-9}
3	D9.E6	IgG2b	$2.11 \cdot 10^{-8}$	1.45×10^{-8}
4	C2.B5	IgG1	–	7.30×10^{-9}

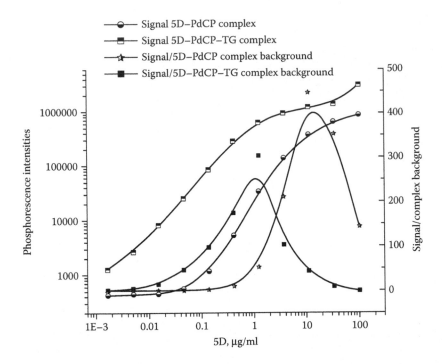

FIGURE 8.19 5D antibody concentration dependence of the 5D–PdCP complex (circles) and 5D–PdCP–TG complex (half squares) phosphorescence intensities and their ratios to background (stars and filled squares, respectively).

To determine the optimal concentrations of PdCP and monoclonal antibody 5D, an antibody–antigen complex was prepared in advance by mixing the antibody (100 μg/ml) and PdCP (10^{-6} M), then the complex was added in different concentrations to the antispecies antibodies adsorbed on the surface of a polystyrene microplate (Figure 8.19). The maximal signal measured after washing and desorption was several million units and the signal:background ratios were 450 for the titration of antibody 5D–PdCP complex and 300 for the antibody 5D–PdCP–thyroglobulin (TG) complex. It is important that the optimum for the protein conjugate is reached at lower concentrations of the complex.

PdCP and monoclonal antibodies obtained against PdCP allow one to develop a simple and universal method for the detection of different antigens: the universal phosphorescence immunoassay (uni-PhIA). Antigens adsorbed on a solid phase bind antibodies that interact with antibodies against the label through the "bridge" antispecies antibodies. This scheme of immunoassay is simple because there is no need to synthesize label-containing conjugates. When it is necessary to change the antigen to be determined, only one component — antigen-specific antibodies — needs to be changed; therefore, this scheme is universal. The potentialities of this method were demonstrated in an immunoassay for insulin.[13] The optimized composition of buffer for the uni-PhIA is basically the same as that for other types of analysis; thus, it is possible to conclude that the label itself (PdCP) does not introduce any additional changes to the standard protocols for carrying out immunochemical reactions.

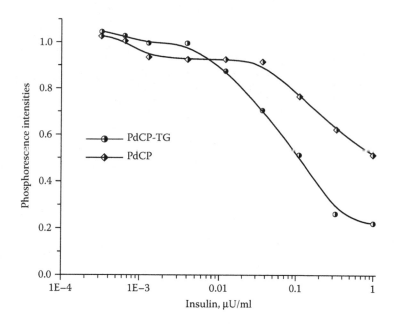

FIGURE 8.20 Insulin concentration dependence of the phosphorescence intensities of PdCP alone (diamonds) and conjugated to thyroglobulin (circles). Competitive assay of insulin by the uni-PhIA method. Insulin (1 μg/ml) was adsorbed in the wells of a polystyrene microplate. The sample, anti-insulin antibodies C4 (1 μg/ml), rabbit anti-mouse antibodies (RAM, 5 μg/ml), 5D antibodies (1 μg/ml), and PdCP–TG conjugate (1 μg/ml) were sequentially added to the microplate wells. Standard MS–PhIA protocol.

The minimal detected concentration of insulin, calculated in this scheme as a reference sample minus standard deviation, is 37 ng/ml using the free label. If a thyroglobulin conjugate (Figure 8.20) is used, the sensitivity of the detection is increased and it is possible to determine insulin at concentrations as low as 4 ng/ml. The increase in sensitivity results from the fact that the protein surface contains up to 77 label molecules. The molecule of monoclonal antibody still interacts with one PdCP molecule, but after desorption, the phosphorescence of all label molecules that are covalently bound with protein is measured. In spite of the almost 5-fold decrease in the quantum yield (see Figure 8.14), the specific signal per binding site increases almost 15-fold, which explains the improvement in detection limits by almost an order of magnitude from 37 to 4 ng/ml.

8.5.5 BISPECIFIC ANTIBODIES AND THEIR USE IN PHOSPHORESCENCE IMMUNOASSAY

Universal phosphorescence immunoassay may be significantly simplified for carrying out one specific type of analysis. Using bispecific antibodies, it is possible to avoid the use of antispecies antibodies, which serve as a connecting link between antibodies against antigen under study and antilabel antibodies. These IgG-type antibodies,

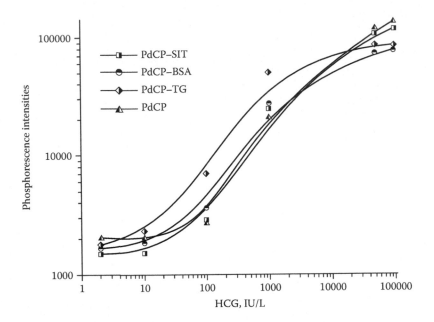

FIGURE 8.21 Sandwich assay of human chorionic gonadotropin using MS–PhIA with bispe-cific antibodies. Antibodies against the α-subunit of human chorionic gonadotropin (10 μg/ml, Diagnostika, Sweden) were adsorbed in the wells of a polystyrene microplate. The sample, affinity-purified bispecific antibodies 6D12, and PdCP conjugates were simultaneously added to the microplate wells. Standard MS–PhIA protocol. Human chorionic gonadotrophin-depen-dent phosphorescence intensities for PdCP alone (triangles) and conjugated to thyroglobulin (diamonds), serum albumin (circles), and SIT (squares).

with one binding site for the α-subunit of human chorionic gonadotropin and another site for PdCP, are described in Demcheva et al.[28]

The comparison of free PdCP and its conjugates with proteins in 10^{-8} M concentration with respect to porphyrin under conditions of sandwich analysis indicated that the calibration curves are practically identical for PdCP and PdCP-soybean trypsin inhibitor (SIT). When PdCP–BSA was used in the experiment, the maximal signal was lower due to the steric hindrance of interaction with the larger protein molecule. The same effect is observed for PdCP–TG at high concentrations of the hormone; in contrast, the sensitivity is higher at lower concentrations (Figure 8.21). The sensitivity of the detection of human chorionic gonadotropin by the sandwich method with bispecific antibodies is 10 IU/L.[28]

8.6 HOMOGENEOUS PHOSPHORESCENCE IMMUNOASSAY

A homogeneous immunoassay for dioxins was developed using a dioxin derivative conjugated to PtCP.[29] When the antibodies against dioxin were bound to the phos-phorescent conjugate, the signal from PtCP was quenched. The interaction between

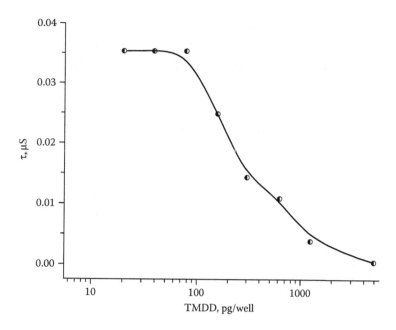

FIGURE 8.22 Tetramethyl dibenzo-*p*-dioxin concentration (TMDD) dependence of short lifetime component determined by homogeneous immunoassay.

the antibody against dioxin (TMDD) and the conjugate was studied by time-resolved luminescence spectroscopy and it was demonstrated that the phosphorescence decay was multiexponential. At the same time, the dependence of the short-lived component of the phosphorescence signal in the presence of antibodies against TMDD on concentrations of the dioxin standard can be clearly seen (Figure 8.22). Varying the concentration of dioxin standards from 78 pg per well to 1.25 ng per well led to a decrease in the lifetime of the short-lived component from 35 to 3.85 μs. The sensitivity obtained is similar to that reported for the enzyme-based immunoassay, but in this case, data were obtained in minutes instead of hours.

8.7 TIME-RESOLVED MICROSCOPY

Phosphorescence immunoassay with measurements on a dry surface is most effective for applications in microscopy.[16] Cell monolayers grown on a glass slide were washed off the culture medium, MAb 242 antibodies (10 μg/ml) were applied, and the slides were incubated for 1 h. After that, the cells were again washed, biotinylated rabbit anti-mouse antibodies were applied to the cells (Biot RAM, 10 μg/ml), and the slides were again incubated for 1 h. After washing, the PdCP–streptavidin conjugate (10 μg/ml) was applied and incubated for 15 min. The cells were then washed and covered with a thin layer of Merckoglas without a coverslip. Measurements were carried out on a Nikon inverted microscope (Japan) with an additional pulsed light source (xenon lamp with 100-Hz repetition rate), excitation interference filter with 400-nm transmission band, and 656-nm filter at the detector.[30] The delay time was 200 μs and the measuring

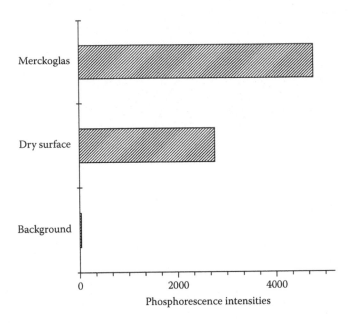

FIGURE 8.23 Intensity of phosphorescence of single cells of the cell line A380 using different methods of detection from the surface on a Nikon inverted microscope (Japan). Indirect immunochemical test using MAb 242 antibodies (10 μg/ml), biotinylated RAM (10 μg/ml), and PdCP–streptavidin (10 μg/ml). PdCP phosphorescence intensities immobilized in Merckoglas or dried on a surface along with a porphyrin-free control.

time was 700 μs. The signal was recorded with a photomultiplier tube in a single-photon counting mode and the accumulation time was 10 s. The phosphorescence intensity of cells of the A380 tumor line carrying the antigen for Mab 242 antibodies, measured from the surface in a precisely focused light beam, is almost two times higher if the sample is dried and covered with Merckoglas as compared with the micellar solution (Figure 8.23). Liquid and semiliquid samples may also be used in time-resolved microscopy. PtCP–streptavidin conjugate and biotinylated probes were successfully used for fluorescence *in situ* hybridization (FISH) of the centromere of chromosome 1 or of 28S RNA.[31] The influence of various embedding media on the intensity of phosphorescence[16,31] and photobleaching[22,31] is different compared with PdCP–streptavidin; this may be the result of different lifetimes and protocols for sample preparation. Phosphorescent labels are suitable for microscopic detection of antigens and DNA in autofluorescent specimens.[32] Soini et al.[33] have described in more detail the role of mounting (embedding) media for phosphorescent measurements.

8.8 HYBRIDIZATION ASSAYS

Phosphorescent labels may be efficiently used in different types of nucleic acid hybridization assays. As mentioned above, streptavidin conjugates and biotinylated probes, as well as direct conjugates of PtCP with oligonucleotides, may be used for FISH.[32] PtCP or PdCP may be attached to oligonucleotides by the carbodiimide method using an amine linker.[32,34] For *p*-isothiocyanatophenyl derivatives of Pt(II)-coproporphyrin I, conjugates with oligonucleotides containing 3'- and 5'-amino modification were synthesized.[35] A 5'-PtCP conjugate with an oligonucleotide was used as a primer for a standard polymerase chain reaction (PCR) protocol. The detection limit of the PCR product or DNA probe was 0.3 pM.[35] Proximity quenching of PtCP oligonucleotide phosphorescence by a dark quencher conjugated to the complementary sequence was demonstrated for homogeneous hybridization assays similar to the "molecular beacon" system.[35,36]

8.9 CONCLUSION

The recently developed methods for measuring phosphorescence at room temperature permit the efficient use of phosphorescing labels in analyses with labeled compounds. Because phosphorescence at room temperature is itself a rather rare phenomenon, both in general and for biological compounds in particular, the background signals of the samples can be reduced to a minimum. Using a time discrimination method for the collection of phosphorescence signals, it is possible to almost completely exclude the instrument noise and greatly improve the signal:background ratio. Phosphorescing metalloporphyrins, especially palladium and platinum complexes with water-soluble porphyrins, are at the present time the most efficient phosphorescence labels. Their detection limit under optimal conditions reaches 10^{-13} M. Complexes of these metals with coproporphyrin are thermodynamically very stable and, more importantly very stable from a kinetic point of view. Platinum or palladium may be removed from the macrocycle only under very harsh chemical conditions that never occur in biological systems. This means that, in spite of the fact that these substances are complex compounds, they behave very homogeneously, even in complex biological specimens. The methods of measuring phosphorescence in micellar and cyclodextrin solutions allow one to use these compounds in any microplate-based method of immunochemical analysis or DNA diagnostics. The methods for measuring phosphorescence from dry surfaces are easily adapted for measurements in the microscope. PtCP and PdCP have absorption bands in the visible region that are especially convenient for existing plate readers and microscopes. Thus, the phosphorescence labels described above are suitable for applications throughout the spectrum of biochemical methods of analysis.

REFERENCES

1. Soini, E. and Hemmila, I., Fluoroimmunoassay: present status and key problems. *Clin Chem*, 25, 353–361, 1979.
2. Diamandis, E.P., Immunoassays with time-resolved fluorescence spectroscopy: principles and applications. *Clin Biochem*, 21, 139–150, 1988.
3. Sidki, A.M., Smith, D.S., and Landon, J., Direct homogeneous phosphoroimmunoassay for carbamazepine in serum. *Clin Chem*, 32, 53–56, 1986.
4. Miller, J.N., Room temperature phosphorimetry—a promising trace analysis method. *Trends Anal Chem*, 1, 31–34, 1981.
5. Savitsky, A.P., Papkovskii, D.B., Ponomarev, G.V., and Berezin, I.V., Phosphorescent immunoanalysis. Metalloporphyrins—an alternative to rare earth fluorescent measurements? *Dokl Akad Nauk SSSR*, 304, 1005–1008, 1989.

6. Tuan V.-D., Room temperature phosphorimetry for chemical analysis. In: *Chemical Analysis*, Elving, P.J. and Winefordner, J.D., eds. New York: John Wiley & Sons, 1983.

7. Garland, P.B. and Moore, C.H., Phosphorescence of protein-bound eosin and erythrosin. A possible probe for measurements of slow rotational mobility. *Biochem J*, 183, 561–572, 1979.

8. Eastwood, D. and Gouterman, M., Luminescence of (Co), (Ni), Pd, Pt complexes. *J Mol Spectrosc*, 35, 359–375, 1970.

9. Vanderkooi, J.M., Maniara, G., Green, T.J., and Wilson, D.F., An optical method for measurement of dioxygen concentration based upon quenching of phosphorescence. *J Biol Chem*, 262, 5476–5482, 1987.

10. Savitsky, A.P., Ponomarev G.V., Lobanov, O.I., and Sadovsky, N.A., Room-temperature phosphorescence of metalloporphyrins and its application to immunoassay. In: *Biomedical Diagnostic Instrumentation*, Bonner, R.F., Cohn, G.F., Laue, T.M., and Priezzhev, A.V., eds. *Proc SPIE*, 2136, 285–298, 1994.

11. Savitsky, A.P., Savitskaja, A.V., Lukyanets, E.A., Dashkevich, S.N., and Makarova, E.A., Near-infrared phosphorescent metalloporphyrins. In: *Advances in Fluorescence Sensing Technology III*, Thompson, R.B., ed. *Proc SPIE*, 2980, 352–357, 1997.

12. O'Riordan, T., Soini, A., Soini, J., and Papkovsky, D., Performance evaluation of the phosphorescent porphyrin label: solid-phase immunoassay of α-fetoprotein. *Anal Chem*, 74, 5845–5850, 2002.

13. Mantrova, E.Y., Demcheva, M.V., and Savitsky, A.P., Universal phosphorescence immunoassay. *Anal Biochem*, 219, 109–114, 1994.

14. Englander, S.W., Calhoun, D.B., and Englander, J.J., Biochemistry without oxygen. *Anal Biochem*, 161, 300–306, 1987.

15. Banks, D.D. and Kerwin, B.A., A deoxygenation system for measuring protein phosphorescence. *Anal Biochem*, 324, 106–114, 2004.

16. Savitsky, A.P., Mantrova, E.Y., Demcheva, M.V., Ponomarev, G.V., Hemmila, I., and Soini, E.J., Solid surface measurement of room temperature phosphorescence of Pd-coproporphyrin and its application for time-resolved microscopy. In: *Microscopy, Holography, and Interferometry in Biomedicine*, Fercher, A.F., Lewis, A., Podbielska, H., Schneckenburger, H., and Wilson, T., eds. *Proc SPIE*, 2083, 49–53, 1994.

17. Savitsky, A.P., Vorobyova, E.V., Berezin, I.V., and Ugarova, N.N., Acid-base properties of protoporphyrin IX: its dimethyl ester and heme solubilized on surfactant micelles: spectrophotometric and fluorometric titration. *J Coll Interface Sci*, 84, 175–181, 1981.

18. Savitsky, A.P., Biomedical application of metalloporphyrins room-temperature phosphorescence. In: *Advances in Fluorescence Sensing Technology*, Lakowicz, J.R. and Thompson, R.B., eds. *Proc SPIE*, 1885, 138–148, 1993.

19. Siegel, B. and Breslow, R., Lyophobic binding of substrates by cyclodextrins in nonaqueous solvents. *J Am Chem Soc*, 97, 6869–6870, 1975.

20. Hurtubise, R.J., Solid-surface luminescence spectrometry. *Anal Chem*, 61, 889A–895A, 1989.

21. Dubrovsky, T.B., Demcheva, M.V., Savitsky, A.P., Mantrova, E.Y., Yaropolov, A.I., Savransky, V.V., and Belovolova L.V., Fluorescent and phosphorescent study of Langmuir–Blodgett antibody films for application to immunosensors. *Biosens Bioelectron*, 8, 377–385, 1993.

22. Savitsky, A.P., Ponomarev, G.V., and Soini, E., Phosphorescent immunostaining. In: *Optical and Imaging Techniques in Biomedicine*, Foth, H.-G., Luis, A., Podbelska, G., Nicould, M.-R., Shnekenburger, H., and Wilson, A., eds. *Proc SPIE*, 2329, 252–257, 1995.

23. Martsev, S.P., Preygerzon, V.A., Mel'nikova, Y.I., Kravchuk, Z.I., Ponomarev, G.V., Lunev, V.E., and Savitsky, A.P., Modification of monoclonal and polyclonal IgG with palladium (II) coproporphyrin I: stimulatory and inhibitory functional effects induced by two different methods. *J Immunol Methods*, 186, 293–304, 1995.

24. Soini, A.E., Yashunsky, D.V., Meltola, N.J., Ponomarev, G.V., Preparation of monofunctional and phosphorescent palladium(II) and platinum(II) coproporphyrin labeling reagents. *J Porphyrins Phthalocyanines*, 5, 735–741, 2001.

25. Savitsky, A.P., Phosphorescent immunoassay. *Yspehi Biologicheskoy Khimii*, 40, 309–328, 2000.

26. O'Riordan, T.C., Soini, A.E., Papkovsky, D.B., Monofunctional derivatives of coproporphyrins for phosphorescent labeling of proteins and binding assays. *Anal Biochem*, 290, 366–375, 2001.

27. Savitsky, A.P., Demcheva, M.V., Mantrova, E.Y., and Ponomarev, G.V., Monoclonal antibodies against metalloporphyrins. Specificity of interaction with structurally different metalloporphyrins. *FEBS Lett*, 355, 314–316, 1994.

28. Demcheva, M.V., Mantrova, E.Y., Savitsky, A.P., Berzing, O., Michel, B., and Hammila, I., Micelle stabilised phosphorescent based on bispecific antibodies against label and antigen. *Anal Lett*, 28, 249–258, 1995.

29. Matveeva, E.M., Gribkova, E.V., Sanborn, J.M., Gee, S.J., Hammock, B.D., and Savitsky, A.P., Development of a homogeneous phosphorescent immunoassay for the detection of polychlorinated dibenzo-p-dioxins. *Anal Lett*, 34, 2311–2320, 2001.

30. Seveus, L., Väisälä, M., Syrjänen, S., Sandberg, M., Kuusisto, A., Harju, R., Salo, J., Hemmilä, I., Kojola, H., and Soini, E., Time-resolved fluorescence imaging of europium chelate label in immunohistochemistry and *in situ* hybridization. *Cytometry*, 13, 329–338, 1992.

31. de Haas, R.R., van Gijlswijk, R.P.M., van der Tol, E.B., Zijlmans, H.J.M.A.A., Bakker-Schut, T., Bonnet, J., Verwoerd, N.P., and Tanke, H.J., Platinum porphyrins as phosphorescent label for time-resolved microscopy. *J Histochem Cytochem*, 45, 1279–1292, 1997.

32. de Haas, R.R., van Gijlswijk, R.P.M., van der Tol, E.B., Veuskens, J., van Gijssel, H.E., Tijdens, R.B., Bonnet, J., Verwoerd, N.P., and Tanke, H.J., Phosphorescent platinum/palladium coproporphyrins for time-resolved luminescence microscopy. *J Histochem Cytochem*, 47, 183–196, 1999.

33. Soini, A.E., Seveus, L., Meltola, N.J., Papkovsky, D.B., and Soini, E., Phosphorescent metalloporphyrins as labels in time-resolved luminescence microscopy: effect of mounting on emission intensity. *Microsc Res Tech*, 58, 125–131, 2002.

34. Fedorova, O.S., Savitsky, A.P., Shoikhet, K.G., and Ponomarev, G.V., Palladium(II)-coproporphyrin I as a photoactivable group in sequence-specific modification of nucleic acids by oligonucleotide derivatives. *FEBS Lett*, 259, 335–337, 1990.

35. O'Sullivan, P.J., Burke, M., Soini, A.E., and Papkovsky, D.B., Synthesis and evaluation of phosphorescent oligonucleotide probes for hybridisation assays. *Nucleic Acids Res*, 30, e114, 2002.

36. Burke, M., O'Sullivan, P.J., Soini, A.E., Berney, H., and Papkovsky, D.B., Evaluation of the phosphorescent palladium(II)-coproporphyrin labels in separation-free hybridization assays. *Anal Biochem*, 320, 273–280, 2003.

37. Savitsky, A.P., Phosphorescent labels in analytical biochemistry. Room-temperature phosphorescence of metalloporphyrins. *Yspehi Biologicheskoy Khimii*, 37, 293–314, 1997.

9 Development of Fluorescent Dipyrrolylquinoxaline-Based Anion Sensors

Pavel Anzenbacher, Jr., Ph.D.
and Karolina Jursíková, M.S.

CONTENTS

9.1 INTRODUCTION

The importance of anions in nature, as agricultural fertilizers and industrial raw materials, and the corresponding environmental concerns, necessitates the development of highly sensitive anion sensors.[1] Here, sensors based on anion-induced changes in fluorescence are particularly attractive, as they offer the potential for high sensitivity at a low analyte concentration.[2] In the last two decades, numerous compounds and materials have been prepared and utilized for the fluorescence-based sensing of anions.[3]

There are several reasons why reliable sensing of anions is a particularly challenging area of research. Anions are larger than isoelectric cations and therefore have a lower charge:radius (surface) ratio, a feature which makes the electrostatic binding of anions to the receptors less effective.[3b,3d] Anions have a wide range of geometries and are often present in delocalized forms, which results in greater design complexity of receptors and sensors required for successful recognition and binding. Anions have very high free energies of solvation, and the receptors must compete more effectively with the medium. This aspect is particularly important in water. For example, fluoride with an ionic radius of 1.33 Å has $\Delta G_{hydration} = -465$ kJ/mol, which is significantly greater than potassium cations with almost the same size (1.38 Å) and a $\Delta G_{hydration} = -295$ kJ/mol.[3d] All of these factors make sensing of anions a difficult task, and, as a result, effective sensors for anions are quite rare.

Among the materials capable of anion binding, compounds utilizing pyrrole moieties as hydrogen-bond donors have been extensively used as a building element of choice, giving rise to a wide variety of sensor types. Examples include porphyrins and expanded porphyrins such as sapphyrins,[4] calixpyrroles,[5] calixphyrins,[6] and cyclopyrroles.[7] Although some of these materials are very effective anion sensors, their practical use is in many instances prohibited by the high cost of their synthesis. The recently rediscovered anion binding and sensing ability of 2,3-di(pyrrol-2-yl)quinoxaline (DPQ),[8] a compound known in the literature since 1911,[9] gave this simple and easy-to-make material a new and brightly illuminated future.

9.2 ANION SENSING BY DPQ AND ITS DERIVATIVES

The structure of DPQ is shown in Figure 9.1. We presume that in the resting state both of the pyrrole units are arranged to achieve maximum conjugation between the pyrrole moieties and the quinoxaline chromophore, creating one large conjugated chromophore. The x-ray structure of the benzo-analog 2 seems to be in an agreement with this theory.

In the presence of an anion, the hydrogen bonding between the pyrrole NHs and the anion enforces a change in the conformation, specifically the rotation around the pyrrole-quinoxaline bond, which in turn results in a dramatic decrease in the size of the conjugated chromophore as well as an order of magnitude decrease in

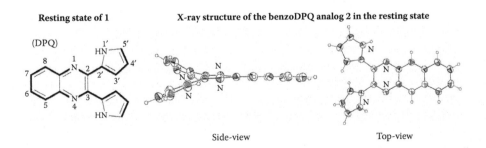

FIGURE 9.1 The structure of DPQ and the resting state conformation of the benzo-DPQ analog 2, which we presume represents the resting state conformation for all of the DPQ derivatives.

the fluorescence quantum yield from 13% to less than 1%. The engagement of the pyrrole–NH protons in the formation of the anion–DPQ complex and the increase in the pyrrole N-H distance result in a partial charge concentration in the pyrrole moieties and a partial charge transfer to the quinoxaline ring. This is, most likely, the reason for the yellow to orange (red) color transition observed during the anion binding process (Figure 9.2).[9]

Although the parent DPQ sensor shows some promise for sensing a fluoride anion in organic media,[9] it is clear that optimization aimed at improving the fluorescence output, anion affinity, and solvent (medium) compatibility of this potential sensor is necessary to meet the requirements of potential applications. The initial efforts were focused on implementation of extended aromatic chromophores (e.g., sensor 2) and implementation of electron-withdrawing substituents such as fluoro (e.g., sensor 3) and nitro groups (sensor 4) to the quinoxaline chromophore and the pyrrole moieties (Figure 9.3).

Although the extension of the quinoxaline chromophore in sensor 2 resulted in an almost twofold increase in the fluorescence quantum yield and enabled us to use a lower sensor concentration, the fact that the anion-binding properties were virtually unchanged compared to the parent DPQ and the extensive π–π stacking precluded any practical use of sensor 2. Interestingly, the introduction of fluorine substituents to the pyrrole moieties in sensor 3 resulted not only in a dramatic increase in the affinity for phosphate anions[10] but also in a 20% increase in the fluorescence quantum yield of sensor 3 ($\phi = 0.16$) compared to DPQ. In addition, the introduction of a strong electron acceptor such as a nitro group in sensor 4 was shown to encourage the partial charge transfer during the sensor–anion complex formation, which resulted in an intensive color change. Unfortunately, the nitro group drastically reduces the fluorescence quantum yield of sensor 4, thus making this material useful only for colorimetric sensing. Two basic trends emerged from these initial studies. First, the increase in fluorescence efficiency (improved signaling) may allow for lowering the sensor concentration, thus making these materials able to detect lower concentrations of anionic analytes (the binding equilibrium and the corresponding sensor–analyte proportion described by the binding constant are of the same magnitude regardless of the sensor concentration). Second, the introduction of halogen to the pyrrole moiety results in the change in relative selectivity for anions. Specifically, DPQ derivatives with halogenated pyrroles showed stronger affinity for phosphate-related anions, a feature that may be useful in the design of DPQ-based sensors for biologically important phosphate anions.

9.3 IMPROVING SENSOR PERFORMANCE

In general, sensor performance can be improved either by increasing the receptor–substrate affinity or by improved signal transduction. So far the focus has been mainly on the supramolecular chemistry of a receptor. In the case of DPQ, this approach has yielded several very interesting sensors[11]; however, it still suffers from shortcomings such as the high cost of synthesis and the potential loss of real-time response due to slow dissociation of the substrate–receptor complex.

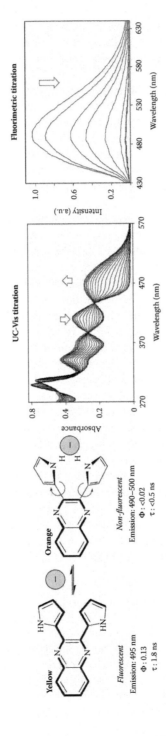

FIGURE 9.2 Anion binding by DPQ results in a change in both the conformation and charge localization that are observed as changes in color or fluorescence intensity.

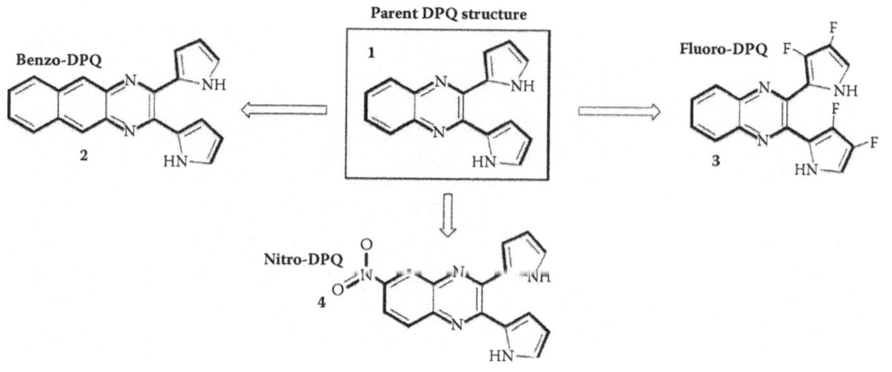

FIGURE 9.3 Examples of modifications of the DPQ structure aimed at improving DPQ sensor performance.

9.3.1 AMPLIFICATION OF SIGNAL TRANSDUCTION VIA EXTENSION OF CONJUGATION

The sensitivity of fluorescence spectroscopy, current advances in the miniaturization of fluorimeters and their components, and the ever-increasing sensitivity inspired us to focus on the signal transduction process as a strategy that may provide fluorescent DPQ-based sensors for anions. Toward this end, we have designed DPQ derivatives composed of extended conjugated chromophores with conjugated fragments attached in the 5- and 8-positions of the quinoxaline ring (Figure 9.4). Sensors 5 through 8 show bright emission (the fluorescence quantum yields were found to be more than twice the quantum yield of the parent DPQ) as well as emission-color tuning (525 nm to 620 nm), depending on the electronic nature of the conjugated substituent.

9.3.2 AMPLIFICATION OF SIGNAL TRANSDUCTION VIA RESONANCE ENERGY TRANSFER

Reduction of the acetylene bonds in sensor 7 allowed for the facile preparation of sensor 9.[12] Sensors 7 and 9 display fluorescence signal amplification, clearly visible in the graphs in Figure 9.5. In sensor 7, the signal amplification is achieved through effective excited state delocalization, as confirmed by both the shift of the emission wavelength from $\lambda_{max} = 495$ nm to $\lambda_{max} = 550$ nm and the increase in fluorescence intensity and lifetime from 1.8 ns recorded for the parent sensor (sensor 1) to 3.7 ns for sensor 7 (Figure 9.5A). The photophysical properties of sensor 9 are largely determined by resonance energy transfer (RET) from the pyrene (donor) moieties to the DPQ (acceptor) moiety of sensor 9, as confirmed by time-resolved fluorescence spectroscopy and a quantum yield measurement (Figure 9.5C,D). The long-lived fluorescence of pyrene (29.0 ns in air-saturated dichloromethane) is quenched, whereas the lifetime of the DPQ acceptor is extended to 2.2 ns from 1.8 ns recorded for sensor 1. Also, the excitation spectrum of sensor 9 shows a remarkable resemblance

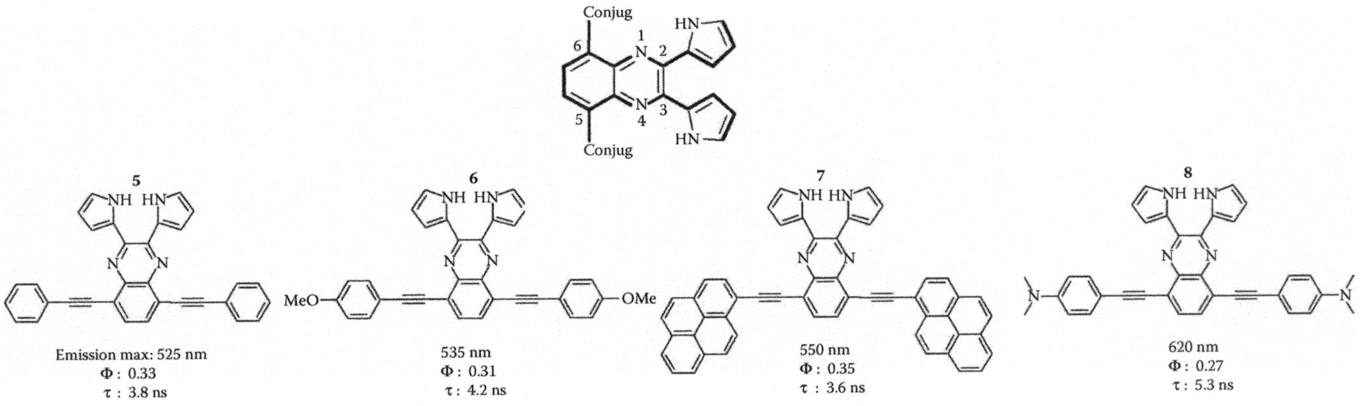

FIGURE 9.4 The 5,8-aryl-substituted DPQ sensors and their photophysical properties.

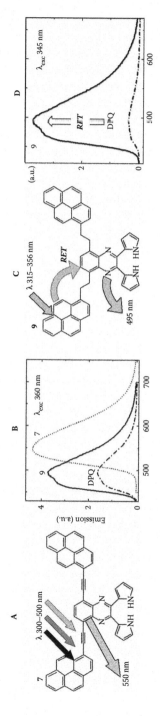

FIGURE 9.5 (A) The excited-state delocalization in sensor 7 results in amplified emission with a significant red shift. (B) Sensors 7 and 9 show a more-than-twofold increase in intensity, even at the excitation wavelength optimal for the parent DPQ sensor. (C) Signal amplification in sensor 9 achieved through RET amplification (excitation in the pyrene-donor moiety). (D) Excitation at the wavelength optimal for pyrene antennae provides more than a 10-fold amplified output signal. (From Pohl et al.[12])

to the absorption spectrum of sensor 2, including the well-resolved bands of pyrene. Figure 9.5B shows that excitation at 360 nm and the presence of pyrene antennae moieties, despite relatively low absorption at 365 nm by the pyrene donors, results in twofold amplification of the signal output recorded for sensors 7 and 9 compared to sensor 1. Figure 9.5D shows more than a 10-fold increase in the fluorescence output signal when a wavelength optimal for fluorescence resonance energy transfer (FRET)-mediated signal amplification (365 nm) is used for excitation.

In addition to the amplified fluorescence output signal, sensors 7 and 9 display increased affinity for anions compared to sensor 1 (DPQ). This effect is particularly strong in the case of sensor 7 and pyrophosphate, with a twofold increase in binding affinity compared to sensor 1. In practice, this means that sensor 9 can be used in an amount 10 times lower compared to the parent DPQ. Also, the detection limit for sensor 9 for fluoride and pyrophosphate was found to be 40 µM and 30 µM, respectively (determined using a 0.2-µM solution of sensor 9). Such a detection limit is not accessible using the parent DPQ, where the practical detection limits for fluoride and pyrophosphate are 0.5 mM or greater.

9.3.3 AMPLIFICATION OF SIGNAL TRANSDUCTION WHILE INCREASING AFFINITY FOR ANIONS

Although the attachment of extended conjugated chromophores provided sensors with a significantly improved fluorescence signal output, given the electron-rich nature of the aryl-ethynylene substituents, these sensors did not show significantly improved affinity for anions compared to the parent DPQ. Therefore, the next design step was to prepare a similar sensor with 5,8-conjugated substituents attached without the electron-rich acetylene bridge (Figure 9.6). Sensors 10 through 14 showed twice the fluorescence quantum compared to the parent DPQ but, most importantly, a significantly enhanced affinity for anions (see table of affinity constants in Figure 9.6).[13] Sensors 10 and 12 showed particularly high anion-binding constants; however, self-association of sensor 10 prevented further development of this material. We have therefore concentrated on sensor 12 and converted to its dichloro analog 13, which shows a dramatic increase in affinity for phosphate-related anions.

The propensity of sensors 12 and 13 to form conductive polymers was utilized in the design of polymeric materials capable of pyrophosphate and phosphate sensing. Thus, electropolymerization of sensor 12 or 13 on a transparent indium tin oxide (ITO) electrode (Figure 9.7) resulted in the formation of conductive polymers poly-12 or poly-13.[14] These materials proved to be efficient anion sensors for aqueous pyrophosphate and (to a lesser extent) for phosphate, providing a colorimetric and conductivity-based output. Unfortunately, poly-12 and poly-13 are not fluorescent.

9.3.4 LIFETIME-BASED CHEMICAL SENSING

The last example of the development of a fluorescence-based anion sensor utilizing the DPQ moiety is sensor 15 (Figure 9.8).[15] The Ru(II) metal bound to the DPQ-type ligand serves two purposes. First, it depletes the DPQ of electronic density, thus making the pyrrole moieties more effective anion binders; second, it creates a

	Affinity constants (M^{-1}) for anion[a]				
	F	Cl	$H_2P_2O_7^{2-}$	$H_2PO_4^-$	
DPQ[9]	18,200	<50	14,300	<100	
10	51,300	<100	93,700	<200	
11	24,700	<100	58,900	<100	
12	25,600	<100	57,300	<100	
13	>10^6	2,000	>10^6	150,000	
14	27,500	<50	39,000	<50	

[a] Tetrabutylammonium salts of anions were used; 1:1 sensor-anion stoichiometry was confirmed by job plots; fitting errors were lower than 15%.

10
Emission: 495 nm
Φ: 0.20
τ: 2.0 ns

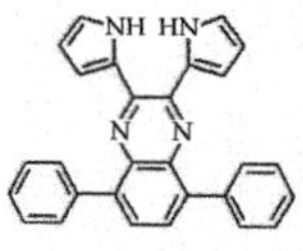

11
505 nm
Φ: 0.24
τ: 2.5 ns

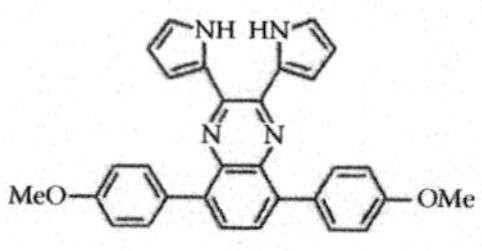

12
590 nm
Φ: 0.22
τ: 10.1 ns

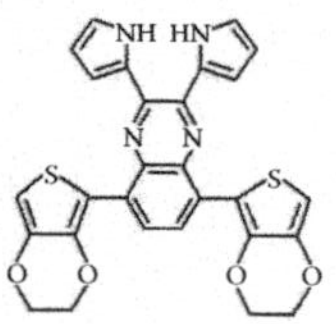

13
598 nm
Φ: 0.15
τ: 12.0 ns

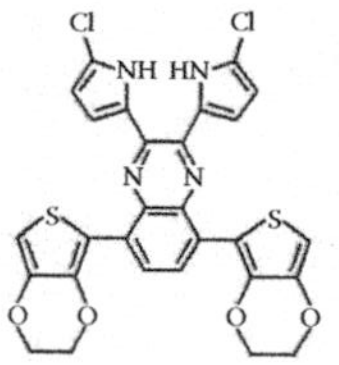

14
610 nm
Φ: 0.26
τ: 3.1 ns

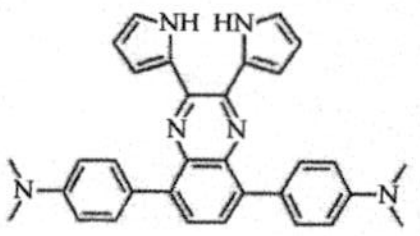

FIGURE 9.6 The structures, photophysical properties, and anion affinities of 5,8-aryl-ethynylene-substituted DPQ sensors. (From Aldakov and Anzenbacher.[13,14])

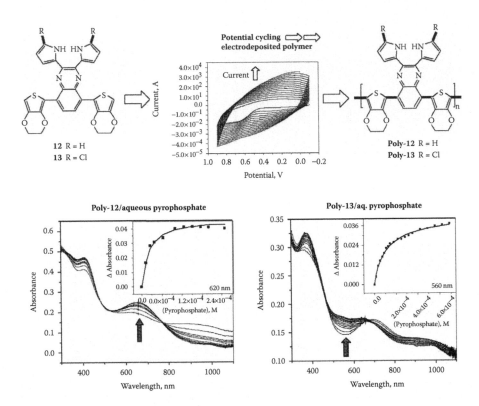

FIGURE 9.7 The electrodeposition of poly-12 or poly-13 on a transparent electrode yields colorimetric sensor devices capable of pyrophosphate sensing in aqueous samples. (From Aldakov and Anzenbacher.[14])

long-lived luminescent chromophore in addition to the DPQ ligand. Sensor 15 was deemed particularly attractive because it not only possesses bright luminescence but also offers the possibility to utilize the changes in long-lived luminescence ($\tau = 380$ ns) as an alternative means of signal transduction. This is a particularly attractive aspect of the development of fluorescence sensors, where the performance and accuracy of sensors based on fluorescence intensity measurements can be limited by many factors, including excitation source intensity fluctuations, light loss in the optical path, light scattering, and detector drift.[16] Signal transduction utilizing emission lifetimes (lifetime-based chemical sensing) is unaffected by the sample's emission intensity, thereby circumventing many limitations of intensity-based methods.[16] Sensor 15 exhibits a long-lived metal-to-ligand charge transfer (MLCT) excited state with a lifetime of up to 380 ns.[15] Similar to previous sensors utilizing DPQ structural features, sensor 15 is an efficient sensor for fluoride and cyanide anions, which are bound to the sensor via hydrogen bonding to the pyrrole NHs.

Figure 9.8 shows the luminescence lifetime quenching of sensor 15 as a function of the cyanide concentration. Prior to the addition of cyanide, sensor 15 exhibited a single exponential lifetime ($\tau = 380$ ns in an air-saturated solution). At low cyanide concentrations, the decay profiles can be adequately fit by a single exponential

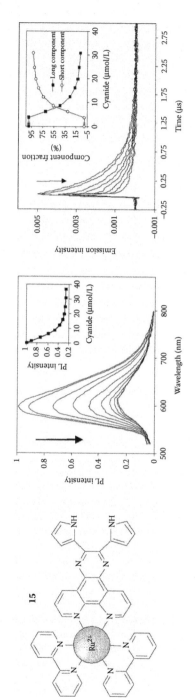

FIGURE 9.8 (Left) A structure of sensor 15. (Center) Changes in luminescence intensity of sensor 1 with the addition of cyanide. (Inset: Binding isotherm monitored by the integrated luminescence intensity.) (Right) Changes in the time-resolved photoluminescence decay of sensor 15 upon the addition of cyanide. (Inset: Shifts in fractional intensity of the two lifetime components obtained from biexponential analysis.) (From Anzenbacher et al.[15])

function, with a slightly reduced lifetime (340 ns). At higher cyanide concentrations, the time profiles exhibited biexponential kinetics. Both exponential components remained effectively constant, ranging from 320 ns to 370 ns (long τ) and 13 ns to 17 ns (short τ). The average lifetimes shorten as a function of the CN concentration, because the fractional intensity shifts from an initial dominant long-lifetime component to a short-lifetime component as the cyanide concentration increases (Figure 9.8). These data suggest that there are at least two distinct luminescent species, consisting of an anion-bound sensor 15 (short τ) and free sensor 15, the sum of which results in the observed lifetime quenching. In this case, the shift in the fractional intensity makes sensor 15 a suitable lifetime-based sensor for anions.

9.4 CONCLUSION

We have demonstrated by several approaches how a relatively simple fluorescent sensor can be successfully upgraded to yield compounds and materials that may be useful in sensor applications. These approaches that utilize well-known features such as excited-state delocalization, RET, and substituent-based electronic modulation of the fluorescence wavelength/quantum yield are, in fact, general and may be used to improve the performance of almost any potential sensor that binds analyte tightly but whose application in sensing is precluded by the lack of effective signal transduction. Our results suggest that the improvements in signal transduction, a road significantly less traveled in the anion-sensing field, may result in dramatic improvement of overall sensor performance and is therefore a very important part of sensor development research efforts.

REFERENCES

1. *Ullmann's Encyclopedia of Industrial Chemistry,* 6th ed. New York: Wiley-VCH, 1999.
2. Lakowicz, J.R., *Principles of Fluorescence Spectroscopy,* 2nd ed. New York: Plenum Press, 1999.
3. (a) Martínez-Máñez, R. and Sancenón, F., Fluorogenic and chromogenic chemosensors and reagents for anions. *Chem Rev,* 103, 4419–4476, 2003; (b) Beer, P.D. and Gale, P.A., Anion recognition and sensing: the state of the art and future perspectives. *Angew Chem Int Ed Engl,* 40, 486–516, 2001; (c) de Silva, A.P., Gunaratne, H.Q.N., Gunnlaugsson, T., Huxley, A.J., McCoy, C.P., Rademacher, J.T., and Rice, T.E., Signaling recognition events with fluorescent sensors and switches. *Chem Rev,* 97, 1515–1566, 1997; (d) Steed, J.W. and Atwood, J.L., *Supramolecular Chemistry.* Chichester, UK: Wiley, 2000.
4. (a) Sessler, J.L. and Weghorn, S.J., *Expanded, Contracted and Isomeric Porphyrins.* Oxford: Elsevier, 1997; (b) Sessler, J.L. and Davis, J.M., Sapphyrins: versatile anion binding agents. *Acc Chem Res,* 34, 989–997, 2001.
5. (a) Gale, P.A., Anzenbacher, P., Jr., and Sessler, J.L., Calixpyrroles II. *Coord Chem Rev,* 222, 57–102, 2001; (b) Gale, P.A., Sessler, J.L., and Král, V., Calixpyrroles. *Chem Commun,* 1, 1–8, 1998.

6. (a) Sessler, J.L., Zimmerman, R.S., Bucher, C., Král, V., and Andrioletti, B., Calix-phyrins. Hybrid macrocycles at the structural crossroads between porphyrins and calixpyrroles. *Pure Appl Chem,* 73, 1041–1057, 2001; (b) Bucher, C., Zimmerman, R.S., Lynch, V., Král, V., and Sessler, J.L., Synthesis of novel expanded calixphyrins: anion binding properties of a calix[6]phyrin with a deep cavity. *J Am Chem Soc,* 123, 2099–2100, 2001.

7. Seidel, D., Lynch, V., and Sessler, J.L., Cyclo[8]pyrrole: a simple-to-make expanded porphyrin with no meso bridges. *Angew Chem Int Ed Engl,* 41, 1422–1425, 2002.

8. Black, C.B., Andrioletti, B., Try, A.C., Ruiperez, C., and Sessler, J.L., Dipyrrolylqui-noxalines: efficient sensors for fluoride anion in organic solution. *J Am Chem Soc,* 121, 10438–10439, 1999.

9. Oddo, B., Syntheses in the pyrrole group. Dipyrroyl and its derivatives. *Gazz Chim Ital,* 41, 248–255, 1911.

10. Anzenbacher, P., Jr., Try, A.C., Miyaji, H., Jursíková, K., Lynch, V.M., Marquez, M., and Sessler, J.L., Fluorinated calix[4]pyrrole and dipyrrolylquinoxaline: neutral anion receptors with augmented affinities and enhanced selectivities. *J Am Chem Soc,* 122, 10268–10272, 2000.

11. Sessler, J.L., Maeda, H., Mizuno, T., Lynch, V.M., and Furuta, H., Quinoxaline-oligopyrroles: improved pyrrole-based anion receptors. *Chem Commun,* 8, 862–863, 2002.

12. Pohl, R., Aldakov, D., Kubát, P., Jursíková, K., Marquez, M., and Anzenbacher, P., Jr., Strategies toward improving the performance of fluorescence-based sensors for inorganic anions. *Chem Commun,* 11, 1282–1283, 2004.

13. Aldakov, D. and Anzenbacher, P., Jr., Dipyrrolyl quinoxalines with extended chromophores are efficient fluorimetric sensors for pyrophosphate. *Chem Commun,* 12, 1394–1395, 2003.

14. Aldakov, D. and Anzenbacher, P., Jr., Sensing of aqueous phosphates by polymers with dual modes of signal transduction. *J Am Chem Soc,* 126, 4752–4753, 2004.

15. Anzenbacher, P., Jr., Tyson, D.S., Jursíková, K., and Castellano, F.N., Luminescence lifetime-based sensor for cyanide and related anions. *J Am Chem Soc,* 124, 6232–6233, 2002.

16. Lakowicz, J.R., ed., *Topics in Fluorescence Spectroscopy,* vol. 4, *Probe Design and Chemical Sensing.* New York: Plenum Press, 1994.

10 Lab-on-a-Chip and Fluorescence Sensing on the Microscale

Hugh N. Chang, M.S.E., Anson V. Hatch, Ph.D., Kenneth R. Hawkins, B.S., and Paul Yager, Ph.D.

CONTENTS

10.1 INTRODUCTION

This chapter provides a brief overview of the issues encountered in lab-on-a-chip or micro total analysis systems (µTAS) that employ fluorescence detection. µTAS are composed of multiple microfabricated devices (which may or may not be monolithic) that are combined as components of a system that performs chemical and biochemical analysis.[1] Often μTAS consist of components similar to those found in a conventional macrolaboratory (e.g., pumps, valves, tubes, filters, chromatographic columns, and detectors), but some of the devices are unique to the microscale. To date, there has been more work on microdevices than on microsystems, but that balance is rapidly shifting toward systems. This pace of work is being driven by economic pressure to capitalize on the potentials of small volumes and highly parallel processing possible with μTAS. Developments in microfluidics and μTAS devices are strongly influenced by one another because microfluidic channels constitute the basic plumbing used to create and integrate μTAS components.

Typically, devices incorporated into μTAS utilize microfluidic flow channels in which at least one dimension is less than 1 mm in length. The geometry of these devices typically results in submicroliter inspection volumes, and occasionally the volumes are a million times smaller. The inherent sensitivity in fluorescence detection (combined with the use of many transparent materials in microdevices) makes it the primary signal measurement method used.

This brief review complements more in-depth reviews of the widely varying applications of μTAS that highlight the increasing potential of microfluidic technologies for biochemical analysis.[2-5] The unique optical challenges and opportunities encountered while working at the microscale will be discussed and examples of how these challenges can be resolved are given. As this field is growing rapidly, no survey can remain definitive for long, so we have provided some general principles, followed by examples taken from our own research.

10.1.1 LAB-ON-A-CHIP APPLICATIONS USING FLUORESCENCE DETECTION

μTAS devices have been developed for use in many different applications, including immunoassays, nucleic acid sequencing and detection, and enzyme assays. Among the wide range of detection techniques employed, fluorescence is currently the most common.[6-11]

10.1.1.1 Immunoassays

Lab-on-a-chip immunoassays employ a number of different fluorescence detection methods. The most commonly employed technique uses laser-induced fluorescence (LIF) to measure the separation of reaction products. Separation techniques include capillary electrophoresis (CE) and immunosorbent chromatography—both examples of miniaturization of a process that works well at a larger scale.

In CE-based immunoassays,[12-14] the capture molecule and its conjugate are incubated in a reaction channel. After incubation, electro-osmotic flow is employed to transport the sample from the reaction chamber to an electrophoretic separation chamber. A competitive immunoassay format employs the use of a labeled antigen that competes with unlabeled antigen for the antibody-binding sites. The changes in electrophoretic mobility of the labeled species due to antibody binding result in a separation of the labeled complex and the free labeled conjugate. The ratio of the fluorescence intensities of these two species allows quantitative measurement of the amount of unlabeled antigen in the sample of interest.

Dodge et al.[15] devised a microfabricated immunoassay that uses principles of immunoaffinity chromatography. In this case, an antigenic molecule (protein A) is immobilized on the surface of the microfluidic channel walls. When labeled antibody (rabbit IgG) is introduced to the channel, the antibody binds to the immobilized antigen. The high surface:volume ratio typical of microfluidic channels reduces the distance the antibody in solution must diffuse to reach the immobilized antigen, resulting in a corresponding reduction in incubation time. This type of immunoassay, performed in a traditional microtiter well, requires incubation times on the order of a few hours. The much smaller diffusion distances in μTAS devices (tens of microns vs. millimeters), enables incubation times on the order of minutes. After incubation, the bound antibody is washed and eluted using a glycine–hydrogen chloride buffer (pH 2.0). The bound, labeled antibody is quantified by the intensity of the fluorescence measured in the eluent. Linder et al.[16] integrated on-chip fluorescence detection with an immunoaffinity-based assay and constructed a heterogeneous immunoassay using polydimethylsiloxane (PDMS) as the primary channel material.

Recently, Roos and Skinner[17] described a two-bead immunoassay performed in a microfluidic device. This assay used capture beads of one emission wavelength (λ_1) conjugated to antibodies specific to the antigen of interest. Detection beads of a different emission wavelength (λ_2) were conjugated to an antibody specific to a different epitope of the same antigen. When mixed, three different fluorescent species exist: the λ_1 capture beads, the λ_2 detection beads, and the λ_1–λ_2 complex of capture beads conjugated to detection beads. Separation techniques may be employed to isolate the two-bead conjugates from the unbound beads, and the fluorescence from the complex may be used to quantitate the amount of antigen in the sample. However, another interesting technique employed by the authors to measure the complex was the use of multiple optical detectors tuned to the specific wavelengths of the capture and detection beads. By using a flow cytometer, individual events representing an unbound capture sphere, an unbound detection sphere, or a capture/detection complex could be analyzed. Using the flow cytometer, quantitation became a count-based measurement rather than a separation/intensity ratio measurement.

These immunoassays all exploit the small working volumes and diffusion distances in μTAS devices to reduce reagent volume requirements and minimize incubation times. Analyte quantitation was achieved through intensity measurements associated with a specific aliquot of sample. A novel immunoassay that exploits differences in the diffusivity of an antibody/antigen pair—the "diffusion immunoassay"—was recently developed in this laboratory.[18] This assay will be discussed in greater detail later in the chapter.

10.1.1.2 Nucleic Acid Analysis

Nucleic acids are analyzed for sequence information, gene regulation, and pathogen identification, among other things. Many different techniques have been employed for nucleic acid analysis in μTAS devices, including CE-based separations in sequence analysis, TaqMan (Applied Biosystems, Foster City, CA) detection for sequence identification, and pattern-based detection of arrays in gene identification and regulation studies.

A number of microfluidic-based DNA analysis devices use CE-based separation techniques. In one of the earliest demonstrations of a complete µTAS, a group at the University of Michigan created a microfabricated DNA analysis device that integrates sample preparation, reagent mixing, DNA amplification, electrophoretic separation, and fluorescence-based detection of separated products onto a single silicon substrate.[19–23] The fluorescently labeled DNA fragments were detected using photodiode arrays that had been fabricated underneath the separation channel[24]; the fragments were excited with a light-emitting diode (LED) external to the device. An interference filter was integrated into the substrate to prevent the excitation light from saturating the photodiodes. A different group developed a four-color confocal rotary scanning detection system to analyze DNA sequences in an array of CE channels microfabricated on a single wafer.[25–29] The confocal configuration rejects scattered light and substrate fluorescence and results in significant enhancement of the signal:noise ratio. In both of these applications, the underlying signal being detected is that of fluorescent bands separated as a result of differences in electrophoretic mobility within a small capillary.

A fluorescence detection method developed for DNA analysis called TaqMan polymerase chain reaction (PCR) uses PCR primers that contain both a fluorescent reporter and a molecule that quenches that label. During the PCR process, the polymerase that facilitates the hybridization of the primer with the target strand removes the reporter molecule from the primer, creating a fluorescent signal directly proportional to the number of primers hybridized to the target DNA. The fluorescent signal generated by the TaqMan probes grows in intensity with the amplification of the sequence of interest. This technique has been used to identify many different pathogens.[30–35] Recently, a microfluidic cartridge was developed that uses the TaqMan reporting mechanism to detect group B streptococcus.[36] TaqMan detection enables the simultaneous amplification and detection of specific markers for group B streptococcus. As with the University of Michigan DNA device, this cartridge integrates all the functionality to process raw sample, lyse bacteria, and amplify bacterial DNA. However, the cartridge is manufactured from a polymer rather than silicon.

Integrated microarrays have recently been developed that incorporate sample processing, probe hybridization, and detection.[37,38] A µTAS incorporating microarrays and PCR chambers was recently developed.[38] In this example, four different PCR/microarray pairs were microfabricated from silicon to study genetic differences in medicinal plants. The results were analyzed by examining the fluorescence pattern generated by the hybridization of fluorescently labeled primers to the different probes immobilized on the surface of the microarray.

10.1.1.3 Enzyme Assays

Enzyme assays in µTAS typically have employed one of two different systems to facilitate fluorescence detection. For example, one system utilized a fluorescently labeled substrate,[39] whereas another used a nonfluorescent but fluorogenic substrate that, when catalytically converted to product, released a fluorescent product.[40–42]

Starkey et al.[39] devised an assay that used extracellular signal-regulated protein kinase (ERK) to catalyze the transfer of a phosphate group from adenosine triphosphate (ATP) to the threonine residue of a fluorescently labeled nonapeptide (APRTPGGRR). When analyzed using microchip-based CE, the substrate and product separate due to the additional phosphate group on the product. Using the system, the activity of endogenous ERK was measured in response to vascular endothelial growth factor.

Two different enzyme-substrate models that release fluorescent products as a result of enzyme-catalyzed reactions have been employed in µTAS-based enzyme assays. The enzyme β-galactosidase hydrolyzes the substrate resorufin βD-galactopyranoside (RBG) releasing resorufin, a fluorescent product. Hadd et al.[40] used this system in a microchannel-based device to assay for β-galactosidase and measure the enzyme reaction kinetics. Schilling et al.[42] also employed this system to determine the extent to which a microfluidic device successfully lysed bacterial cells and extracted β-galactosidase from the cells. Fluorescein mono-βD-glucuronide (FMG) is a substrate for the enzyme β-glucuronidase. When hydrolyzed by β-glucuronidase, FMG releases a fluorescein molecule. Starkey et al.[41] employed this system in a different assay than described above to measure the degree to which D-saccharic acid-1,4-lactone inhibits the activity of β-glucoronidase on FMG.

10.1.2 Microscale Challenges

Microscale devices hold the potential to reduce the cost of biochemical analysis by reducing reagent volumes, reaction times, and instrumentation requirements. Additional cost reductions may be achieved by using lower-cost polymeric materials. Both the scale of the devices and the device materials pose inherent challenges for fluorescence detection.

The small channel volumes inherent in µTAS devices allow the use of small reagent volumes and a reduction of reaction times due to shorter diffusion lengths between reaction components. However, the smaller volumes also result in low numbers of fluorophores in any given inspection volume; consequently, the intensity of the emitted light may be very small. The very high surface:volume ratio is an additional problem. Because biological samples often adsorb to the channel walls, the number of signal-reporting fluorophores free in solution may be further reduced. Another complicating aspect of the device geometry lies in the light path used to inspect the fluids. Because the channel dimensions are usually on the order of a few hundred microns or less, the measured fluorescence is often a composite of the fluorescence generated by the signal of interest and the fluorescence generated by the device materials. By using different materials and different optical interrogation methods, the issues related to background fluorescence may be ameliorated.

10.2 FLUORESCENCE-BASED DETECTION METHODS FOR µTAS

The basic elements of fluorescence detection for microscale devices are the same as on the macroscale. A light source illuminates a fluorescently labeled sample. Upon appropriate excitation, the fluorophores emit light and the emitted light is collected and quantified to determine the precise amount of fluorophore in the sample of interest. Unique considerations for microscale devices include methods to overcome the autofluorescence of device materials and the ability to miniaturize detection systems. This section discusses the general principles underlying some commonly used detection systems and provides specific examples of implementations of these systems for microscale devices.

10.2.1 SINGLE-FOCUS TECHNIQUES

Fluorescence microscopy is commonly employed to interrogate fluorescent signals in microfluidic channels. When a broad-spectrum light source such as a mercury arc lamp is used for illumination, a filter is placed in the excitation light path. If a narrowband light source such as a laser or LED is employed, the filters for conditioning the excitation light may be unnecessary. Emission filters condition the collected light so that only wavelengths emitted by the fluorophore are presented to the photodetectors and as much stray excitation light as possible is rejected in the collection path. When the signal generated in the volume of interest has high fluorescence intensity relative to the background, fluorescence microscopy is an effective detection mechanism. µTAS applications that have employed fluorescence microscopy include enzyme assays,[40-42] immunoassays,[18,43] and protein adsorption studies.[44]

10.2.2 CONFOCAL OPTICS

When background noise is of the same order of intensity as the fluorescence generated by the interrogation volume, different optical techniques may be employed to reduce the undesirable sources of fluorescence. These techniques include the use of confocal optics and multiphoton fluorescence excitation.

Confocal detection uses point illumination and point detection of a sample to exclude most fluorescence emission generated outside the focal point of the objective lens. In general, the objective lens in the system is used for both focusing the excitation source and collecting the emitted fluorescence (epifluorescence illumination). A pinhole is employed as a spatial filter that effectively allows only light generated at the focal point of the objective lens to reach the detector. A typical configuration for the optical components is shown in Figure 10.1. Detailed explanations of the operating principles for confocal microscopes are available elsewhere.[45-47] The use of confocal detectors to measure fluorescence generated in microfluidic channels can improve the limit of detection in these devices by minimizing the scattered light and fluorescence generated by the device materials. Obviously, the thickness of the microfluidic channel must be greater than the depth of field of the optical system for this to work well. The rejection of light out of the focal point of the objective also enables greater packing densities by reducing cross talk between adjacent channels.

Applications of confocal detection systems include the development of a four-color confocal scanner for genome analysis in a highly parallel microcapillary array electrophoresis device,[29] single molecule detection of double-stranded DNA in a polymeric device,[48] and the detection of an immunoassay capillary electrophoresis separation in a glass device.[13]

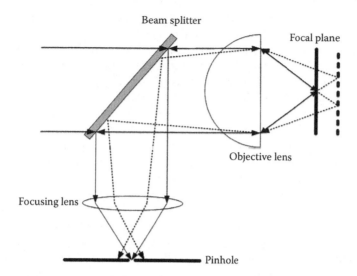

FIGURE 10.1 Schematic of a simple confocal system for capturing fluorescence intensity in microdevices. The excitation light is focused to a point by the objective lens at its focal plane. Light from the focal plane (represented by solid lines) is collected by the objective directed through the pinhole, under which would be placed a single-channel detector such as a photomultiplier tube. Light generated out of the focal plane (represented by dotted lines) is blocked by the pinhole and is not detected.

10.2.3 OTHER METHODS

Multiphoton excitation exploits the ability of fluorophores to be excited by the near-simultaneous absorption of multiple low-energy photons. By combining diffraction-limited optics with powerful pulsed laser light sources, the high photon intensity needed to achieve multiphoton excitation is limited to the focal point of the objective lens in the system. No pinholes are required to spatially filter the emitted fluorescence because excitation and subsequent emission occurs only at the focal point of the system. Denk et al.[49] developed the first multiphoton microscope, and a wide variety of applications in the biosciences have been developed using this underlying principle.[50] The technique has been used for bioassays in capillaries[51] and on planar microfluidic devices.[52]

Miniaturization of the optical components necessary to integrate fluorescence detection systems into μTAS devices requires the microfabrication of lenses, light sources, and detectors. These devices may be constructed from many different materials, including plastic laminates, silicon, PDMS, and glass. Each material exhibits unique machining, electrical, optical, and biological properties that must be balanced. For example, glass is a much better electrical insulator than silicon and may have performance advantages when high voltages are applied for electrophoresis. μTAS devices may also be a composite of these materials,[24] and a variety of fabrication steps may be necessary to achieve appropriate isolation between material layers.

A number of microfabricated fluorescence detection systems have successfully balanced the different design considerations mentioned above. For example, a PDMS device integrated a microavalanche photodetector and used optic fiber to direct the excitation source close to the microfluidic channel.[53] A composite device integrated a plastic electrophoresis channel with a silicon/silicon dioxide–based fluorescence detection system.[24] In these cases, the signals from the samples of interest were much greater than the background fluorescence of the materials.

The miniaturization of confocal detection systems requires the microfabrication of lenses in addition to photodetectors. Lenses have been microfabricated in many different materials,[54–58] including silicon, glass, PDMS, and plastic polymers. A microfabricated two-dimensional confocal scanning array has been fabricated in silicon.[59] This microelectromechanical system (MEMS) structure incorporates the lenses and scanning actuators to enable raster scanning in two directions of a microfluidic channel.

The design of fluorescence detection systems for microscale devices entails making a wide range of design decisions. Bench-scale fluorescence, confocal, and multiphoton microscopes are commercially available and are used extensively in research settings. These instruments often are customized by choosing different photodetectors (charge-coupled device [CCD] cameras, photomultiplier tubes, etc.) and different light sources (broad-spectrum lamps, lasers, etc.). During the development phase of μTAS devices, these instruments are particularly useful. The choice of device materials and the limit of detection for the analytes of interest may favor one instrument type over another. The widespread deployment of these devices may depend on the availability of low-cost, portable detectors, and the cost of the devices themselves may force specific choices in device materials. Miniaturization of detection systems is possible, and decisions about the components integrated into the device and the components integrated into the detection instrument will impact the performance and cost of the overall system.

10.3 BACKGROUND FLUORESCENCE OF MATERIALS EMPLOYED FOR LAB-ON-A-CHIP DEVICES AND SYSTEMS

Background luminescence is an important consideration when performing fluorescence detection on any scale. Such background luminescence may be fluorescence or phosphorescence and is often spectrally broad and therefore poorly characterized. For the purposes of this chapter, we will refer to it as fluorescence, although the issues described refer to all possible sources of luminescence.

Background fluorescence may come from the solvent, the optical elements, or the device holding the sample. However, we will not address background due to the optics or solutions used in the experiments because they are not specific to microdevices. In microfluidic applications, the total number of fluorescent photons emitted may be very small due to the small volumes of fluorescent samples excited, so there is generally a need for care (if not expense, as well) in the design of optical systems. Moreover, other design parameters of lab-on-a-chip devices frequently compromise the minimization of the background fluorescence. In these cases, despite the use of materials that are acceptable at the macroscale, the background fluorescence of the microdevices may be significant compared to the fluorescence signal of interest.

10.3.1 SILICON AND GLASS

The first generation of microfluidic "lab-on-a-chip" devices was generally made using conventional microelectronics processes. Therefore, most such devices consisted of channels etched in single-crystal silicon; closed channels were usually formed either by bonding to the first silicon part either another polished silicon wafer or a glass such as Pyrex capable of being anodically bonded.[60–62] Channels may also be etched in glass and sealed to other glass pieces or silicon. The advantages of silicon and glass are similar:

1. Photolithographic fabrication was developed by the integrated circuit industry and therefore is well characterized.
2. Small feature sizes (down to submicron in silicon) are easily obtained.
3. The parallel nature of the lithographic fabrication method enables mass production of numerous identical devices.
4. Both materials have many attractive chemical and mechanical properties, including low intrinsic fluorescence.

Wet-etching and dry-etching techniques can produce a wide variety of features of use in chemical detection.[60,62] As a result of these advantages, the microfluidics literature is filled with examples of glass and silicon devices, many that use fluorescent labels as reporters.[63,64] For applications using electro-osmotic pumping, glass is still preferred for optimal flow control.[65] Single-crystal silicon is essentially opaque to light, with a wavelength of less than 400 nm or greater than 1100 nm, and is not very transparent to visible wavelengths,[60] so a glass "lid" is usually bonded to the wafer as a window when fluorescence detection is desired. This effectively eliminates the possibility of excitation below about 350 nm.

10.3.2 POLYMERS

As the field has matured, researchers have begun to recognize the limitations of silicon and glass as substrates for lab-on-a-chip devices. Fabrication processes for silicon and glass are time consuming and relatively expensive for the first few prototypes; polished wafer substrates are expensive, and obtaining the first devices requires either the services of an expensive fabrication facility or the laborious process of working out the fabrication protocols in-house. The finished products are fragile and not particularly well suited to connection to external devices with commercial tubing (although some new commercially available connector kits have gone a long way toward solving this problem). Therefore, glass and silicon are poorly suited for rapid prototype development or use in small research projects; they are also particularly poorly suited to the production of very large numbers of devices at low cost.

Lately, many investigators have turned to polymeric substrates for constructing microfluidic devices and µTAS. Several recent reviews are devoted to this topic.[66–68] Polymers have numerous desirable characteristics as substrates for microfluidic devices: the raw materials are relatively inexpensive, several fabrication techniques can yield nearly straight channel walls,[18,69,70] and methods exist for rapid prototyping at low cost.[18,69,70] Lamination of multiple transparent layers is a straightforward approach to producing devices and systems with complex three-dimensional features that are amenable to optical interrogation.[71–74] There are also several disadvantages that affect some polymers more than others, including poor chemical resistance (although both glass and silicon are susceptible to etching by alkaline solutions that do not affect most polymers), permeability to gases and many organic solvents,[75] poorly controlled surface chemistry (due to slow surface remodeling),[76] and relatively high intrinsic fluorescence.[66,77] Polymers are also, by their nature, less defined materials than glass and silicon and can vary significantly between manufacturers and even between batches from the same vendor.[77] Nevertheless, the economic advantage of using polymers is so great that polymers are certain to be the dominant materials for microfluidic devices in the future.

It has long been known that many polymers are fluorescent, particularly when excited with ultraviolet (UV; less than 400 nm), blue (400–500 nm), or green (500–550 nm) light.[77,78] In some cases, the fluorescence is intrinsic to the polymer chain itself (usually because of the presence of aromatic rings), although in most cases other components of the final material are the culprits, including mechanical or optical additives, impurities, or degradation products.[79] To optimize a fluorescence-based detection system, the fluorescence must be characterized and then avoided or eliminated. Formulating a custom polymer is usually not an option for microfluidics research groups, so they are generally limited to purchasing what is available off the shelf. Most investigators have minimized confounding background by employing confocal optics,[48] two-photon excitation,[51] long-wavelength fluorophores,[80] modulating the signal of interest and employing lock-in detection,[78] or simply working at a high-enough fluorophore concentration that the signal:noise ratio is acceptably high. However, in many microfluidic applications, these solutions are not practical.

The proliferation of materials and fabrication techniques makes generalizations about the fluorescence properties of plastic lab-on-a-chip devices difficult. Photolithographic patterning of the photoresist SU-8 has been employed to rapidly prototype high-resolution thin and thick devices with nearly vertical channel sidewalls. SU-8 is itself strongly fluorescent, however, so its use for optical detection is generally restricted to forming well-defined molds for other less fluorescent polymers. PDMS replica molding (using molds from either silicon or SU-8) has gained widespread popularity,[69,70] and because of its low intrinsic luminescence, many PDMS devices that use fluorescence detection have been described.[9–11] Extensive development of PDMS has led to surface modification methods that allow permanent sealing of PDMS to oxide-coated surfaces, including glass.[69,71,81,82] The primary limitations of the most useful of the PDMS materials, such as Sylgard 184 (Dow Corning, Midland, MI), are high mechanical compliance, permeability to organics, and the tendency to absorb hydrophobic compounds (including fluorophores).

Polymethylmethacrylate (PMMA) and other rigid polymers produce stiffer, more resistant structures where required[83,84]; they can be formed using laser ablation, hot embossing, "soft embossing,"[85] and injection molding. Several investigators report that PMMA and polycarbonate have background fluorescence only slightly higher than glass of a similar thickness.[66,67,77] Hybrid chips using several materials are common.[7,8,10,18,75,84] Unfortunately, the fabrication techniques above usually result in devices that are much thicker than required for the depth of the microfluidic channels—a disadvantage when simple illumination and detection systems are used. Direct-write carbon dioxide laser cutting of polymer thin films and adhesive lamination enable the very rapid assembly of three-dimensional microfluidic devices that are only as thick as a credit card.[18,73,74] This method has the advantage of minimizing the volume of polymer in the light path when confocal optics or two-photon excitation is not possible, thus minimizing the background observed. Optical-grade Mylar (polyethylene terephthalate [PET]) polyester film is often employed in these laminated devices because of its mechanical toughness, although its intrinsic fluorescence is relatively high.[18,73,74] Recently, PMMA films have been used for the outermost "capping" layers in laminated devices to reduce background fluorescence.[86] In general, when optimal design is desired, the fluorescence properties of these polymers will have to be weighed against their other properties of interest and evaluated in the context of the application.

10.3.3 PLASTIC THIN-FILM AUTOFLUORESCENCE

A recent publication[77] surveyed the fluorescence behavior of some commonly available polymer films under standardized conditions; the aim was to determine which materials are most suitable for high-sensitivity fluorescence detection with conventional illumination and detection schemes (Table 10.1). When illuminated with a

TABLE 10.1
A Listing of Polymer Thin Films Compared in Terms of Their Intrinsic Luminescence (Presumably Fluorescence) and Its Rate of Burn Off

	Material	Thickness (µm)	Source
1	Mylar (PET polyester)	100	Fralock, Santa Clara, CA
2	Mylar (PET polyester)	50	DuPont, Wilmington, DE
3	Lexan (polycarbonate)	2500	GE Plastics, Pittsfield, MA
4	Polypropylene	60	ExxonMobil, Houston, TX
5	Trycite (polystyrene)	50	Dow Chemicals, Midland, MI
6	Zeonor (polyolefin)	110	Zeon Chemicals, Louisville, KY
7	Topas (polyolefin)	90	Ticona, Summit, NJ
8	Topas (polyolefin)	210	Ticona, Summit, NJ
9	Rohaglas (PMMA)	340	Cyro Industries, Orange, CT
10	Glass microscope slide	960	Corning Glass, Corning, NY
11	Glass microscope slide	1970	Corning Glass, Corning, NY
12	UV-grade fused silicon	1030	Esco Products, Oak Ridge, NJ

From Hawkins and Yager.[77]

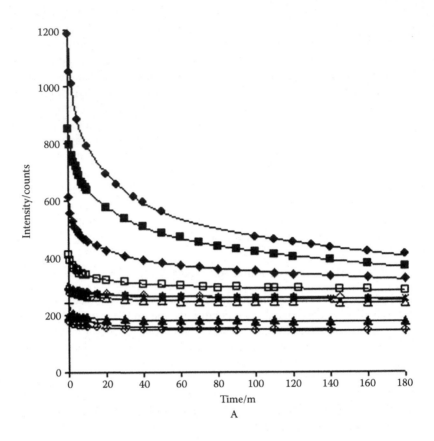

FIGURE 10.2A (See color insert following page 142.) Plots of the decrease in fluorescence of most of the materials from Tables 10.1 and 10.2[77]. For the two materials where two different thicknesses were evaluated (glass, Topas™) a single representative plot is shown for clarity. Symbol size approximates the error bars. Each series of discrete observations is accompanied by a best fit curve (see Table 10.2 for equation of best fit curve). Curves were fit using non-linear least squared regression by a generalized-gradient method (Excel Solver function, Microsoft, Redmond, WA, USA). A) Fluorescence intensity (CCD counts) versus time. (●)=Fralock Mylar™ data, (▲)=Glass (thin) data, (■)=polycarbonate data, (◆) = Dupont Mylar™ data, (◇) = Zeonor data, (□) = Rohaglas data, (△) = Fused Si data, (*) = polystyrene data, (+) = polypropylene data, (○) = Topas™ (thin) data.

mercury arc lamp conditioned by an epifluorescence microscope filter set for fluorescein, the most commonly used fluorophore, the initial fluorescence intensities of many of the materials surveyed (as detected with a CCD camera) were significantly higher than glass and fused silica controls. The intensity for all materials decreased over a 3 h "burn-off" period with complex kinetics (Figure 10.2). The other materials had initial intensities similar to glass and fused silica but also burned off, as did the glass and fused silica (Figure 10.2A and Table 10.2). The observations were normalized to the individual initial intensities to facilitate comparison of the kinetics of burn off (Figure 10.2B and Table 10.2).

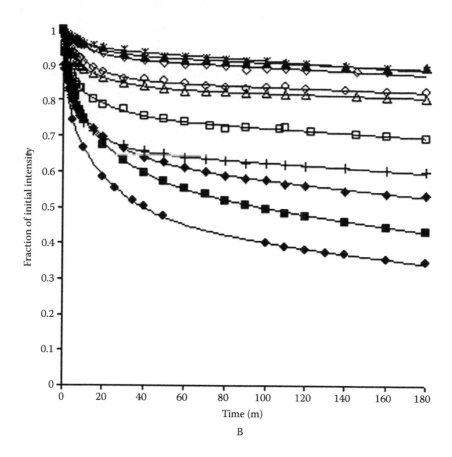

FIGURE 10.2B (**See color insert following page 142.**) Fluorescence intensity as a fraction of initial intensity versus time. Symbols and fitted lines are identical to "A" (The multiplicative and additive coefficients are re-scaled to reflect the normalization, the time constants in the exponent remain the same.)

Because the samples of materials surveyed were not identical in thickness, normalization was required to correct for the different volumes illuminated. Applying a linear normalization to the initial intensities (Table 10.2) changes the ranking of the materials significantly (Figure 10.3). These data also highlight the desirability of minimizing the amount of material in the illumination path. It is also notable that different samples of the same polymer did not always display the same thickness-normalized initial intensity; however, glass controls gave essentially identical results upon normalization (Figure 10.3). The differences in magnitudes of the initial fluorescence for similar polymers could be explained by two possible mechanisms. First, differences in processes between vendors may result in real material differences that manifest themselves as differences in fluorescence. Second, the postproduction age and handling of material from the same vendor may vary (this was not a controlled variable in the study). The decrease in fluorescence observed over time (Figure 10.2)

TABLE 10.2
Initial Fluorescence Intensity, Best-Fit Multiple Exponential Models, and Coefficients of Determination for Fitted Exponential Decay Models for All Materials Surveyed

	Material	Initial intensity (counts)	"Best-fit" model	$at + b$	$ae^{bt} + c$	$ae^{bt} + de^{ft} + c$	$ae^{bt} + de^{ft} + ge^{ht} + c$
						R² by model	
1	Fralock Mylar	1184	$y = 257e^{-0.462t} + 355e^{-0.044t} + 535e^{-0.002t} + 37.5$	0.791	0.796	0.994	0.999
2	DuPont Mylar	613	$y = 104e^{-0.641t} + 119e^{-0.048t} + 353e^{-0.001t} + 37.5$	0.697	0.778	0.988	0.997
3	Polycarbonate	849	$y = 135e^{-0.343t} + 217e^{-0.040t} + 458e^{-0.002t} + 37.5$	0.600	0.863	0.981	0.999
4	Polypropylene	242	$y = 38e^{-0.731t} + 44e^{-0.075t} + 1221e^{-0.0007t} + 37.5$	0.877	0.641	>0.999	>0.999
5	Polystyrene	286	$y = 14e^{-0.113t} + 234e^{-0.0005t} + 37.5$	0.826	0.886	0.988	0.989
6	Zeonor	293	$y = 5e^{-0.654t} + 20e^{-0.048t} + 231e^{-0.0003t} + 37.5$	0.729	0.829	0.938	0.999
7	Topas (thin)	176	$y = 6e^{-0.645t} + 19e^{-0.068t} + 1139e^{-0.0003t} + 37.5$	0.710	0.717	0.992	0.999
8	Topas (thick)	294	$y = 18e^{-0.646t} + 14e^{-0.054t} + 225e^{-0.0003t} + 37.5$	0.732	0.742	0.976	0.996
9	Rohaglas	410	$y = 52e^{-0.427t} + 49e^{-0.052t} + 271e^{-0.0005t} + 37.5$	0.673	0.713	0.989	0.998
10	Glass (thin)	198	$y = 14e^{-0.075t} + 147e^{-0.003t} + 37.5$	0.431	0.838	0.988	0.996
11	Glass (thick)	419	—	—	—	—	—
12	UV-grade fused silicon	301	$y = 20e^{-0.425t} + 30e^{-0.058t} + 214e^{-0.0002t} + 37.5$	0.660	0.678	0.988	0.999

The fitted constants in front of the exponential terms are expressed in the units of "counts," whereas the time constants of the exponential decay are expressed in the units "per minute." Only the initial fluorescence of the thick glass slide was evaluated.

From Hawkins and Yager.[77]

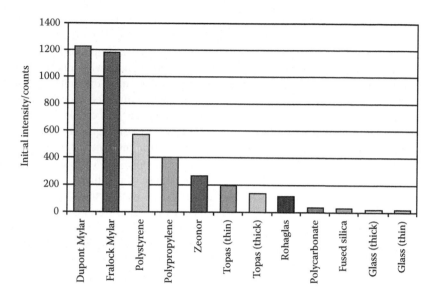

FIGURE 10.3 (See color insert following page 142.) Initial fluorescence intensities of polymeric sheets from Figure 10.2 but normalized to 200-μm thickness.[77] The color scheme is the same as that in Figure 10.2.

suggests a likely mechanism for some of the variation seen in two samples of material from the same vendor—material that had been exposed to the ambient light of a warehouse during prolonged storage could have lost a significant fraction of its intrinsic fluorophores through photobleaching.

The decrease in polymer fluorescence with illumination time must be caused by photobleaching of fluorophores in the polymer. Simple models of photobleaching predict an exponential decrease in intensity.[87] In systems where multiple fluorophores are present, linear combinations of exponential decay functions (if there are no interactions between fluorophores) or more complex nonlinear combinations may result (when energy transfer between fluorophores occurs).[88] Multiple decay rates may be due to multiple fluorophores or multiple local chemical environments for the same fluorophore.[79] Exponential decay functions can be fitted to a dataset using nonlinear least squares (NLS) regression to enable the extraction of the individual initial magnitudes and time constants of the decay processes of interest.[89,90] Several model decay functions were fitted to the experimental results above (Table 10.2). Many of the materials appear to have three exponential decay processes operating over the 3-h experiment—all appear to have at least two (Table 10.2). The time constants of these processes range from about 1 min to more than 80 h, complicating potential background correction algorithms. (Note that the numerical constants in the exponential terms have units per minute; for example, the numerical constant 0.02 corresponds to a time constant of 50 min.) This suggests that simple photobleaching of two or three species is producing the observed behavior or that there are differences in the rates of photobleaching of the polymer surface and interior.

Device fluorescence will impact the on-chip detection of fluorescently labeled reporters primarily through an increase in the intensity of background fluorescence.[67,78] If the background noise does not overwhelm the signal of interest, a simple difference algorithm may be sufficient to correct for the background if the device fluorescence is relatively constant over time. Robust correction algorithms are more difficult to formulate when the device fluorescence changes systematically on the same time scale as collection of the signal of interest. Signal integration over several minutes may enable the detection of small signals, although some form of deconvolution is then required to avoid confounding of the results by the background. If, as is often the case, the polymer emission is spectroscopically distinguishable from that of the sample, use of multiwavelength detection may partially overcome this limitation of the use of polymers.

In general, the significance of an observed change in background depends on the magnitude of the fluorescence signal of interest. For example, using a microscope-based illumination system,[77] the change in intensity over 3 min for a 200-μm-thick Mylar layer produced a change in intensity that was as great as, or greater than, the signal produced by a 20-nM fluorescein solution (Figure 10.4) throughout the entire 3 h the change was monitored. A glass slide never produced a change that was as great as the 20-nM fluorescein signal at any time during the 3 h (Figure 10.4). Thus, a device that includes 200 μm of Mylar in the light path does not perform well in an assay that uses a 20-nM fluorescein reporter, but a glass device does. Topas (cyclic polyolefin) samples produced

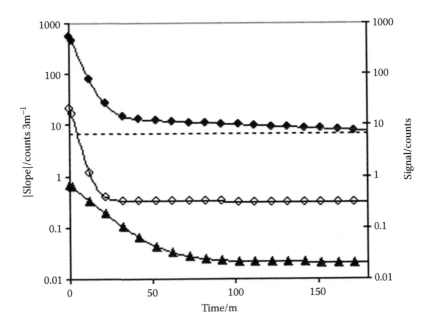

FIGURE 10.4 (See color insert following page 142.) Slope of decay curves for a three minute analysis time (♦)=Fralock Mylar™, (○) = Topas (thin), (▲) = Glass (thin), compared to the fluorescence signal (not slope) produced by 20 nM fluorescein (---). All materials normalized to 200 μm thickness.

a change that slightly exceeded the signal of 20-nM fluorescein for the first few minutes after exposure but rapidly dropped below the 20-nM fluorescein signal (Figure 10.4).

Prior exposure of the thin polymer films to the illumination source to be used in the experiment can improve the quality of detection by partial burn off, which both reduces the total background and reduces its rate of change. A device made with Topas would require a short burn off of this type if it were to be employed for an assay that used a 20-nM fluorescein reporter. This can be difficult to control precisely. All the materials investigated, including glass and fused silica, have some lower bound to their limit of detection caused by the fluorescence change over time, as long as they are in the light path. Only the adoption of a more complex illumination and detection scheme (confocal optics or multiphoton excitation) will relieve this constraint. This kind of burn-off analysis should be part of the evaluation of materials for lab-on-a-chip fluorescence detection applications.

One approach under study is to work with polymeric laminates of the least fluorescent materials, such as Rohaglas (PMMA) sheets.[91] This material can be laminated using a combination of solvent pretreatment, heat, and pressure, producing functional microfluidic devices with very low intrinsic fluorescence in the visible spectrum (Figure 10.5).

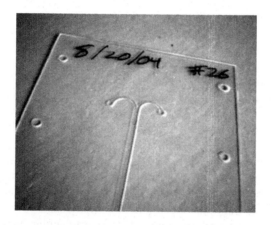

FIGURE 10.5 A photograph of a T-sensor device fabricated using solvent-assisted thermal welding of PMMA. Microfluidic channels cut in films of Rohaglas PMMA (175-µm thickness) are laminated by first immersing them in ethanol and then pressing in an assembly jig at a temperature close to the bulk PMMA glass-transition temperature. The ethanol disrupts a surface layer of PMMA sufficiently to lower the local glass-transition temperature and the elevated temperature and pressure create a weld that spans only the activated surface layers. Thus, the low-autofluorescence properties of PMMA are combined with a glueless, monolithic lamination process that retains a high degree of dimensional fidelity. This process has been optimized to produce the T-sensors shown[91] and has also been successfully employed in the fabrication of a more complicated device that combines the T-sensor with three-dimensional sheath flow to prevent solute adsorption the surfaces and absorption into the glue layers.[104]

10.4 SPECIFIC EXAMPLE: THE DIFFUSION IMMUNOASSAY

Fluorescence detection methodologies used for diffusion-based binding assays are of growing importance and present some illustrative challenges. The assays may be conducted in a T-sensor (Figure 10.6), a simple microfluidic device element that exploits laminar flow conditions prevalent in microchannels[92] or in a hydrogel.[93] Both microfluidic devices take advantage of the short timescale of diffusive transport at the microscale.

The T-sensor operates by injection of different fluids into a common channel by pressure-driven flow. With steady laminar flow conditions, a stable and predictable interface is formed between the injected fluids as they flow side by side through the main channel. Mixing of solute between incoming fluid streams (along d) occurs only by diffusion. The extent of solute transport along d is then determined by molecular diffusion coefficients and binding interactions (if any). The T-sensor converts time-based experiments to space-based experiments as long as continuous delivery of reaction species can be supplied. This puts a premium on the spatial resolution of the fluorescence imaging system.

Interdiffusion of components is illustrated in Figure 10.6 by the diffusive transport of species M, initially contained in the sample fluid stream, into the neighboring fluid stream. In T-sensor analysis, the fluorescence signal is spatially resolved across d, referred to as a "diffusion profile," to infer the distribution of analyte (Figure 10.6). Interdiffusion of chemical species is driven by the formation of a concentration gradient at the interface between incoming fluid streams. The "interdiffusion zone" for each species is the region in which the concentration gradient is appreciable. As fluid travels along the axis of flow (z), the interdiffusion zone spreads laterally across d. Spreading of the interdiffusion zone for each molecular species depends on its diffusivity and potential for binding interactions during the time of transport.

Optical measurements of diffusive transport are generally made using a CCD camera and a fluorescence microscope imaging a fixed position L downstream using fluorescence microscopy. This location is chosen so that sufficient inter-action time between fluids has been allowed for measurable diffusive transport along the d dimension and for interaction of molecular analytes. Molecular diffusion coefficients have been rapidly and accurately determined using a con-figuration like that shown in Figure 10.6.[94,95] Other uses of the T-sensor include determining sample viscosity,[96] determining sample solute concentrations (pH, human serum albumin [HSA], calcium, and phenytoin),[97–99] and evaluating molecular binding interactions (extracting kinetics and equilibrium constants of binding).[93,100]

The time required for representative molecules to traverse three different length scales is shown in Table 10.3, based on the Einstein equation of diffusive transport in one direction:

$$x_{RMS} = \sqrt{2Dt} \qquad\qquad (10.1)$$

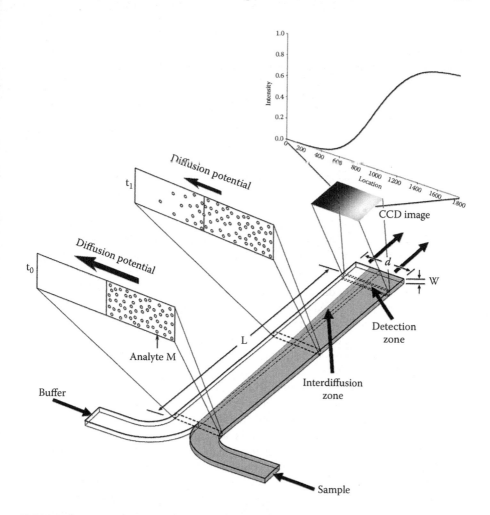

FIGURE 10.6 Illustration of a T-sensor device. A concentration gradient is generated by interfacing two different fluids in the main channel. Fluids are injected by pressure-driven flow through two inlets. Under laminar flow conditions, the fluid streams meet and flow side by side down the main channel. Mixing of the two streams is dominated by diffusion rather than convection. Optical measurements are typically made at a distance, L, downstream of the inlet junction where sufficient interdiffusion and reaction of molecular components have occurred. The CCD image shows fluorescence of the analyte, fluorescein, at the detection area. Pixel values across d are used for analysis, as shown by the graph of normalized fluorescence intensity values. With the simple configuration shown here, it is possible to measure molecular diffusion coefficients as reported by Kamholz et al.[95]

where x_{RMS} is the root mean square distance traversed by a solute species, t is the time allowed for diffusion, and D is the molecular diffusion coefficient. Smaller molecules generally have higher diffusion coefficients than large molecules and the time required for diffusive transport scales roughly with molecular weight to the

TABLE 10.3
Diffusion Coefficients of Representative Analytes in Aqueous Buffer at 25°C

Molecule	Molecular Weight (Da)	D (cm²/s)	Time for Transport		
			x_{RMS} 10 μm	x_{RMS} 100 μm	x_{RMS} 1 cm
Phenytoin[a]	252	5.8×10^{-6}	0.086	8.6	1.0 days
Fluorescein[b]	376	4.3×10^{-6}	0.116	11.6	1.4 days
Bovine serum albumin[b]	66,000	6.3×10^{-7}	0.794	79.4	8.9 days
IgG antibody[a]	150,000	4.3×10^{-7}	1.163	116.3	13.7 days

[a] Diffusion coefficient estimated based on molecular weight.

[b] Diffusion coefficient from Culbertson et al.[105]

power; larger molecules require more time than smaller molecules to traverse the same distance. Days to years would be required for many biomolecules to traverse macro-scale containers such as a test tube or beaker by diffusion alone because the time required for diffusive transport scales with distance squared. For the rapid measurements conducted in a T-sensor, diffusive transport on the order of microns is observed for most species of interest.

The signal along d that must be spatially resolved is the portion where gradients in analyte concentration are appreciable (interdiffusion zone). The extent of this zone is related to the time allowed and the molecular diffusion coefficient of the species of interest. Accurate detection is also aided by internally referencing fluorescence emissions in regions outside the interdiffusion zone for each image. Internal referencing of regions within the fluid channel where fluorescent reporter molecules are either undiluted or absent is useful. This also imposes a design constraint on the fabrication and operation of the T-sensor. Internal referencing can greatly increase accuracy by accounting for possibilities of systematic drift in the detection system, changing levels of background fluorescence, and shifts in sample fluorescence and its use will be discussed in more detail in the next section. Reference regions are typically resolved when using standard CCD detection, leaving the inner one half to one third of data points useful for quantifying the diffusion-based response.

In reported T-sensor work, at least 10 and up to several hundred data points are typically resolved within interdiffusion zones that range from a few to several hundred microns. To a certain point, higher spatial resolution increases accuracy. The acceptable level of resolution depends on the application, assay format, experimental conditions, and the signal:noise ratio, and no general criteria have been established. However, Fourier analysis of typical T-sensor experiments, followed by application of the Nyquist sampling frequency, suggests that the spatial resolution typically achieved with a CCD camera is far greater than required. Because it is the concentration gradient along d that is used for analysis, only a narrow region of CCD image data along L is typically needed to quantify a response; a single row of photodetection elements along d could be used for analysis. With CCD image data, intensity values along L are often averaged (typically from 1 to 50 pixels) to increase the signal:noise ratio, but the extent to which the diffusion profile may be changing along L must be considered, particularly when using computational model predictions to quantify the response.

10.4.1 Binding Assays

Assessment of binding interactions between molecular species is important as an analytical tool for determining concentrations of biological analytes and to evaluate molecular function. Using the T-sensor, fluorescent analysis of binding reactions may be subdivided into classes based on the fluorescence properties of the reporter molecule. In some cases, the reporter molecule exhibits a dramatic increase in quantum yield or a shift in emission wavelength upon a specific binding event. For interactions between the indicator albumin blue 580 (AB580) with HSA and the indicator Fluo3FF with calcium ions, fluorescence emission was dramatically increased upon binding. In such cases, the peak signal intensity (where maximum binding had occurred) has been used to determine concentrations of albumin or calcium based on an empirical calibration curve.[92] Kamholz et al.[96] also demonstrated that computational modeling of transport accounting for convection, diffusion, conversion of species through binding interactions, and fluorescence response can predict the observed fluorescence intensity profiles.

Baroud et al.[100] recently reported using a T-sensor to determine the association rate of calcium ions with the fluorescent indicator calcium green. Calcium green exhibits a dramatic increase in fluorescence upon binding. Numerical predictions of transport and species-dependent fluorescence were matched to experimental findings to infer the rate of complex association.[100] Numerical predictions were based on a one-dimensional equation of transport with binding treated as irreversible (complex dissociation was neglected and the validity of this assumption was not discussed in the article). The value of k_{on} determined ($1.0 \pm 0.47 \times 10^6$ $M^{-1}s^{-1}$) was reportedly too fast for traditional stopped-flow determinations, an apparent advantage of this type of analysis, and the equilibrium constant and dissociation rate were not evaluated.[100]

For most binding interactions, a fluorescent reporter molecule that responds to specific binding events with a shift in emission wavelength or quantum efficiency will not be available. In such cases, a change in the diffusive transport of indicator molecules due to binding interactions may be exploited to evaluate binding. When molecules interact to form a larger, slower diffusing complex, their diffusive transport is reduced. This also constrains analysis to binding pairs where the diffusivity of the fluorescent reporter is sufficiently reduced (approximately twofold or greater) by complex formation. Diffusion-based analysis has been exploited to infer different parameters associated with specific binding interactions, including antibody/antigen and protein/ligand interactions. In these cases, signal quantification relies more heavily on spatial resolution of the diffusion profile and changes in the diffusivity of reporter molecules due to binding.[93,99] With this class of analytes, computational modeling is still useful for determining analyte concentrations and binding kinetics, but simple peak detection may not be possible. An alternative method of generating empirical calibration curves has been first-derivative analysis of the intensity profile used to infer concentrations of phenytoin in blood samples.[93,99]

It is important to note that nonuniform flow velocities associated with pressure-driven flow in a T-sensor can impact analysis. The flow effects must be considered, particularly when attempting to analyze results using simplified models of transport that do not account for nonuniformities in flow velocities. Ismagilov et al.[101] observed the effects on transport with confocal microscopy. Both Ismagilov et al. and Kamholz et al. have provided useful insights for minimizing these velocity effects, and Kamholz et al. demonstrated cases, in which diffusive transport across d matches one-dimensional approximations.[94,95,101–103]

10.4.2 IMAGE PROCESSING

Several processing steps are necessary to convert raw image data to a diffusion profile that can be used for analysis. The first step is to extract pixel values from an image. Often the primary image format is specific to the camera or frame grabber used. To generate a diffusion profile, pixel values must be collected across the image corresponding to the d dimension. At each location across the d dimension, a window of pixels along the L dimension may be averaged to increase the signal:noise ratio.

Issues related to the imaging system (temporal and spatial nonuniformities in illumination and signal collection), microfluidic device (variations in transparency and autofluorescence), and fluorophore properties (photobleaching, quenching due to binding) necessitate the use of signal correction before interpreting image data. An example dataset processed in this manner is shown in Figure 10.7. For the first signal correction step, reference conditions were imaged to correct for spatial nonuniformities associated with the native fluorescence and system response (termed here background/flood correction [bfc]). Two reference measures were captured: "flood" images, where the device was filled with fluorescent analyte, and "blank" images, where the device was rinsed thoroughly and loaded with (putatively nonfluorescent) buffer. Images of experimental conditions were then corrected at each data point across the diffusion profile by subtracting background fluorescence and dividing by the flood response at the corresponding location according to the following equation:

$$I_{bfc}(x) = \frac{I_0(x) - I_{blank}(x)}{I_{flood}(x) - I_{blank}(x)} \qquad (10.2)$$

where x denotes the pixel number or x location across the d dimension.

Temporal variations affecting signal as a whole were corrected for by normalizing data to a scale from zero to one using internal reference points in each image. The internal reference points were signals from two regions outside the interdiffusion zone: I_{min}, where no fluorescent analyte is present, and I_{max}, where sample

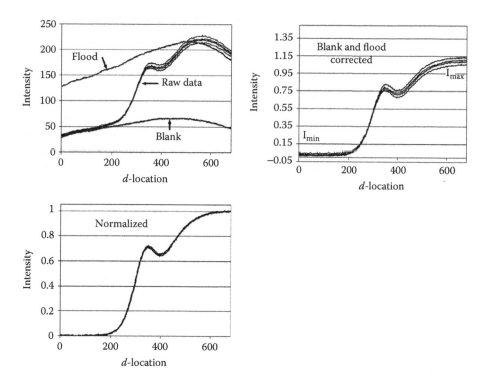

FIGURE 10.7 An example of processing of fluorescence images for analysis of the diffusion immunoassay. Plotted are (upper left) a set of raw image data and reference images (blank and flood), (upper right) a blank/background corrected dataset, and (lower) a fully normalized dataset as described in the text.

analyte is essentially undiluted. This normalization step is described by the following equation:

$$I_{normalized}(x) = \frac{I_{bfc}(x) - I_{min}}{I_{max} - I_{min}} \qquad (10.3)$$

This allowed all images of samples having equivalent fluorescent signals to be normalized to the same relative scale. It was also convenient to use normalized data for comparison with numerical simulations.

Normalization must be used with caution. Care must be taken to ensure that two regions of the image can accurately serve as internal references. To this end, channels with sufficiently large d and a sufficiently large field were imaged so that the entire interdiffusion zone was contained within the image. It is desirable to have the reference region sufficiently far from the channel walls that light scattering does not affect the signal. The signal level of fluorescing analyte may also vary, depending on the initial sample conditions (due to quenching or shifts in emission spectra that depend on the concentrations of fluid constituents), and in these cases the normalization step is not justified.

10.5 CONCLUSION

Fluorescence has been demonstrated to be one of the better methods for detecting biochemical species in microfludic devices; it is particularly well suited for analyzing the small volumes inherent in the use of microfluidics. The current rapid growth in the field of microfluidics for chemical analysis guarantees that there will be continuing interest in optimizing fluorescence detection in such devices for the foreseeable future. It is interesting to note that very few of the many types of fluorescence detection of analytes have yet been applied to microfluidic systems; thus, there will be a need for a second edition of this chapter. The greatest interest in microfluidic systems in the future will probably involve the use of polymers to reduce the inherent cost of microfluidic disposables; polymers are much more economical choices than silicon and glass for products that must be produced in very large numbers. This presents a continuing challenge—to minimize the inherent luminescence of the devices themselves in the presence of inherent luminescence in the polymers and to use detection methods that are minimally sensitive to that luminescence.

REFERENCES

1. Ramsey, J.M., Proceedings of the µTAS 2001 Symposium. In: *Micro Total Analysis Systems 2001,* Ramsey, J.M. and van den Berg, A., eds. Monterey, CA: Kluwer Academic, 2001:vii.
2. Weigl, B.H., Bardell, R.L., and Cabrera, C.R., Lab-on-a-chip for drug development. *Adv Drug Delivery Rev,* 55, 349–377, 2003.
3. Reyes, D.R., Iossifidis, D., Auroux, P.-A., and Manz, A., Micro total analysis systems. 1. Introduction, theory, and technology. *Anal Chem,* 74, 2623–2636, 2002.
4. Auroux, P.-A., Iossifidis, D., Reyes, D.R., and Manz, A., Micro total analysis systems. 2. Analytical standard operations and applications. *Anal Chem,* 74, 2637–2652, 2002.
5. Verpoorte, E., Microfluidic chips for clinical and forensic analysis. *Electrophoresis,* 23, 677–712, 2002.
6. Schwarz, M.A. and Hauser, P.C., Recent developments in detection methods for microfabricated analytical devices. *Lab Chip,* 1, 1–6, 2001.
7. Zhang, C.X. and Manz, A., High-speed free-flow electrophoresis on chip. *Anal Chem,* 75, 5759–5766, 2003.
8. Kwakye, S. and Baeumner, A., A microfluidic biosensor based on nucleic acid sequence recognition. *Anal Bioanal Chem,* 376, 1062–1068, 2003.
9. Yan, L.S., Liang, N., Luo, G.A., Wang, Y.M., and Wang, J.Y., Rapid fabrication of monolithic PDMS microchips and highly sensitive chemiluminescence detection for amino acids. *Chem J Chin Univ,* 24, 1193–1197, 2003.
10. Sato, Y., Irisawa, G., Ishizuka, M., Hishida, K., and Maeda, M., Visualization of convective mixing in microchannel by fluorescence imaging. *Measure Sci Technol* 14, 114–121, 2003.

11. Monahan, J., Gewirth, A.A., and Nuzzo, R.G., Indirect fluorescence detection of simple sugars via high-pH electrophoresis in poly(dimethylsiloxane) microfluidic chips. *Electrophoresis,* 23, 2347–2354, 2002.
12. Chiem, N.H. and Harrison, D.J., Microchip systems for immunoassay: an integrated immunoreactor with electrophoretic separation for serum theophylline determination. *Clin Chem,* 44, 591–598, 1998.
13. Jiang, G.F., Attiya, S., Ocvirk, G., Lee, W.E., and Harrison, D.J., Red diode laser induced fluorescence detection with a confocal microscope on a microchip for capillary electrophoresis. *Biosens Bioelectron,* 14, 861–869, 2000.
14. Cheng, S.B., Skinner, C.D., Taylor, J., Attiya, S., Lee, W.E., Picelli, G., and Harrison, D.J., Development of a multichannel microfluidic analysis system employing affinity capillary electrophoresis for immunoassay. *Anal Chem,* 73, 1472–1479, 2001.
15. Dodge, A., Fluri, K., Verpoorte, E., and de Rooij, N.F., Electrokinetically driven microfluidic chips with surface-modified chambers for heterogeneous immunoassays. *Anal Chem,* 73, 3400–3409, 2001.
16. Linder, V., Verpoorte, E., Thormann, W., de Rooij, N.F., and Sigrist, M., Surface biopassivation of replicated poly(dimethylsiloxane) microfluidic channels and application to heterogeneous immunoreaction with on-chip fluorescence detection. *Anal Chem,* 73, 4181–4189, 2001.
17. Roos, P. and Skinner, C.D., A two bead immunoassay in a micro fluidic device using a flat laser intensity profile for illumination. *Analyst,* 128, 527–531, 2003.
18. Hatch, A., Kamholz, A.E., Hawkins, K.R., Munson, M.S., Schilling, E.A., Weigl, B.H., and Yager, P., A rapid diffusion immunoassay in a T-sensor. *Nat Biotechnol,* 19, 461–465, 2001.
19. Sammarco, T.S., Burke, D.T., Mastrangelo, C.H., and Burns, M.A., Thermocapillary pumping in microfabricated liquid analysis devices. *Abstr Papers Am Chem Soc,* 213, 331-PMSE, 1997.
20. Sammarco, T.S., Johnson, B.N., Burke, D.T., Mastrangelo, C.H., and Burns, M.A., A microfluidic pumping and reaction system for chemical analysis devices. *Abstr Papers Am Chem Soc,* 211, 219-BIOT, 1996.
21. Burns, M.A., Johnson, B.N., Brahmasandra, S.N., Handique, K., Webster, J.R., Krishnan, M., Sammarco, T.S., Man, P.M., Jones, D., Heldsinger, D., Mastrangelo, C.H., and Burke, D.T., An integrated nanoliter DNA analysis device. *Science,* 282, 484–487, 1998.
22. Mastrangelo, C.H., Burns, M.A., and Burke, D.T., Microfabricated devices for genetic diagnostics. *Proc IEEE,* 86, 1769–1787, 1998.
23. Brahmasandra, S.N., Ugaz, V.M., Burke, D.T., Mastrangelo, C.H., and Burns, M.A., Electrophoresis in microfabricated devices using photopolymerized polyacrylamide gels and electrode-defined sample injection. *Electrophoresis,* 22, 300–311, 2001.
24. Webster, J.R., Burns, M.A., Burke, D.T., and Mastrangelo, C.H., Monolithic capillary electrophoresis device with integrated fluorescence detector. *Anal Chem,* 73, 1622–1626, 2001.
25. Scherer, J.R., Kheterpal, I., Radhakrishnan, A., Ja, W.W., and Mathies, R.A., Ultrahigh throughput rotary capillary array electrophoresis scanner for fluorescent DNA sequencing and analysis. *Electrophoresis,* 20, 1508–1517, 1999.
26. Lagally, E.T., Emrich, C.A., and Mathies, R.A., Fully integrated PCR-capillary electrophoresis microsystem for DNA analysis. *Lab Chip,* 1, 102–107, 2001.
27. Lagally, E.T., Medintz, I., and Mathies, R.A., Single-molecule DNA amplification and analysis in an integrated microfluidic device. *Anal Chem,* 73, 565–570, 2001.

28. Lagally, E.T., Simpson, P.C., and Mathies, R.A., Monolithic integrated microfluidic DNA amplification and capillary electrophoresis analysis system. *Sensors Actuators B,* 63, 138–146, 2000.

29. Medintz, I.L., Paegel, B.M., Blazej, R.G., Emrich, C.A., Berti, L., Scherer, J.R., and Mathies, R.A., High-performance genetic analysis using microfabricated capillary array electrophoresis microplates. *Electrophoresis,* 22, 3845–3856, 2001.

30. Higgins, J.A., Nasarabadi, S., Karns, J.S., Shelton, D.R., Cooper, M., Gbakima, A., and Koopman, R.P., A handheld real time thermal cycler for bacterial pathogen detection. *Biosens Bioelectron,* 18, 1115–1123, 2003.

31. Borg, I., Rohde, G., Loseke, S., Bittscheidt, J., Schultze-Werninghaus, G., Stephan, V., and Bufe, A., Evaluation of a quantitative real-time PCR for the detection of respiratory syncytial virus in pulmonary diseases. *Eur Respir J,* 21, 944–951.

32. Brinkman, N.E., Haugland, R.A., Wymer, L.J., Byappanahalli, M., Whitman, R.L., and Vesper, S.J., Evaluation of a rapid, quantitative real-time PCR method for enumeration of pathogenic candida cells in water. *Appl Environ Microbiol,* 69, 1775–1782, 2003.

33. Emanuel, P.A., Bell, R., Dang, J.L., McClanahan, R., David, J.C., Burgess, R.J., Thompson, J., Collins, L., and Hadfield, T., Detection of *Francisella tularensis* within infected mouse tissues by using a hand-held PCR thermocycler. *J Clin Microbiol,* 41, 689–693, 2003.

34. Mackay, I.M., Jacob, K.C., Woolhouse, D., Waller, K., Syrmis, M.W., Whiley, D.M., Siebert, D.J., Nissen, M., and Sloots, T.P., Molecular assays for detection of human metapneumovirus. *J Clin Microbiol,* 41, 100–105, 2003.

35. Kim, S.G., Shin, S.J., Jacobson, R.H., Miller, L.J., Harpending, P.R., Stehman, S.M., Rossiter, C.A., and Lein, D.A., Development and application of quantitative polymerase chain reaction assay based on the ABI 7700 system (TaqMan) for detection and quantification of *Mycobacterium avium* subsp paratuberculosis. *J Vet Diagn Invest,* 14, 126–131, 2002.

36. Handique, K., Advanced development and manufacturing engineering of a microfluidic multiplexed molecular diagnostic device. In: *BioMEMs and Microfluidics.* San Diego: IBC USA Conferences, 2003.

37. Liu, Y. and Rauch, C.B., DNA probe attachment on plastic surfaces and microfluidic hybridization array channel devices with sample oscillation. *Anal Biochem,* 317, 76–84, 2003.

38. Trau, D., Lee, T.M.H., Lao, A.I.K., Lenigk, R., Hsing, I.M., Ip, N.Y., Carles, M.C., and Sucher, N.J., Genotyping on a complementary metal oxide semiconductor silicon polymerase chain reaction chip with integrated DNA microarray. *Anal Chem,* 74, 3168–3173, 2002.

39. Starkey, D.E., Abdelaziez, Y., Ahn, C.H., Tu, J., Anderson, L., Wehmeyer, K.R., Izzo, N.J., Carr, A.N., Peters, K.G., Bao, J.J., Halsall, H.B., and Heineman, W.R., Determination of endogenous extracellular signal-regulated protein kinase by microchip capillary electrophoresis. *Anal Biochem,* 316, 181–191, 2003.

40. Hadd, A.G., Raymond, D.E., Halliwell, J.W., Jacobson, S.C., and Ramsey, J.M., Microchip device for performing enzyme assays. *Anal Chem,* 69, 3407–3412, 1997.

41. Starkey, D.E., Han, A., Bao, J.J., Ahn, C.H., Wehmeyer, K.R., Prenger, M.C., Halsall, H.B., and Heineman, W.R., Fluorogenic assay for beta-glucuronidase using microchip-based capillary electrophoresis. *J Chromatogr B,* 762, 33–41, 2001.

42. Schilling, E.A., Kamholz, A.E., and Yager, P., Cell lysis and protein extraction in a microfluidic device with detection by a fluorogenic enzyme assay. *Anal Chem,* 74, 1798–1804, 2002.

43. Schultz, N.M. and Kennedy, R.T., Rapid immunoassays using capillary electrophoresis with fluorescence detection. *Anal Chem,* 65, 3161–3165, 1993.
44. Locascio, L.E., Perso, C.E., and Lee, C.S., Measurement of electroosmotic flow in plastic imprinted microfluid devices and the effect of protein adsorption on flow rate. *J Chromatogr A,* 857, 275–284, 1999.
45. Corle, T.R. and Kino, G.S., *Confocal Scanning Optical Microscopy and Related Imaging Systems.* San Diego: Academic Press, 1996.
46. Pawley, J.B., ed., *Handbook of Biological Confocal Microscopy.* New York: Plenum Press, 1995.
47. Wilson, T., *Confocal Microscopy.* London: Academic Press, 1990.
48. Wabuyele, M.B., Ford, S.M., Stryjewski, W., Barrow, J., and Soper, S.A., Single molecule detection of double-stranded DNA in poly(methylmethacrylate) and polycarbonate microfluidic devices. *Electrophoresis,* 22, 3939–3948, 2001.
49. Denk, W., Strickler, J.H., and Webb, W.W., Two-photon laser scanning fluorescence microscopy. *Science,* 248, 73–76, 1990.
50. Zipfel, W.R., Williams, R.M., and Webb, W.W., Nonlinear magic: multiphoton microscopy in the biosciences. *Nat Biotechnol,* 21, 1368–1376, 2003.
51. Baker, G.A., Munson, C.A., Bukowski, E.J., Baker, S.N., and Bright, F.V., Assessment of one- and two-photon excited luminescence for directly measuring O_2, pH, Na^+, Mg^{2+}, or Ca^{2+} in optically dense and biologically relevant samples. *Appl Spectrosc,* 56, 455–463, 2002.
52. Zugel, S.A., Burke, B.J., Regnier, F.E., and Lytle, F.E., Electrophoretically mediated microanalysis of leucine aminopeptidase using two-photon excited fluorescence detection on a microchip. *Anal Chem,* 72, 5731–5735, 2000.
53. Chabinyc, M.L., Chiu, D.T., McDonald, J.C., Stroock, A.D., Christian, J.F., Karger, A.M., and Whitesides, G.M., An integrated fluorescence detection system in poly(dimethylsiloxane) for microfluidic applications. *Anal Chem,* 73, 4491–4498, 2001.
54. Naessens, K., Ottevaere, H., Van Daele, P., and Baets, R., Flexible fabrication of microlenses in polymer layers with excimer laser ablation. *Appl Surf Sci,* 208, 159–164, 2003.
55. Ong, N.S., Koh, Y.H., and Fu, Y.Q., Microlens array produced using hot embossing process. *Microelectron Eng,* 60, 365–379, 2002.
56. Roulet, J.C., Volkel, R., Herzig, H.P., Verpoorte, E., de Rooij, N.F., and Dandliker, R., Fabrication of multilayer systems combining microfluidic and microoptical elements for fluorescence detection. *J Microelectromech Syst,* 10, 482–491, 2001.
57. Camou, S., Fujita, H., and Fujii, T., PDMS 2D optical lens integrated with microfluidic channels: principle and characterization. *Lab Chip,* 3, 40–45, 2003.
58. Llobera, A., Wilke, R., and Buttgenbach, S., Poly(dimethylsiloxane) hollow Abbe prism with microlenses for detection based on absorption and refractive index shift. *Lab Chip,* 4, 24–27, 2004.
59. Kwon, S. and Lee, L.P., Stacked two dimensional microlens scanner for micro confocal imaging array. *Proc 15th Annu IEEE Int MEMS Conf,* Las Vegas, NV, 2002.
60. Kovacs, G.T.A., *Micromachined Transducers Sourcebook.* Boston: WCB-McGraw-Hill, 1998.
61. Madou, M., *Fundamentals of Microfabrication.* Boca Raton, FL: CRC Press, 1997.
62. Maluf, N., *An Introduction to Microelectromechanical Systems Engineering.* Boston: Artech House, 2000.
63. Kohler, J.M., Mejevaia, T., and Saluz, H.P., eds., *Microsystem Technology: A Powerful Tool for Biomolecular Studies.* Basel, Switzerland: Birkhauser Verlag, 1999.

64. Xing, W.-L. and Cheng, J., eds., *Biochips: Technology and Applications*. Berlin: Springer, 2003.

65. Lacher, N.A., de Rooij, N.F., Verpoorte, E., and Lunte, S.M., Comparison of the performance characteristics of poly(dimethylsiloxane) and Pyrex microchip electrophoresis devices for peptide separations. *J Chromatogr A*, 1004, 225–235, 2003.

66. Becker, H. and Locascio, L.E., Polymer microfluidic devices. *Talanta*, 56, 267–287, 2002.

67. Boone, T., Fan, Z.H., Hooper, H., Ricco, A., Tan, H.D., and Williams, S., Plastic advances microfluidic devices. *Anal Chem*, 74, 78A–86A, 2002.

68. deMello, A., Plastic fantastic. *Lab Chip*, 2, 31N–36N, 2002.

69. McDonald, J.C. and Whitesides, G.M., Poly(dimethylsiloxane) as a material for fabricating microfluidic devices. *Acc Chem Res*, 35, 491–499, 2002.

70. Ng, J.M.K., Gitlin, I., Stroock, A.D., and Whitesides, G.M., Components for integrated poly(dimethylsiloxane) microfluidic systems. *Electrophoresis*, 23, 3461–3473, 2002.

71. Wu, H.K., Odom, T.W., Chiu, D.T., and Whitesides, G.M., Fabrication of complex three-dimensional microchannel systems in PDMS. *J Am Chem Soc*, 125, 554–559, 2003.

72. Leclerc, E., Sakai, Y., and Fujii, T., Cell culture in 3-dimensional microfluidic structure of PDMS (polydimethylsiloxane). *Biomed Microdev*, 5, 109–114, 2003.

73. Munson, M.S. and Yager, P., A novel microfluidic mixer based on successive lamination. In: *Micro Total Analysis Systems 2003*, Northrup, M.A., Jensen, K.F., and Harrison, D.J., eds. *Mesa Monographs*, 2003:495–498.

74. Munson, M.S. and Yager, P., Simple quantitative optical method for monitoring the extent of mixing applied to a novel microfluidic mixer. *Anal Chim Acta*, 501, 63–71, 2004.

75. Chang, W.J., Akin, D., Sedlak, M., Ladisch, M.R., and Bashir, R., Poly(dimethylsiloxane) (PDMS) and silicon hybrid biochip for bacterial culture. *Biomed Microdev*, 5, 281–290, 2003.

76. Wang, B., Chen, L., Abdulali-Kanji, Z., Horton, J.H., and Oleschuk, R.D., Aging effects on oxidized and amine-modified poly(dimethylsiloxane) surfaces studied with chemical forced titrations: effects on electroosmotic flow rate in microfluidic channels. *Langmuir*, 19, 9792–9798, 2003.

77. Hawkins, K.R. and Yager, P., Nonlinear decrease of background fluorescence in polymer thin-films—a survey of materials and how they can complicate fluorescence detection in μTAS. *Lab Chip*, 3, 248–252, 2003.

78. Wang, S.C. and Morris, M.D., Plastic microchip electrophoresis with analyte velocity modulation. Application to fluorescence background rejection. *Anal Chem*, 72, 1448–1452, 2000.

79. Soutar, I., Application of luminescence spectroscopy in polymer science. In: *Multidimensional Spectroscopy of Polymers: Vibrational, NMR, and Fluorescence Techniques*, Urban, M.W. and Provder, T., eds. Washington, DC: American Chemical Society, 1995.

80. Owens, C.V., Davidson, Y.Y., Kar, S., and Soper, S.A., High resolution separation of DNA restriction fragments using capillary electrophoresis with near-IR, diode-based, laser-induced fluorescence detection. *Anal Chem*, 69, 1256–1261, 1997.

81. McDonald, J.C., Duffy, D.C., Anderson, J.R., Chiu, D.T., Wu, H., Schueller, O.J., and Whitesides, G.M., Fabrication of microfluidic systems in poly(dimethylsiloxane). *Electrophoresis*, 21, 27–40, 2000.

82. Deng, T., Wu, H.K., Brittain, S.T., and Whitesides, G.M., Prototyping of masks, masters, and stamps/molds for soft lithography using an office printer and photographic reduction. *Anal Chem,* 72, 3176–3180, 2000.

83. Fiorini, G.S., Jeffries, G.D.M., Lim, D.S.W., Kuyper, C.L., and Chiu, D.T., Fabrication of thermoset polyester microfluidic devices and embossing masters using rapid prototyped polydimethylsiloxane molds. *Lab Chip,* 3, 158–163, 2003.

84. Ko, J.S., Yoon, H.C., Yang, H.S., Pyo, H.B., Chung, K.H., Kim, S.J., and Kim, Y.T., A polymer-based microfluidic device for immunosensing biochips. *Lab Chip,* 3, 106–113, 2003.

85. Carvalho, B.L., Schilling, E.A., Schmid, N., and Kellogg, G.J., Soft embossing of microfluidic devices. In: *The 7th International Conference on Micro Total Analysis Systems,* Squaw Valley, CA, 2003.

86. Hawkins, K.R., A diffusion immunoassay for a model protein analyte. Manuscript.

87. Moshrefzadeh, R.S., Misemer, D.K., Radcliffe, M.D., Francis, C.V., and Mohapatra, S.K., Nonuniform photobleaching of dyed polymers for optical waveguides. *Appl Phys Lett,* 62, 16–18, 1993.

88. Song, L.L., vanGijlswijk, R.P.M., Young, I.T., and Tanke, H.J., Influence of fluorochrome labeling density on the photobleaching kinetics of fluorescein in microscopy. *Cytometry,* 27, 213–223, 1997.

89. Lakowicz, J.R., *Principles of Fluorescence Spectroscopy.* New York: Plenum Press, 1983.

90. Straume, M., Frasier-Cadoret, S.G., and Johnson, M.L., Topics in fluorescence spectroscopy. In: *Topics in Fluorescence Spectroscopy,* Lakowicz, J.R., ed. New York: Plenum Press, 1991.

91. Hawkins, K.R., Markel, D., and Yager, P., Rapid fabrication of microfluidic laminate devices using PMMA. *Lab Chip,* in preparation.

92. Weigl, B. and Yager, P., Microfluidic diffusion-based separation and detection. *Science,* 283, 346–347, 1999.

93. Hatch, A., Garcia, E., and Yager, P., Diffusion-based analysis of molecular interactions in microfluidic devices. *Proc IEEE,* 92, 126–139, 2004.

94. Kamholz, A.E., Quantitative analysis of diffusion and chemical reaction in pressure-driven microfluidic channels. Ph.D. dissertation. Seattle: University of Washington, 2002.

95. Kamholz, A.E., Schiling, E.A., and Yager, P., Optical measurement of transverse molecular diffusion in a microchannel. *Biophys J,* 80, 1967–1972, 2001.

96. Kamholz, A.E., Weigl, B.H., Finlayson, B.A., and Yager, P., Quantitative analysis of molecular interaction in a microfluidic channel: the T-sensor. *Anal Chem,* 71, 5340–5347, 1999.

97. Weigl, B.H., Yager, P., Brody, J.P., Kenny, M., Schutte, D., Hixson, G., Zebert, D.M., Kamholz, A., Wu, C., and Altendorf, E., Microfabricated diffusion-based chemical sensor. U.S. patent no. 5716852, Feb. 10, 1998.

98. Weigl, B.H. and Yager, P., Silicon-microfabricated diffusion-based optical chemical sensor. *Sensors Actuators B,* B39, 452–457, 1997.

99. Hatch, A. and Yager, P., Diffusion immunoassay in polyacrylamide hydrogels. In: *Micro Total Analysis Systems 2001,* Ramsay, J.M. and van den Berg, A., eds. Boston: Kluwer Academic, 2001:571–572.

100. Baroud, C.N., Okkels, F., Menetrier, L., and Tabeling, P., Reaction-diffusion dynamics: confrontation between theory and experiment in a microfluidic reactor. *Phys Rev E,* 67, 060104(R), 2003.

101. Ismagilov, R.F., Stroock, A.D., Kenis, P.J.A., and Whitesides, G., Experimental and theoretical scaling laws for transverse diffusive broadening in two-phase laminar flows in microchannels. *Appl Phys Lett,* 76, 2376–2378, 2000.

102. Kamholz, A.E., and Yager, P., Theoretical analysis of molecular diffusion in pressure-driven laminar flow in microfluidic channels. *Biophys J,* 80, 155–160, 2001.

103. Kamholz, A.E. and Yager, P., Molecular diffusive scaling laws in pressure-driven microfluidic channels: deviation from one-dimensional Einstein approximations. *Sensors Actuators B,* 82, 117–121, 2002.

104. Munson, M.S., Hawkins, K.R., Hasenbank, M.S., and Yager, P., Sheath flow T-sensor. *Lab Chip,* submitted.

105. Culbertson, C.T., Jacobson, S.C., and Ramsey, J.M., Diffusion coefficient measurements in microfluidic devices. *Talanta,* 56, 365–373, 2002.

11 The Array Biosensors

Frances S. Ligler, D.Phil., D.Sc.,
and Chris R. Taitt, Ph.D.

CONTENTS

11.1 INTRODUCTION

A combination of forces is converging to promote the development of biosensors that are capable of measuring multiple analytes simultaneously. First is customer demand; no longer are most users satisfied with analyzing each sample for one target at a time. Second, the development of microarrays for genomic applications has demonstrated the potential to measure hundreds or even thousands of targets simultaneously in laboratory systems. Third, biosensor technology itself has become more familiar both to scientists and to users as an inexpensive alternative to centralized laboratories and offers both fast response and on-site analysis. An ideal biosensor that detects multiple analytes would also be portable and fully automated to facilitate on-site use by a variety of users. Array biosensors offer the most promising approach to achieving this goal.

Many biosensor development efforts are currently addressing the need to detect multiple analytes simultaneously. A perusal of commercially available optical biosensors reveals a number of laboratory systems that are partially automated and can assay for more than one target. These include the Biacore surface plasmon resonance (SPR) system, the ORIGEN electrochemiluminescence system, the IAsys resonant mirror system, the IMPACT and FAST 6000 displacement flow immunosensors, and the RAPTOR fiber-optic biosensor. Each of these is unique in both fluidics system design and method of analyte detection.

The Biacore SPR system (Biacore International, Upsala, Sweden) detects binding of target molecules to recognition elements on the surface by monitoring changes in the conditions required for excitation of surface plasmons; these changes reflect changes in the refractive index at the surface caused by biomolecular interactions. Measurements are made continuously and without the need for a molecular tag; thus, such SPR instruments are termed "reagentless." For detection of multiple analytes, the commercial system performs four analyses using parallel channels on a single surface (www.biacore.com/lifesciences/products/systems_overview/3000/systems_information/index.html; July 1, 2005). Homola et al.[1] have extended the SPR approach to multiplexed assays such that multiple analytes may be detected on a single chip. Using a beveled prism coupler to serially couple light at different areas of the sensing element, multiple loci on a single chip may be individually interrogated. This group has also integrated a high refractive index overlay to shift the coupling wavelength in a portion of the sensing chip,[2] allowing spatial resolution of the chip in angular-modulated systems. Although these latter methods to detect multiple binding events on a single chip have not been incorporated into commercial systems, they are paving the way for incorporation of multianalyte chips into future SPR systems.

The IAsys resonant mirror system (Affinity Sensors, Cambridge, UK) is another reagentless system that detects target molecules by monitoring changes in refractive index near the sensing surface (www.thermo.con/eThermo/CMA/PDFs/products/product PDF-15191.pdf, July 1, 2005). These refractive index changes are detected as a peak in intensity of light coupled into and out of a resonant structure and can be measured continuously. Like the commercial Biacore system, the IAsys system requires that analyses be performed in parallel, using the two flow-through cuvettes provided, and it is limited in the number of simultaneous measurements that can be accomplished.[3]

Two flow-through systems, the IMPACT (Lifepoint, Rancho Cucamonga, CA) and FAST 6000 (Research International, Monroe, WA) systems, use displacement immunoassays to detect analytes of interest (www.lifepointinc.com, Feb. 1, 2004; www.resrchintl.com/fast6000.html, Feb. 1, 2004). Injected sample flows over beads or filters coated with antibodies specific for targets. These systems require that the target be a relatively small molecule and have limited binding affinity for its cognate antibodies; the IMPACT system detects drugs, whereas the FAST 6000 detects explosives. In the presence of the target, a fluorescent analog of the target (drug/explosive) is displaced from the immobilized antibody and measured downstream. Both of these systems can test for multiple targets by parallel operation of displacement assays in 6 (FAST 6000) or 10 (IMPACT) flow paths. In contrast to the other sensors described above, the FAST 6000 system is designed for field use rather than as a laboratory instrument.[4,5]

IGEN (Gaithersburg, MD) commercialized a biosensor based on electrochemiluminescence (www.bioveris.com/products_services.html, July 1, 2005). The ORIGEN system uses a voltage potential to power a luminescent redox reaction. Immunomagnetic beads are used to recognize and capture the target onto a magnetized anode. Detection is accomplished by addition of a tracer antibody labeled with $Ru(bpy)_3^{2+}$; excitation of the $Ru(bpy)_3^{2+}$ label results in emission of photons. Inclusion of an immunomagnetic separation (IMS) step renders this system particularly suitable for analysis of foods and environmental samples where a target preconcentration step significantly lowers the limits of detection.[6–9] For detection of multiple analytes, parallel analyses are performed, using up to eight separate substrates.

The RAPTOR fiber-optic biosensor (Research International, Monroe, WA) uses standard sandwich fluoroimmunoassays performed on the surface of optical fibers to detect targets present in the sample (www.resrchintl.com/raptor.html, Feb. 1, 2004). This instrument uses four fiber-optic probes coated with antibodies to extract target from samples and generates a signal when a fluorescent tracer antibody binds to target captured by the antibody-coated probes. This system uses an electromagnetic component of the light launched in the optical fibers, the evanescent wave, to selectively excite fluorescent immunocomplexes bound to the surface of the optical fibers. This surface selectivity confers the same advantage shared by SPR, interferometric, and resonant mirror sensors, namely real-time detection. Furthermore, evanescent wave excitation allows analysis of nonhomogeneous or dirty samples with minimal optical interference or contributions from the sample matrix.[10-13] The RAPTOR system is designed and manufactured for use in the field and may be integrated with a computer-controlled air sampler for remote, fully automated use. The RAPTOR system is designed for detection of up to four analytes in a single sample, using four optical fibers.[14] However, it has recently been demonstrated that the multianalyte detection capabilities may be doubled by a simple modification of the biochemical assay protocols.[15]

In comparison to the sensors described above, use of a planar array format eliminates the need for assays to be performed in parallel on multiple substrates. Not only does this feature enable detection of many more target species, but it also provides the opportunity for inclusion of positive and negative controls on each sensing surface, which is more reliable than having controls located on parallel but separate sensing surfaces. Recognition species used in array-based systems include oligonucleotides, proteins (antibodies, enzymes, other proteins), peptides, carbohydrates, haptens, and cells.[16]

DNA array technology has led device development efforts with two systems employing fluorescence technology and optical waveguides produced and marketed by Zeptosens (Witterswil, Switzerland)[17,18] and Illumina (San Diego, CA).[19] Although both of these systems are highly sensitive and incorporate thousands of recognition elements, neither has been fully automated or adapted for on-site use by inexpert operators (www.zeptosens.com, Feb. 1, 2004; www.illumina.com, Feb. 1, 2004). Another elegant laboratory system has been developed by Zyomyx (Hayward, CA). This instrument is capable of performing about 35 immunoassays simultaneously on six samples and is based on planar antibody arrays (www.zyomyx.com, Feb. 1, 2004).[20]

In this chapter we will limit the consideration of optical array biosensors to systems where the readout system is integrated with the sensing surface in a single, preferably portable, unit. Therefore, we will exclude discussion of microarrays that are imaged using a large reader (usually confocal), scanning tunneling microscope, or optical microscope.

Planar array technology has been pursued in parallel with the development of SPR systems, resonant mirror sensors, and high-resolution fluorescent microarrays by investigators interested in portable devices. However, the only optical array biosensors using this planar array technology reported to date are based on interferometry and fluorescence.

Researchers at the University of Tubingen (Tubingen, Germany) have developed an elegant interferometric system based on reflectometric interference spectroscopy.[21-23]

This technique uses the reflection of white light at a thin waveguide placed on a glass substrate to produce an interference pattern over a wide spectrum. A separate sensing layer, formed by recognition molecules (e.g., antibodies) immobilized on the surface of the waveguide, is also used as an "interface" to reflect the incoming light. A change in optical thickness of this sensing layer occurs when target binds to its cognate recognition molecule and the resulting shift of the interference pattern is then detected by a diode array. What makes this reflectance system particularly interesting is that measurements are performed in multiple locations on the sensing surface and at multiple wavelengths.

Some of the early work in the area of fluorescence-based readout systems for planar array biosensors used evanescent illumination of the waveguide, fluorescence sandwich immunoassays, and image capture using a charge-coupled device (CCD).[24,25] A variation on this approach used direct illumination and a complementary metal oxide semiconductor (CMOS) chip.[26] The fluorescence array sensors have proven to be sensitive and amenable to miniaturization and relatively immune to interference from complex sample components. The remainder of this chapter focuses on the chemistry for fabricating the planar arrays, their wide variety of applications, and miniaturization of the hardware.

11.2 THE BIOMOLECULAR RECOGNITION ELEMENT

The choice of biological recognition elements used in array biosensors is largely dependent on availability of appropriate binding molecules and the intended application.[27] To date, the majority of array-based biosensors have utilized antibody–antigen interactions, nucleic acid hybridization (DNA/RNA), and (to a lesser extent) receptor–ligand binding to detect, identify, and quantify targets of interest. Although many biomolecules typically contain some type of intrinsic fluorescence, extrinsic labels are generally used in these array-based systems. These extrinsic labels, typically introduced into the target itself or into a tracer molecule, are most often fluorescent dyes, such as rhodamine, coumarin, cyanine, or fluorescein, and allow the use, through spectral selection, of readily available excitation sources such as laser diodes.

To date, antibody–antigen binding interactions are the most well-characterized systems employed in planar array biosensors, in large part because the binding reactions may be performed under ambient conditions in aqueous (nontoxic) solutions. Immunoassay formats may be divided into four main categories: direct, competitive, displacement, and sandwich immunoassays.[28] The direct assay is the simplest method to perform, but it requires that the antigen be intrinsically fluorescent. In the absence of a fluorescent antigen, the preferred formats are competitive and sandwich assays. Competitive formats are particularly useful in the detection of small molecules that do not possess two distinct epitopes (e.g., haptens), as required for the sandwich assays.[24,29–35] Use of displacement assays has only recently been demonstrated using a planar waveguide biosensor.[33] Table 11.1 shows the limits of detection obtained for a variety of analytes using the U.S. Naval Research Laboratory (NRL; Washington, DC) array biosensor (described later) and a variety of antibodies.

TABLE 11.1
Current Limits of Detection Using the NRL Array Biosensor

Proteins

Toxins	Allergens
Staphylococcal enterotoxin B (0.5 ng/ml)	Ovalbumin (0.025 ng/ml)
Cholera toxin (1.6 ng/ml)	
Botulinum toxoid A (40 ng/ml)	**Physiological markers**
Botulinum toxoid B (200 ng/ml)	*Yersinia pestis* F1 (25 ng/ml)
Ricin (8 ng/ml)	D-dimer (25 ng/ml)

Bacteria

Gram negative	Gram positive
Erwinia herbicola (10^4 Colony-forming units [CFU]/ml)	*Bacillus globigii* (10^5 CFU/ml)
Brucella abortus (3×10^3 CFU/ml)	*Bacillus anthracis* (623 CFU/ml)
Francisell tularensis LVS (10^5 CFU/ml)	*Listeria* (10^4 CFU/ml)
Salmonella typhimurium (4×10^4 CFU/ml, 8×10^3 CFU/g excreta)	
Campylobacter (10^3 CFU/ml)	
Escherichia coli O157:H7 (10^3 CFU/ml)	**Low-molecular-weight toxins**
Shigella (5×10^4 CFU/ml)	Trinitrotoluene (1 ng/ml)
	Ochratoxin (0.2 ng/ml)
Virus	Deoxynivalenol (≤ 1 ng/ml)
MS2 (10^7 plaque-forming units [PFU]/ml)	Fumonisin (<62 ng/ml)

Nucleic acids are becoming increasingly popular for use as recognition molecules in systems using planar waveguides.[36-38] Although descriptions of automated, portable biosensors using DNA arrays have not as yet been published, most likely due to the demands of sample preparation and the exacting conditions required for typical nucleic acid hybridizations, the use of nucleic acid aptamers on a planar waveguide sensor for detection of a non-nucleic acid-based target has been described.[39] The binding of thrombin to a fluorescently labeled, immobilized DNA aptamer was monitored using evanescent wave-induced fluorescence anisotropy. Aptamers have also been employed to measure bacteria in an electrochemiluminescence system.[40] As DNA and RNA aptamers, as well as catalytic DNAs,[41] may be synthesized for a number of target analytes with specificities comparable to the corresponding antibodies,[42] aptamers clearly have potential as recognition elements in array-based sensors.

Use of receptors or other molecules as recognition species in array-based biosensors is limited to a handful of reports. Green fluorescent protein and serotonin receptor,[42,43] sodium-potassium ATPase (Na,K-ATPase),[44] and gangliosides[45,46] have all been incorporated into planar array sensors and used to detect targets of interest. A limitation of using receptors as recognition species is the requirement that the immobilized receptor protein remain active. Preliminary studies using the NRL array biosensor demonstrated that it is also possible to use antibiotics, tethered sugars, and siderophores as recognition elements (Ligler et al., unpublished data). In laboratory systems employing peptide arrays, Salisbury et al.[47] measured the cleavage of fluorescent fragments by proteases.

11.3 IMMOBILIZATION OF THE BIOMOLECULE ONTO THE WAVEGUIDE

A number of the researchers currently involved in developing planar waveguide biosensors previously carried out much of their initial research in the field of fiber optics. Planar waveguides offer a number of advantages compared to fiber-optic technology, including the relative ease of preparation and integration into fluidic systems. As a precedent to patterning arrays, researchers have immobilized capture biomolecules uniformly over the planar surface and monitored the fluorescent signal intensity as a function of either time or the concentration of the labeled binding partner.[32,36,44,48–50]

One obvious prerequisite for all immobilization techniques is that the immobilized recognition species retain its binding or enzymatic activity. Various methods for immobilization of the biological component of a biosensing system include physical adsorption, covalent immobilization, and entrapment in polymer matrices.[51] Physical adsorption and covalent binding to functionalized surfaces are most commonly used in planar array biosensors. Li and Reichert[52] used a robotic microarrayer to print microarrays of capture species. Although high-density microarrays of adsorbed recognition molecules were obtained, the patterned microporous nitrocellulose slides were not appropriate for use as planar waveguides. A checkerboard pattern of two different oligonucleotides was produced by Budach et al.[37] using ink-jet printing of capture biomolecules on a tantalum oxide (Ta_2O_5) waveguide using 3-glycidoxypropyl trimethoxysilane. Using hybridization assays, Budach et al. were able to detect sample concentrations as low as 50 fM in 12-min assays. Knecht et al.[35] used a noncontact arrayer to attach antibiotic haptens to a glycidoxypropyl silane-coated slide for competitive assays.

Photoimmobilization and photolithographic patterning have been used by a number of groups to immobilize recognition species onto planar arrays.[50,53–61] Hofmann et al.,[50] for example, used a dextran-based photoimmobilization procedure to produce a network-like multilayer structure of immobilized rabbit IgG capture antibodies. Photolithographic patterning typically involves light-mediated activation of a surface species to create patterns for immobilization of the capture biomolecule in specific loci. For example, Bhatia et al.[60,61] used ultraviolet (UV) light to convert thiol moieties of mercaptosilane-treated slides into sulfonates; although these latter moieties are protein resistant, the nonilluminated regions of the slides (thiol moieties) were subsequently used for protein immobilization by standard cross-linking chemistry. This proved to be a convenient method for creating high-resolution patterns of immobilized capture biomolecules but had the disadvantage that only a single biomolecule could be patterned.

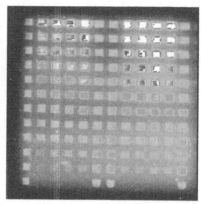

FIGURE 11.1 (See color insert following page 142.) Twelve-channel flow chamber modules used for patterning of recognition elements (left, bottom) and sample analyses (left, top). The resulting array (right) has 144 sensing regions.

Avidin–biotin chemistry has also been extensively used for the immobilization of molecular recognition elements. Typically, this method involves either physical adsorption of avidin onto the surface[24,32,38,48,62] or covalent immobilization of avidin onto the substrate.[30,63,64] Biotinylated capture/recognition species are then applied to specific regions of the avidin-coated surface. For example, Silzel et al.[24] ink-jet printed avidin onto polystyrene waveguides in 200-μm-diameter zones and subsequently immobilized biotinylated antibodies onto the avidin spots. Delehanty and Ligler,[65,66] on the other hand, used a robotic noncontact microarray printer to spot biotinylated capture antibodies onto avidin-coated slides.

Physically isolated patterning, using either polydimethylsiloxane (PDMS) flow channels[67] or a Teflon block fitted with O-rings,[38] has been used to immobilize biotinylated recognition species onto avidin-treated surfaces. As used in the NRL array biosensor, a PDMS flow cell containing a number of channels is temporarily attached to the surface of an avidin-coated waveguide and each channel is filled with a solution of the biotinylated capture biomolecule, as shown in Figure 11.1. After incubation, the flow channels are rinsed and the flow cell removed; this procedure results in a planar waveguide patterned with stripes of immobilized biomolecules. Samples and fluorescently labeled antibody are then passed over the surface using a second flow guide with channels oriented perpendicular to those of the patterning channels, allowing each sample to be interrogated for multiple different targets.

Physically isolated patterning has been used in the absence of avidin–biotin chemistry to attach recognition species to sensor substrates. Wadkins et al.[25,56] used a photopolymerizable polymer to create three-dimensional wells on sensor substrates; upon removal of unpolymerized material from each well, a different capture antibody was immobilized by covalent attachment. Rowe-Taitt et al.[45] used the same PDMS flow guides as Feldstein et al.[67] to create stripes of different ganglioside receptors on planar waveguides; the gangliosides were essentially locked into place by the hydrophobic interactions of their sphingosine tails with the C_{18}-silane used to coat the sensor surface. Plowman et al.[29] adsorbed capture antibodies onto grating-coupled integrated optical waveguides using a series of rubber gaskets to physically separate different species. Bernard et al.[68] also immobilized capture antibodies by adsorption but utilized a silicon microfluidic network to create patterned stripes measuring 20 μm in diameter, separated by 20 μm.

11.4 APPLICATIONS

Signal intensity over background in array biosensors can be increased by the use of fluorophores that are excited between 600 and 700 nm, as there are few components of natural sample matrices that can generate background fluorescence even if non-specific binding occurs. For this reason, samples analyzed at these wavelengths using planar arrays require little, if any, preparation, except as required to eliminate components that interfere with binding. Generally for immunoassays, this sample preparation is limited to homogenization of solids, dilution of highly viscous solutions, and/or removal of particulates by centrifugation or filtration as required. Nearly any aqueous sample with viscosity that is sufficiently low to flow over the sensor surface is appropriate for analysis. The sensing surface is usually subjected to a quick rinse after sample addition to remove any loosely adsorbed target that would bind the labeled tracer, increasing the background. This relative immunity from signal generation by nonspecifically adsorbed sample components, along with miniaturization and automation, extends their use from the laboratory to the point of sample collection.

Although the large majority of publications focus on assay development using well-defined buffers, the use of several systems for analysis of dirty, nonhomogeneous samples has been demonstrated. A planar optical biosensor used by Plowman et al.[29] was able to detect target in serum but showed a decrease in signal. This was attributed to an increase in viscosity and a subsequent decrease in the delivery of the antigen to the surface. The Zyomyx laboratory–based array immunosensor is being used with clinical fluids to measure cytokine and specific antibody responses with the goal of characterizing disease indicators (www.zyomyx.com, Jan. 2004); as important diagnostic targets become better documented, the critical assays may be incorporated into the portable array systems. The NRL array biosensor was used to detect and measure plague F1 antigen, staphylococcal enterotoxin B (SEB), and the physiological marker D-dimer in a variety of clinical fluids, including whole blood, plasma, urine, saliva, and nasal secretions,[30] at physiologically relevant concentrations. This system was also used to detect three different bacteria and three different toxins in the presence of sample components such as sand, clay, pollen, and smoke extracts.[69] Effects of such environmental matrices may become increasingly important for routine monitoring for homeland security.

Most recently the NRL group has focused on the detection of toxins and pathogens in foods. SEB was detected at 0.5 ng/ml in ham extracts, ground beef homogenate, chicken carcass rinse, milk, eggs, and cantaloupe homogenate.[70] *Salmonella typhimurium* was detected at concentrations as low as 8×10^3 colony-forming units (CFU)/ml in a number of foodstuffs, using only homogenization and filtration for sample preparation.[71] The same sensitivity was also demonstrated in

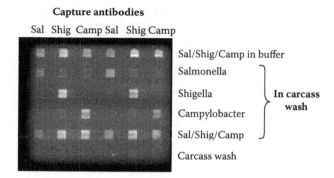

FIGURE 11.2 CCD image of a single sensor surface exposed to multiple pathogens. The slides were patterned with antibodies against salmonella (Sal), shigella (Shig), or campylobacter (Camp) using the physically isolated patterning procedure.[72] Six samples and a mixture of Cy5-labeled tracer antibodies were flowed over the sensor waveguide.

chicken excreta, raising the potential for screening flocks for disease and administering therapy prior to harvest. *Campylobacter jejuni* and *Shigella dysenteriae* were detected in chicken carcass rinse at 3×10^3 CFU/ml and 5×10^4 CFU/ml, respectively, and in ground turkey at 2×10^3 CFU/ml and 8×10^5 CFU/ml, respectively.[72] The ability to detect shigella in milk and lettuce washes was also demonstrated. Organisms were detected individually and in mixtures as shown in Figure 11.2.

One potential advantage of several of these array-based biosensors is due to use of evanescent illumination to excite surface-bound fluorophores. The surface selectivity of the evanescent wave allows these systems to measure binding in real time, even in the presence of free fluorophores in solution. Because the evanescent field decays exponentially with distance from the surface, unbound fluorophores at reasonably low concentrations will not generate a signal. The NRL array biosensor has been used to measure the kinetics of binding for both antibody–antigen interactions and nonspecific adsorption.[73,74] A similar system, also based on evanescent excitation,[75] was used to monitor hybridization rates in real time for discrimination of single nucleotide polymorphisms.

11.5 AUTOMATION AND MINIATURIZATION

Fluorescence-based array biosensor systems have been fabricated by a number of groups for on-site, automated use. Each system has a distinctive feature: the device developed at the University of Utah (Salt Lake City, UT) uses polystyrene waveguides incorporating a hemispheric lens on the proximal end; a biosensor developed at Oak Ridge National Laboratory (Oak Ridge, TN) uses a CMOS camera for signal detection; a system developed by Nanogen (San Diego, CA) employs electrophoretic methods to concentrate analyte at the surface; and the system developed at the NRL uses glass waveguides modified with a reflective cladding.

Herron et al.,[62] at the University of Utah, developed a planar array biosensor based on a molded polystyrene waveguide. After extensive investigation of alternative methods for reproducibly injecting light into the waveguide, they developed a molded waveguide with the cylindrical focusing lens physically integrated with the proximal end of the waveguide. The light is generated using a 638-nm laser and the waveguide is imaged using a CCD.[75] Data using this sensor for multianalyte analysis of coded blind samples[76] and physiological samples[29] have been published. Furthermore, this group recently determined real-time hybridization kinetics in multiple samples using a temperature-controlled flow guide with multiple flow cells.[75] This special flow guide allowed precise measurements of hybridization curves and identification of different single-nucleotide polymorphisms.

Researchers at Oak Ridge National Laboratory have published a series of papers describing an array biosensor built around a CMOS chip.[77-79] Their initial concept of immobilizing the capture molecules directly on the chip was abandoned due to cost concerns in favor of using a porous membrane as the sensing surface. In the most recently published version,[79] the system uses a 635-nm laser beam diffracted into 16 separate beams corresponding to the number of elements in a specially configured CMOS chip that includes application-specific signal amplification and filtering circuitry. The individual photosensors in the 4×4 array are 900 μm^2 in size. The sensitivity of the system for measuring directly labeled IgG was 130 ng/ml,[78] which is not as good as that obtained by the groups using CCD cameras; however, the sensitivity was significantly enhanced with the addition of an enzymic amplification method.[79] Although the increased number of biochemical steps complicates automation of the assays, this system has the advantage that the detector is very small.

The Nanogen array biosensor utilizes a CMOS chip that includes an array of 40×40 electrodes.[80] These electrodes are used to concentrate capture molecules before immobilization and subsequently target molecules from the sample. Signal generation is accomplished optically; fluorescent signal is generated using illumination from a helium–neon (HeNe) laser and imaged using a cooled CCD camera. The fluidics system has been miniaturized and integrated with the stacked biochip structure. More impressively, this instrument is able to perform automated sample preparation for nucleic acid analysis; bacterial targets captured on the array may be lysed, with the released DNA automatically channeled into an adjacent region for amplification and hybridization to a DNA microarray.

In contrast to the University of Utah system, which utilized an integral lens for focusing excitation light into the waveguide, the NRL array biosensor uses a 635-nm diode laser fired into the end of a microscope slide and a mode-mixing region to give uniform lateral and longitudinal excitation in the sensing region. As an early step in selecting the imaging method, photodiode arrays, commercial CMOS cameras, and commercial CCD cameras were compared for their ability to discriminate fluorescence at the waveguide surface.[81] The CCD was shown to have a significantly better signal:noise ratio (Figure 11.3) and thus better discrimination of targets at low concentrations; therefore, a cooled CCD is used in the current prototype. However, CMOS imagers are continually improving and have the advantages that the mechanical shutter is eliminated and the signal from each pixel is already digitized, simplifying data acquisition.

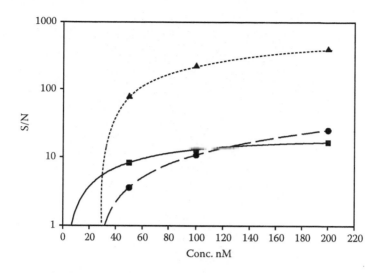

FIGURE 11.3 Signal:noise ratio (S/N) for three detection systems.[81] Log plot shows the mean fluorescence signal (three data points) divided by the system noise (standard deviation of the background signal) for measurements of the fluorescent dye Cy5 in glycerol on a planar waveguide excited using evanescent excitation light at 635 nm. Shown are the photodiode (square, solid line), CMOS (circle, dashed line), and CCD (triangle, dotted line).

The NRL array biosensor has progressed through a series of prototypes to a portable system to which a user may add six samples, with minimal, if any, sample preparation, and test for a variable number of targets.[66,69,82] The current system weighs less than 6 kg and is operated using a laptop computer.[83] The waveguide is coated with a reflective cladding at all points where the six-channel flow guide contacts it; this prevents both stripping of the excitation light out of the waveguide[67,84] and scattering of incident light from the flow cell.[83]

The fluidics system has evolved from a low-tech system using syringe barrels as reservoirs to a fully automated version. The current system uses two modules, each containing six reservoirs for either sample or tracer reagent. Upon insertion of the two modules into the system, a needle pierces the rubber septum at the bottom of each well, connecting each reservoir to the automated fluidics system. Use of separate modules for the six samples and the six tracers gives the system additional flexibility over previous prototypes using a single module for both sample and tracer reagents.[83] In addition, the current reservoir modules are more readily adaptable to manufacture by injection molding. Furthermore, this design also allows the user to prepare a battery of sample and tracer modules to perform a series of experiments in a short time. The ability to lyophilize tracer reagents within the reservoirs and rehydrate them immediately before use,[85] as well as the ability to exchange the sample reservoirs for connections to autonomous sampling devices, results in a plug-and-play methodology.

Although the fluidics system automatically runs the samples over the waveguide, exposes the waveguide to tracer, and washes out excess tracer, image acquisition

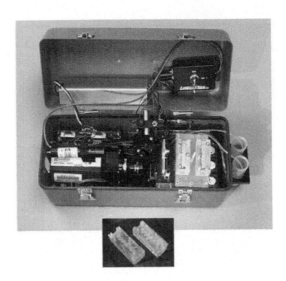

FIGURE 11.4 The most recent version of the NRL array biosensor with integrated optics and fluidics components. Six-reservoir modules for holding samples and fluorescent tracer reagents are shown separately at the bottom.

and data analysis are still performed manually. However, a semiautomated program has been developed to determine the locations, sizes, and intensities of the array elements, as well as the background controls, and to calculate the net increase in fluorescence due to analyte binding. Of the four portable array biosensors described in the literature, the NRL system is the most advanced in terms of integrating all the components into a single portable device and automating the assays (Figure 11.4).

A similar instrument based on chemiluminescence measurements of competitive immunoassays has been automated by Knecht et al.[35] A single sample is flowed over 140 spots to test for 10 different antibiotic residues in milk. Eight computer-controlled syringe pumps add sample, antibody cocktail, rinse buffer, peroxidase-labeled secondary antibody, and chemiluminescence substrate. A Peltier cooled CCD camera is used to measure the signal. Limits of detection ranged from 0.4 to 50 ng/ml for the various antibiotics, with coefficients of variation less than 10%.

11.6 FUTURE DIRECTIONS

Two parallel fields of science and engineering will contribute to the continued and rapid development of array biosensors. First, the rapid engineering progress of optics and microfluidics technologies will provide components that reduce the cost and size of the instrument and simplify automation. Second, advances in proteomics technology will provide enhanced techniques for the immobilization of capture molecules and the identification of new targets for disease diagnosis, therapeutic monitoring, and identification of biohazards.

A strong consumer market for CCD and CMOS cameras and the high-tech market for low-light detection systems are advancing the capabilities of detection components at lightning speeds. Already, the Peltier cooled CCDs used in most array biosensors may be replaced with newer technology chips at much lower cost if they are designed into the systems instead of being inserted as off-the-shelf cameras. Diode lasers and light-emitting diodes (LEDs) are available in an increasing array of wavelengths and power outputs; in addition, virtual pixels (vixels) and organic light-emitting diodes (OLEDs) offer new possibilities for excitation sources. Lens and filter technologies have improved to the point that methods are available for making the appropriate combinations right on the surface of the detector; such integration of the optical components results in more portable, more robust systems. Holographic filters offer the possibility of changing the wavelength of the excitation or emission light as electronically dictated.

Increased flexibility in the wavelengths of operation offers the opportunity to use a wider variety of labels with the tracer molecules. In addition to near-infrared fluorophores, quantum dots and metal colloids may be used with these systems. Quantum dots offer the advantages that multiple colors can be excited with a single excitation wavelength, the signal can be integrated over extended time periods because there is no photobleaching, and they are now commercially available in forms appropriate for attachment of proteins. Preliminary data using an NRL array biosensor with an appropriate light source have demonstrated the utility of the biosensor to measure quantum dot-based assays (K. Sapsford, I. Medintz, and J. Golden, personal communication). The NRL array biosensor has also demonstrated the ability to measure signals from immunoassays and DNA assays using gold colloids as labels (unpublished data). Other types of labels worth investigating include phycobilisomes and dye-labeled dendrimers. Enzymatic signal amplification strategies may also be used to increase assay sensitivity,[79] but the increased complexity extends the assay time and complicates the automation of the assay procedures.

One of the most important features of fluorescence-based array biosensors is the excellent discrimination between the target analyte generating the fluorescent signal and autofluorescence from sample components. The issue of signal vs. background has recently been addressed by several groups involved in using hydrogels to increase the number of capture molecules immobilized on each locus in an array. Researchers have detailed the higher immobilization efficiencies obtainable with highly porous hydrogels,[58,86–88] and Angenendt et al.[89] recently published a comprehensive comparison of various microarray substrates, including several hydrogels. For some applications, the use of such hydrogels on the waveguides using with the array biosensor may enhance the limit of detection.

The proteomics community is producing large arrays of genetically and combinatorially derived molecules in order to explore the interactions between biomolecules and to identify biomarkers for disease progression and therapeutic response. These studies will produce a more limited number of critical targets for clinical use, pathogen detection, and toxicology. As the biomarkers are characterized, the array biosensor will provide an appropriate method for rapid screening of changes in biomarker concentration. The analyses will not be limited to antibody-based methods; protein–protein, receptor–ligand, drug–target, and interactions with intact cells have all been proven feasible.

Finally, the affinity-based recognition schemes that are key to the successful use of the array biosensor for screening and presumptive identification may be married to other technologies. Nanogen reported a lab-on-a-chip system where antibody arrays were used to collect anthrax for subsequent lysis and polymerase chain reaction (PCR) analysis.[80] Targets binding to affinity arrays and detected using SPR imaging are subsequently being evaluated using mass spectroscopy.[90] Kramer and Lim[91] and Tims and Lim[92] reported the affinity capture of bacteria on a waveguide and subsequent cell culture directly on the waveguide. In all of these approaches, the sensing surface not only is used to screen for the target of interest, but it also collects the target and separates it from most of the other sample components. An independent method of analysis may subsequently be used to provide additional information about the target. Performing such procedures in an array format increases the number of targets or target features that may be addressed.

REFERENCES

1. Homola, J., Lu, H.B., Nenninger, G.G., Dostálek, J., and Yee, S.S., A novel multichannel surface plasmon resonance biosensor. *Sensors Actuators B,* 76, 403–410, 2001.
2. Homola, J., Lu, H.B., and Yee, S.S., Dual-channel surface plasmon resonance sensor with spectral discrimination of sensing channels using a dielectric overlayer. *Electron Lett,* 35, 1105–1106, 1999.
3. Kinning, T. and Edwards, P.E., The resonant mirror optical biosensor. In: *Optical Biosensors: Present and Future,* Ligler, F.S. and Taitt, C.R., eds. New York: Elsevier, 2002:253–276.
4. Kusterbeck, A.W., Flow immunosensor. In: *Optical Biosensors: Present and Future,* Ligler, F.S. and Taitt, C.R., eds. New York: Elsevier, 2002:123–142.
5. Shriver-Lake, L.C., Charles, P.T., and Kusterbeck, A.W., Non-aerosol detection of explosives with a continuous flow immunosensor. *Anal Bioanal Chem,* 377, 550–555, 2003.
6. Yu, H. and Bruno, J.G., Immunomagnetic-electrochemiluminescent detection of *Escherichia coli* O157 and *Salmonella typhimurium* in foods and environmental water samples. *Appl Environ Microbiol,* 62, 587–592, 1995.
7. Bruno, J.G. and Yu, H., Immunomagnetic electrochemiluminescent detection of *Bacillus anthracis* spores in soil matrices. *Appl Environ Microbiol,* 62, 3474–3476, 1996.
8. Crawford, C.G., Wijey, C., Fratamico, P., Tu, S.I., and Brewster, J., Immunomagnetic-electrochemiluminescent detection of *E. coli* O157:H7 in ground beef. *J Rapid Meth Automat Microbiol,* 8, 249–264, 2000.
9. Kuczynska, E., Boyer, D.G., and Shelton, D.R., Comparison of immunofluorescence assay and immunomagnetic electrochemiluminescence in detection of *Cryptosporidium parvum* oocysts in karst water samples. *J Microbiol Methods,* 53, 17–26, 2003.
10. Anderson, G.P. and Rowe-Taitt, C.A., Water quality monitoring using an automated portable fiber optic biosensor: RAPTOR. *Proc SPIE,* 4206, 58–63, 2001.
11. Golden, J.P., Saaski, E.W., Shriver-Lake, L.C., Anderson, G.P., and Ligler, F.S., Portable multichannel fiber optic biosensor for field detection. *Opt Eng,* 36, 1008–1013, 1997.

12. DeMarco, D.R., Saaski, E.W., McCrae, D.A., and Lim, D.M., Rapid detection of *Escherichia coli* O157:H7 in ground beef using a fiber-optic biosensor. *J Food Protect,* 62, 711–716, 1999.

13. Jung, C.C., Saaski, E.W., McCrae, D.A., Lingerfelt, B.M., and Anderson, G.P., RAPTOR: a fluoroimmunoassay-based fiber optic sensor for detection of biological threats. *IEEE Sens J,* 3, 352–360, 2003.

14. Anderson, G.P., King, K.D., Gaffney, K.L., and Johnson, L.H., Multianalyte interrogation using the fiber optic biosensor. *Biosens Bioelectron,* 14, 771–778, 2000.

15. Anderson, G.P., Lingerfelt, B.M., and Taitt, C.R., Eight analyte detection using a four-channel optical biosensor. *Sensor Lett,* 2, 18–24, 2004.

16. Weller, M.G., Classification of protein microarrays and related techniques. *Anal Bioanal Chem,* 875, 15–17, 2003.

17. Pawlak, M., Schick, E., Bopp, M.A., Schneider, M.J., Oroszlan, P., and Ehrat, M., Zeptosens' protein microarrays: a novel high performance microarray platform for low abundance protein analysis. *Proteomics,* 2, 383–393, 2002.

18. Neuschäfer, D., Budach, W., Wanke, C., and Chibout, S.-D., Evanescent resonator chips: a universal platform with superior sensitivity for fluorescence-based microarrays. *Biosens Bioelectron,* 18, 489–497, 2003.

19. Epstein, J.R. and Walt, D.R., Fluorescence-based fibre optic arrays: a universal platform for sensing. *Chem Soc Rev,* 32, 203–214, 2003.

20. Peluso, P., Wilson, D.S., Do, D., Tran, H., Venkatasubbaiah, M., Quincy, D., Heidecker, B., Poindexter, K., Tolani, N., Phelan, M., Witte, K., Jung, L.S., Wagner, P., and Nock, S., Optimizing antibody immobilization strategies for the construction of protein microarrays. *Anal Biochem,* 312, 113–124, 2003.

21. Piehler, J., Brecht, A., Geckeler, K.E., and Gauglitz, G., Surface modification for direct immunoprobes. *Biosens Bioelectron,* 11, 579–590, 1996.

22. Jung, A., Stemmler, I., Brecht, A., and Gauglitz, G., Covalent strategy for immobilization of DNA-microspots suitable for microarrays with label-free and time-resolved optical detection of hybridization. *Fresenius J Anal Chem,* 371, 128–136, 2001.

23. Kroger, K., Bauer, J., Fleckenstein, B., Rademann, J., Jung, G., and Gauglitz, G., Epitope-mapping of transglutaminase with parallel label-free optical detection. *Biosens Bioelectron,* 17, 937–944, 2002.

24. Silzel, J.W., Cercek, B., Dodson, C., Tsay, T., and Obremski, R.J., Mass-sensing, multianalyte microarray immunoassay with imaging detection. *Clin Chem,* 44, 2036–2043, 1998.

25. Wadkins, R.M., Golden, J.P., Pritsiolas, L.M., and Ligler, F.S., Detection of multiple toxic agents using a planar array immunosensor. *Biosens Bioelectron,* 13, 407–415, 1998.

26. Vo-Dinh, T., Alarie, J.P., Isola, N., Landis, D., Wintenberg, A.L., and Ericson, M.N., DNA biochip using a phototransitor integrated circuit. *Anal Chem,* 71, 358–363, 1999.

27. Iqbal, S.S., Mayo, M.W., Bruno, J.G., Bronk, B.V., Batt, C.A, and Chambers, J.P., A review of molecular recognition technologies for detection of biological threat agents. *Biosens Bioelectron,* 15, 549–578, 2000.

28. Rabbany, S.Y., Donner, B.L., and Ligler, F.S., Optical immunosensors. *Biomed Eng,* 22, 307–346, 1994.

29. Plowman, T.E., Durstchi, J.D., Wang, H.K., Christensen, D.A., Herron, J.N., and Reichert, W.M., Multiple-analyte fluoroimmunoassay using an integrated optical waveguide sensor. *Anal Chem,* 71, 4344–4352, 1999.

30. Rowe, C.A., Scruggs, S.B., Feldstein, M.J., Golden, J.P, and Ligler, F.S., An array immunosensor for simultaneous detection of clinical analytes. *Anal Chem,* 71, 433–439, 1999.

31. Rowe, C.A., Tender, L.M., Feldstein, M.J., Golden, J.P, Scruggs, S.B., MacCraith, B.D., Cras, J.J., and Ligler, F.S., Array biosensor for simultaneous identification of bacterial, viral, and protein analytes. *Anal Chem,* 71, 3846–3852, 1999.

32. Schult, K., Katerkamp, A., Trau, D., Grawe, F., Cammann, K., and Meusel, M., Disposable optical sensor chip for medical diagnostics: new ways in bioanalysis. *Anal Chem,* 71, 5430–5435, 1999.

33. Sapsford, K.E., Charles, P.T., Patterson, C.H., Jr., and Ligler, F.S., Demonstration of four immunoassay formats using the array biosensor. *Anal Chem,* 74, 1061–1068, 2002.

34. Ligler, F.S., Taitt, C.R., Shriver-Lake, L.C., Sapsford, K.E., Shubin, Y., and Golden, J.P., Array biosensor for detection of toxins. *Anal Bioanal Chem,* 377, 469–477, 2003.

35. Knecht, B.G., Strasser, A., Dietrich, R., Märtlbauer, E., Niessner, R., and Weller, M.G., Automated microarray system for the simultaneous detection of antibiotics in milk. *Anal Chem,* 76, 646–654, 2004.

36. Duveneck, G.L., Pawlak, M., Neuschafer, D., Bar, E., Budach, W., Pieles U., and Ehrat, M., Novel bioaffinity sensors for trace analysis based on luminescence excitation by planar waveguides. *Sensors Actuators B,* 38–39, 88–95, 1997.

37. Budach, W., Abel, A.P., Bruno, A.P., and Neuschafer, D., Planar waveguides as high performance sensing platforms for fluorescence-based multiplexed oligonucleotide hybridization assays. *Anal Chem,* 71, 3347–3355, 1999.

38. Schuderer, J., Akkoyun, A., Brandenburg, A., Bilitewski, U., and Wagner, E., Development of a multichannel fluorescence affinity sensor system. *Anal Chem,* 72, 3942–3948, 2000.

39. Potyrailo, R.A., Conrad, R.C., Ellington, A.D., and Hieftje, G.M., Adapting selected nucleic acid ligands (aptamers) to biosensors. *Anal Chem,* 70, 3419–3425, 1998.

40. Bruno, J.G. and Kiel, J.L., *In vitro* selection of DNA aptamers to anthrax spores with electrochemiluminescence detection. *Biosens Bioelectron,* 14, 457–464, 1999.

41. Li, J. and Lu, Y., A highly sensitive and selective catalytic DNA biosensor for lead ions. *J Am Chem Soc,* 122, 10466–10467, 2000.

42. Wang, L., Carrasco, C., Kumar, A., Stephens, C.E., Bailly, C., Boykin, D.W., and Wilson, W.D., Evaluation of the influence of compound structure on stacked-dimer formation in the DNA minor groove. *Biochemistry,* 40, 2511–2521, 2001.

43. Schmid, E.L., Tairi, A.-P., Hovius, R., and Vogel, H., Screening ligands for membrane protein receptors by total internal reflection fluorescence: the 5-HT3 serotonin receptor. *Anal Chem,* 70, 1331–1338, 1998.

44. Pawlak, M., Grell, E., Schick, E., Anselmetti, D., and Ehrat, M., Functional immobilization of biomembrane fragments on planar waveguides for the investigation of site-directed ligand binding by surface-confined fluorescence. *Faraday Discuss,* 111, 273–288, 1998.

45. Rowe-Taitt, C.A., Cras, J.J., Patterson, C.H., Golden, J.P., and Ligler, F.S., A ganglioside-based assay for cholera toxin using an array biosensor. *Anal Biochem,* 281, 123–133, 2000.

46. Fang, Y., Frutos, A.G., and Lahiri, J., Ganglioside microarrays for toxin detection. *Langmuir,* 19, 1500–1505, 2003.

47. Salisbury, C.M, Maly, D.J., and Ellman, J.A., Peptide microarrays for the determination of protease substrate specificity. *J Am Chem Soc,* 124, 14868–14870, 2002.

48. Herron, J.N., Caldwell, K.D., Christensen, D.A., Dyer, S., Hlady, V., Huang, P., Janatova, V., Wang, H.-K., and Wei, A.-P., Fluorescent immunosensors using planar waveguides. *Proc SPIE*, 1885, 28–39, 1993.

49. Brecht, A., Klotz, A., Barzen, C., Gauglitz, G., Harris, R.D., Quigley, G.R., Wilkinson, J.S., Sztajnbok, P., Abuknesha, R., Gascon, J., Oubina, A., and Barcelo, D., Optical immunoprobe development for multiresidue monitoring in water. *Anal Chim Acta*, 362, 69–79, 1998.

50. Hofmann, O., Voirin, G., Niedermann, P., and Manz, A., Three-dimensional micro-fluidic confinement for efficient sample delivery to biosensors surfaces. Application to immunoassays on planar optical waveguides. *Anal Chem*, 74, 5243 5250, 2002.

51. Hall, E.A.H., *Biosensors*. Buckingham, UK: Open University Press, 1990.

52. Li, Y. and Reichert, W.M., Adapting cDNA microarray format to cytokine detection protein arrays. *Langmuir*, 19, 1557–1566, 2003.

53. Conrad, D.W., Davis, A.V., Golightley, S.K., Bart, J.C., and Ligler, F.S., Photoactivatable silanes for the site-specific immobilization of antibodies. *Proc SPIE*, 2978, 12–21, 1997.

54. Conrad, D.W., Golightley, S.K., and Bart, J.C., Photoactivatable o-nitrobenzyl polyethylene glycol-silane for the production of patterned biomolecular arrays. U.S. patent no. 5,773,308, June 30, 1998.

55. Guschin, D., Yershov, G., Zaslavsky, A., Gemmell, A., Shick, V., Proudnikov, D., Arenkov, P., and Mirzabekov, A., Manual manufacturing of oligonucleotide, DNA, and protein chips. *Anal Biochem*, 250, 203–211, 1997.

56. Wadkins, R.M., Golden, J.P., and Ligler, F.S., A patterned planar array immunosensor for multianalyte detection. *J Biomed Opt*, 2, 74–79, 1997.

57. Schwarz, A., Rossier, J.S., Roulet, E., Mermod, N., Roberts, M.A., and Girault, H.H., Micropatterning of biomolecules on polymer substrates. *Langmuir*, 14, 5526–5531, 1998.

58. Arenkov, P., Kuhktin, A., Gemmell, A., Voloshchuk, S., Chupeeva, V., and Mirzabekov, A., Protein microchips: use for immunoassay and enzymatic reactions. *Anal Biochem*, 278, 123–131, 2000.

59. Liu, X.H., Wang, H.K., Herron, J.N., and Prestwich, G.D., Photopatterning of antibodies on biosensors. *Bioconjug Chem*, 11, 755–761, 2002.

60. Bhatia, S.K., Hickman, J.J., and Ligler, F.S., A new approach to engineering patterned biomolecular assemblies. *J Am Chem Soc*, 114, 4432–4433, 1992.

61. Bhatia, S.K., Teixeira, J.L., Anderson, M., Shriver-Lake, L.C., Calvert, J.M., Georger, J.H., Hickman, J.J., Ducley, C.S., Schoen P.E, and Ligler, F.S., Fabrication of surfaces resistant to protein adsorption and application to two-dimensional protein patterning. *Anal Biochem*, 208, 197–205, 1993.

62. Herron, J.N., Christensen, D.A., Wang, H.-K., Caldwell, K.D., Janatova, V., and Huang, S.-C., Apparatus and methods for multi-analyte homogeneous fluoroimmunoassays. U.S. patent no. 5,512,492, October 14, 1997.

63. Birkert, O., Haake, H.-M., Schutz, A., Mack, J., Brecht, A., Jung, G., and Gauglitz, G., A streptavidin surface on planar glass substrates for the detection of biomolecular interactions. *Anal Biochem*, 282, 200–208, 2000.

64. Wilson, D.S. and Nock, S., Recent developments in protein microarray technology. *Angew Chem Int Ed Engl*, 42, 494–500, 2003.

65. Delehanty, J.B. and Ligler, F.S., A method for printing functional protein microarrays. *Biotechniques*, 34, 380–385, 2002.

66. Delehanty, J.B. and Ligler, F.S., A rapid microarray immunoassay for simultaneous detection of proteins and bacteria. *Anal Chem*, 74, 5681–5687, 2002.

67. Feldstein, M.J., Golden, J.P., Rowe, C.A., MacCraith, B.D., and Ligler, F.S., Array biosensor: optical and fluidics systems. *J Biomed Microdev,* 1, 139–153, 1999.

68. Bernard, A., Michel, B., and Delamarche, E., Micromosaic immunoassays. *Anal Chem,* 73, 8–12, 2001.

69. Rowe-Taitt, C.A., Hazzard, J.W., Hoffman, K.E., Cras, J.J., Golden, J.P., and Ligler, F.S., Simultaneous detection of six biohazardous agents using a planar waveguide array biosensor. *Biosens Bioelectron,* 15, 579–589, 2000.

70. Shriver-Lake, L.C., Shubin, Y., and Ligler, F.S., Detection of staphylococcal enterotoxin B in spiked food samples. *J Food Prot,* 66, 1851–1856, 2003.

71. Taitt, C.R., Shubin, Y.S., Angel, R., and Ligler, F.S., Detection of *Salmonella enterica* serovar typhimurium by using a rapid, array-based immunosensor. *Appl Environ Microbiol,* 70, 152–158, 2004.

72. Sapsford, K.E., Rasooly, A., Taitt, C.R, and Ligler, F.S., Rapid detection of campylobacter and shigella species in food samples using an array biosensor. *Anal Chem,* 76, 433–440, 2004.

73. Sapsford, K.E., Liron, Z., Shubin, Y.S., and Ligler, F.S., Kinetics of antigen binding to arrays of antibodies in different sized spots. *Anal Chem,* 73, 5518–5524, 2001.

74. Sapsford, K.E. and Ligler, F.S., Real-time analysis of protein adsorption to a variety of thin films. *Biosens Bioelectron,* 19, 1045–1055, 2004.

75. Tolley, S.E., Wang, H.-K., Smith, R.S., Christensen, D.A., and Herron, J.N., Single-chain polymorphism analysis in long QT syndrome using planar waveguide fluorescent biosensors. *Anal Biochem,* 315, 223–237, 2003.

76. Sipe, D.M., Schoonmaker, K.P., Herron, J.N., and Mostert, M.J., Evanescent planar waveguide detection of biological warfare simulants. *Proc SPIE,* 3913, 215–222, 2000.

77. Vo-Dinh, T., Development of a DNA biochip: principle and applications. *Sensors Actuators B,* 51, 52–59, 1998.

78. Moreno-Bondi, M.C., Alarie, J.P., and Vo-Dinh, T., Multianalyte analysis system using an antibody-based biochip. *Anal Bioanal Chem,* 375, 120–124, 2003.

79. Stratis-Cullum, D.N., Griffin, G.D., Mobley, J., Vass, A.A., and Vo-Dinh, T., A miniature biochip system for detection of aerosolized *Bacillus globigii* spores. *Anal Chem,* 75, 275–280, 2003.

80. Yang, J.M., Bell, B., Huang, Y., Tirado, M., Thomas, D., Forster, A.H., Haigis, R.W., Swanson, P.D., Wallace, B., Martinsons, B., and Krihak, M., An integrated stacked microlaboratory for biological agent detection with DNA and immunoassays. *Biosens Bioelectron,* 17, 605–618, 2002.

81. Golden, J.P. and Ligler, F.S., A comparison of imaging methods for use in an array biosensor. *Biosens Bioelectron,* 17, 719–725, 2002.

82. Taitt, C.R., Anderson, G.P., Lingerfelt, B.M., Feldstein, M.J., and Ligler, F.S., Nine-analyte detection using an array-based biosensor. *Anal Chem,* 74, 6114–6120, 2002.

83. Golden, J.P., Taitt, C.R., Shriver-Lake, L.C., Shubin, Y.S., and Ligler, F.S., A portable automated multianalyte biosensor. *Talanta,* 65, 1078–1085, 2005.

84. Feldstein, M.J., Golden, J.P., Ligler F.S., and Rowe, C.A., Reflectively coated optical waveguide and fluidics cell integration. U.S. patent no. 6,192,168, February 20, 2001.

85. Taitt, C.R., Golden, J.P., Shubin, Y.S., Shriver-Lake, L.C., Sapsford, K.E., Rasooly, A., and Ligler, F.S., A portable array biosensor for detecting multiple analytes in complex samples. *Microb Ecol,* 47, 175–185, 2004.

86. Barsky, V.E., Kolchinsky, A.M., Lysov, Y.P., and Mirzabekov, A.D., Biological microchips with hydrogel-immobilized nucleic acids, proteins, and other compounds: properties and applications in genomics. *Mol Biol,* 36, 437–455, 2002.

87. Charles, P.T., Goldman, E.R., Rangasammy, J.G., Schauer, C.L., Chen, M-S., and Taitt, C.R., Fabrication and characterization of 3D hydrogel microarrays to measure antigenicity and antibody functionality for biosensor application. *Biosens. Bioelectron.* 20, 753–764, 2004.
88. Rubina, A.Y., Dementie, E.I., Stomakhin, A.A., Darri, E.L., Pan'kov, S.V., Barsky, V.E., Ivanov, S.M., Konovalova, E.V., and Mirzabekov, A.D., Hydrogel-based protein microchips: manufacturing, properties, and applications. *Biotechniques,* 34, 1008–1022, 2003.
89. Angenendt, P., Glokler, J., Murphy, D., Lehrach, H., and Cahill, D.J., Toward optimized antibody microarrays: a comparison of current microarray support materials. *Anal Biochem,* 309, 253–260, 2002.
90. Xinglong, Y., Donsheng, W., Dingxin, W., Hua, O.Y.J., Zibo, Y., Yanggui, D, Wei, L., and Sheng, Z.X., Micro-array detection system for gene expression products based on surface plasmon resonance imaging. *Sensors Actuators B,* 91, 133–137, 2003.
91. Kramer, M.F. and Lim, D.V., A rapid and automated fiber optic-based biosensor assay for the detection of salmonella in spent irrigation water used in the sprouting of sprout seeds. *J Food Prot,* 67, 46–52, 2004.
92. Tims, T.B. and Lim, D.V., Confirmation of viable *E. coli* O156:H7 by enrichment and PCR after rapid biosensor detection. *J Microbiol Methods,* 55, 141–147, 2003.

12 Planar Waveguide Biosensors for Point-of-Care Clinical and Molecular Diagnostics

James N. Herron, Ph.D., Hsu-Kun Wang, M.S., Lyndon Tan, B.S., Stacy Z. Brown, B.S., Alan H. Terry, M.S., Samuel E. Tolley, Ph.D., Jacob D. Durtschi, B.S., Eric M. Simon, M.E., Mark E. Astill, M.S., Richard S. Smith, Ph.D., and Douglas A. Christensen, Ph.D.

CONTENTS

12.1 INTRODUCTION

12.1.1 Point-of-Care Testing

The past decade witnessed the migration of several key *in vitro* diagnostics (IVD) assays from centralized clinical laboratories to patient settings in the home, physician's office, and hospital. The top five such "point-of-care" (POC) assays (in terms of market share) include blood glucose (22%), pregnancy (19%), infectious disease (12%), cholesterol (11%), and critical care (10%).[1] Coagulation, cardiac markers, drugs of abuse, and bilirubin collectively account for an additional 12% market share.[1] Interestingly, the top two assays (blood glucose and pregnancy) are sold over the counter at most pharmacies for home use. Worldwide sales of POC assays are expected to reach $3.2 billion in 2005.[1] Although the POC market segment accounts for only 11% of total worldwide IVD sales ($28.6 billion expected for 2005), it has grown phenomenally (40%) over the past five years, with double-digit annual growth rates for cholesterol, coagulation, cardiac markers, and bilirubin assays.[1] In contrast, annual growth rates of 5% are typical in the conventional IVD market.[1]

12.1.2 Impact of Biosensors in POC Testing

The aforementioned migration was enabled in part by biosensor technology.[2] For example, almost all home-use blood glucose assays are amperometric biosensors in which oxidation of glucose by either glucose oxidase or glucose dehydrogenase generates a measurable current.[3,4] Electrochemical biosensors are also employed in professional-use POC assays for critical care (e.g., blood gases, electrolytes, hemoglobin), glucose, and metabolites such as creatinine, lactate, and urea.[5,6] Pulse oximetry is probably the most ubiquitous POC application of optical biosensor technology to date. In this technique, blood oxygen saturation is determined by absorption of light through tissue (typically a finger, ear, or nose) at two different wavelengths (650 to 700 nm for deoxyhemoglobin and 800 to 940 nm for oxyhemoglobin).[7] Typically, two light-emitting diodes (LEDs) (red and infrared) are placed on one side of the tissue, with a single detector on the other. Hemoglobin concentration within the optical path increases slightly with each pulse, producing a concomitant increase in absorption that is entirely due to hemoglobin.[7] Optical biosensors are also employed in home-use POC coagulation (prothrombin time) assays for patient self-monitoring of anticoagulation therapy,[8] professional-use POC assays for transcutaneous bilirubin in neonates,[9] and cardiac markers (described below).

Professional-use POC immunoassays exist for several different cardiac markers, including (1) creatine kinase isoform MB, troponin I, troponin T, and myoglobin for diagnosis of acute myocardial infarction (AMI)[10–12]; (2) B-type natriuretic peptide for diagnosis of congestive heart failure (CHF)[13]; and (3) C-reactive protein (CRP) for prognosis of cardiovascular disease (CVD).[14–16] AMI tests migrated to near-patient testing during the past five years, whereas CHF and CVD tests are in transition now. Several vendors (e.g., Biosite, Response Medical, Roche Diagnostics) market POC AMI tests based on optical biosensor technology. Each test measures one or more cardiac markers directly from whole blood samples in 20 min or less. All employ similar immunochromatography techniques in which a complex forms between the analyte and an antibody labeled with either colloidal gold or a fluorescent dye and then migrates through a fiber strip or capillary channel until being captured specifically by a second antibody immobilized in the detection zone.[17] In multiple analyte tests, the detection zone is subdivided according to analyte by immobilizing different analyte-specific capture antibodies at defined positions along the zone (referred to as "spatial resolution"). Detection involves either colorimetric determination of colloidal gold labels or front-face fluorescence measurements of fluorescent labels.

Other immunoassays and some molecular diagnostics assays are also poised to transition from centralized laboratory to near-patient testing in the next few years. These include diagnosis, prognosis, and screening assays for diseases such as allergy, anemia, cancer, diabetes, metabolic disease, microbial infections, osteoporosis, sexually transmitted infections, and thyroid disease, and applications such as drug testing, genetic screening, personalized medicine, reproductive endocrinology, therametrics, and therapeutic drug monitoring. As in the past, biosensor technology is expected to be a significant factor in this ongoing migration from central laboratory to near-patient testing.

12.1.3 AFFINITY BIOSENSORS

Both immunoassays and molecular diagnostics assays fall into the broader category of affinity assays, in which analyte binding to a specific recognition molecule (typically an antibody or single-stranded DNA probe) results in detection. In many cases the recognition molecule is immobilized to a solid phase (e.g., beads, particles, and multiwell plates). Solid-phase assays can readily be ported to biosensor formats by immobilizing the recognition molecule to the biosensor's transduction element (electrode, optical element, etc.).

Affinity biosensors have been the subject of intense research in both academia and industry for more than two decades, but aside from the previously mentioned cardiac marker assays (which account for only 3% of worldwide POC market),[1] they have yet to make a significant impact in the POC IVD marketplace.[18] Nevertheless, the predominance of optical immunosensors in the cardiac marker POC market segment, together with this segment's phenomenal 20% annual growth rate,[1] suggests that affinity biosensors will have a major impact in the burgeoning POC market for immunoassays and molecular diagnostics assays. Several key attributes should be considered when designing affinity biosensors for use in such assays. These include

1. Comparable or better sensitivity than conventional IVD assays
2. Short assay times (preferably 5 min or less)
3. Clinical Laboratory Improvement Amendments of 1988 (CLIA '88) moderately complex or waived assay procedure
4. Immunity to nonspecific binding (NSB) and sample matrix effects
5. Multiple tests on a single patient specimen
6. Inexpensive, disposable sensor cartridge with internal calibration

12.1.4 MASS SENSORS

Although not currently a major player in the POC IVD market, affinity sensors enjoy commercial success in both the drug discovery and life science research markets. Most familiar is the Biacore biosensor (Biacore International AB, Uppsala, Sweden), which is based on surface plasmon resonance (SPR) and is used to monitor biomolecular interactions in real time without the use of fluorescent labels or radiolabels.[19,20] Affinity Sensors (Franklin, MA; a division of Thermo Labsystems) makes a similar system called IAsys that uses a slightly different optical geometry called resonant mirrors.[19] Both are examples of a more general optical technique called total internal reflection (TIR). As the name implies, light reflects totally off an internal face of the sensor (see Figure 12.1). The incident and reflected beams interfere constructively, creating a standing electric wave at the reflection point. Although intense at the reflection point, the electric field decays exponentially over a distance of a few hundred nanometers as it enters the sample (which has a lower refractive index than the sensor). This decaying or "evanescent" field may be used to optically interrogate molecules bound to the sensor. For example, analyte binding to an SPR sensor (via an immobilized recognition molecule) produces a local

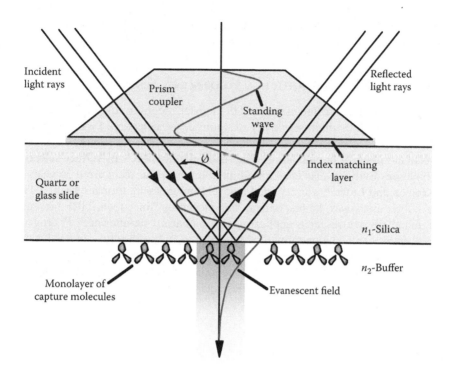

FIGURE 12.1 Total internal reflection spectroscopy. A prism couples light into an optical substrate (either fused quartz or glass) at angle Ø. For coupling angles greater than critical angle $Ø_c$, light rays reflect totally off the interior face at the bottom of the substrate and exit through the other side of the prism. Constructive interference between the incident and reflected rays creates a standing wave at the reflection point. The electric field decays over a few hundred nanometers into the sample, which has a lower refractive index than the optical substrate. This "evanescent" electric field may be used to optically interrogate molecules immobilized to the exterior face of the substrate.

refractive index change in the sensor's evanescent field, shifting the characteristic reflection angle for the SPR effect. Such devices are called mass sensors because their signal changes in proportion to the mass bound within the evanescent field.

Mass sensors in general offer the significant advantage of real-time kinetic measurements without the need for extrinsic labels, wash steps, or additional reagents—features that reduce both assay time and complexity (attributes 2 and 3 above). In addition, the IAsys system in particular employs an inexpensive, disposable sample cuvette (attribute 6). Nevertheless, mass sensors have at least three limitations. First, they respond to any molecule located within the evanescent field, whether bound specifically to the recognition molecule or not. Thus, NSB (attribute 4) may be a significant problem, especially with complex sample matrices such as human serum, plasma, or whole blood (attribute 4). Second, they are less sensitive (attribute 1) than comparable methods using extrinsic labels. Third, they require larger sensing areas than extrinsic label methods, making impractical medium- and high-density array mass sensors (attribute 5). These limitations explain why few, if any, mass sensors have been commercialized in POC IVD applications.

12.1.5 Total Internal Reflection Fluorescence Sensors

Total internal reflection fluorescence (TIRF) spectroscopy weds the surface selectivity of TIR with the high sensitivity of fluorescence detection. First described by Hirschfeld[21] four decades ago, this technique is responsible for significant advancements in the fields of protein adsorption[22] and cell adhesion.[23] Its surface selectivity and high sensitivity are also ideal for affinity biosensors, as recognized 30 years ago by Kronick and Little,[24] who reported a novel immunoassay format based on TIRF. In fact, their assay was the first optical immunosensor. Since then, TIRF biosensors have found diverse research applications in biohazard monitoring,[25–28] drug testing,[29–31] environmental monitoring,[32–34] food testing,[35–37] intracellular bioanalysis,[38,39] IVD,[40–46] molecular diagnostics,[47–53] and pharmacology and toxicology.[54,55] Fiber-optic TIRF sensors constitute the vast majority of these reports, although a few articles describe other optical geometries such as single-bounce TIRF and planar waveguide TIRF.

12.1.6 Fiber-Optic TIRF Sensors

Fabrication of fiber-optic TIRF sensors involves stripping the cladding from a fiber's distal end (typically only a few millimeters are removed), followed by immobilization of recognition molecules on the exposed surface. Intrinsically fluorescent or fluorescently labeled analytes may be detected directly. Unlabeled analytes are detected indirectly using a fluorescently labeled "tracer" molecule (typically an antibody) that binds specifically to the analyte recognition molecule complex. In either case, the sensor's evanescent field both excites and collects analyte fluorescence (such evanescent collection is also known as reciprocity), which is then conducted by the fiber to a remote photodetector at its proximal end. Such remote sensing is ideal for applications such as biohazard monitoring, environmental monitoring, and process control. Interested readers are referred to a recent review of fiber-optic TIRF sensors by Wolfbeis.[56]

The sensitivity and surface selectivity of fiber-optic TIRF sensors make them compatible with the first three attributes listed above. In addition, they are less susceptible to NSB (attribute 4) than mass sensors because measurable NSB only arises from fluorescent analytes or tracers rather than from any molecule within the evanescent field. However, a solitary fiber is inherently a single assay device, necessitating the use of fiber bundles to measure multiple analytes. Interfiber variability is a noise source with such bundles,[57] limiting their utility in multianalyte testing (attribute 5). Internal calibration (attribute 6) is also challenging for the same reason. Nevertheless, there have been a few reports of fiber bundle TIRF sensors[58,59] and a couple of start-up companies marketing them (Illumina, San Diego, CA; and Research International, Inc., Monroe, WA). Illumina markets its products for genotyping and gene expression profiling, whereas Research International's products are for monitoring biological agents, toxins, explosives, and chemical contaminants. Neither company has products in the POC IVD marketplace.

12.1.7 PLANAR WAVEGUIDE TIRF SENSORS

Optical fibers are actually cylindrical waveguides. In concept, the cylinder can be unrolled into a plane and patterned with recognition molecules to form an array biosensor. Thus, planar waveguides are ideal for multianalyte sensing with onboard calibration (attributes 5 and 6). However, the planar format precludes fluorescence collection by reciprocity because signals arising from different analytes would commingle within the waveguide. For that reason, nearly all multianalyte applications of planar waveguide sensors employ free space collection with an imaging photodetector (e.g., a charge-coupled device [CCD] camera) placed alongside the waveguide.[27,28,43,45,52,60–62]

Both single- and multimode planar waveguide sensors have been reported. Single-mode sensors are typically 0.15- to 1.5-μm thin films fabricated from either silicon oxynitride, tantalum pentoxide (Ta_2O_5), or titanium dioxide (TiO_2) and deposited on either silicone or silicon dioxide substrates for support.[45,61–63] Their principal advantages are exquisite sensitivity and high spatial uniformity evanescent fields. For example, Plowman et al.[63] reported low femtomolar sensitivities using 1-μm silicon oxynitride planar waveguide sensors, and Pawlak et al.[62] reported high-performance protein microarrays (maximum density 1600 spots/cm^2 with 250-μm pitch) patterned on Ta_2O_5 planar waveguide sensors for proteomic applications. This latter technology has been commercialized by Zeptosens AG (Witterswill, Switzerland). On the down side, fabrication costs are high ($25 to $50 per chip), making thin-film waveguides prohibitively expensive for one-shot, disposable POC assays.

Multimode planar waveguide sensors are thicker (typically 0.1 to 1 mm) than single-mode sensors, resulting in a less intense evanescent field (intensity scales inversely with waveguide thickness) with periodic intensity fluctuations. On the other hand, they are also easier to fabricate and significantly less expensive, making them suitable for low-density array applications. The simplest design is a 25 mm × 75 mm microscope slide (either glass or quartz) patterned with recognition molecules in a two-dimensional array.[27,28,41,64–72] Fluorescence from the array elements is imaged either collectively with a camera lens or individually with a two-dimensional lens array. A CCD is used for detection in either case.

Our group has developed a more complex injection-molded planar waveguide design specifically for POC IVD applications. This device is shown in Figure 12.2. It has been licensed to BioCentrex, LLC (Culver City, CA), which has developed a product for the POC cardiac marker market.

12.1.8 RATIONALE FOR THE PLANAR WAVEGUIDE APPROACH

We first proposed planar waveguides as fluorescent affinity biosensors in 1987.[60] Since then, our group has used the attributes listed in Section 12.1.3 as a road map for investigating POC applications of planar waveguide TIRF sensor technology.

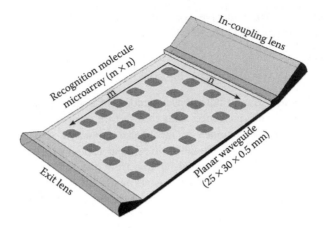

FIGURE 12.2 Polystyrene planar waveguide sensor. This injection-molded piece incorporates the waveguide and lens into the same disposable unit for greater alignment tolerance. The depicted version also contains an exit lens for dumping light out of the sensor. Exciting light (typically generated by a 635-nm diode laser) is collimated into a sheet beam (approximately 20 mm × 2 mm) and steered into the coupling lens. After being coupled, the light propagates down the waveguide, creating an evanescent field. The illustration depicts an $m \times n$ microarray of recognition molecules (e.g. antibodies, oligonucleotides) enabling multiple assays to be performed on a single sensor.

Initially, we focused on immunoassays for human chorionic gonadotropin (hCG) and cardiac markers for AMI,[41,43,45,63,64,65] but we later broadened our scope to include molecular diagnostics assays.[49,51,52] Our approach contains several key elements, including injection-molded planar waveguides as an optical platform for evanescent excitation and free space fluorescent detection of analytes using Cy5, a red-emitting fluorescent label. The following paragraphs present our rationale for this approach.

Injection-molded planar waveguides were employed for three reasons. First, injection molding enables inexpensive manufacture of disposable sensors (attribute 6). Second, patterning of the waveguide's large, planar sensing region with different recognition molecules allows multiple analytes to be assayed in a single specimen (attribute 5). Third, patterning with internal standards enables on-sensor assay calibration (attribute 6).

Evanescent excitation of fluorescence offers at least three advantages over conventional excitation by transmitted light. First, the evanescent field is more intense than transmitted light (all else being equal) due to constructive interference within the sensor, thereby enhancing sensitivity (attribute 1). Second, its surface selectivity eliminates precipitation, filtration, reagent addition, and/or wash steps, thereby reducing both assay time and complexity (attributes 2 and 3). Third, this surface selectivity minimizes sample matrix effects (attribute 4) arising from particles and cells that are significantly larger than the evanescent field's penetration depth (approximately 70 nm).

Fluorescence detection was chosen because fluorescent immunoassays rival both radioimmunoassays (RIAs) and enzyme-linked immunosorbent assays (ELISAs) in terms of sensitivity (attribute 1) but do not require radiolabels or a time-consuming enzyme amplification step, thereby reducing both assay time and complexity (attributes 2 and 3). Free space collection is more sensitive (attribute 1) than reciprocity because only a small fraction (less than 10%) of emitted fluorescence evanescently back-couples into the waveguide.[64] As mentioned previously, fluorescence sensors are less susceptible to NSB (attribute 4) than mass sensors. A red-emitting fluorophore was chosen because whole blood samples are more transparent to red light, again minimizing sample matrix effects (attribute 4).

In summary, our planar waveguide TIRF sensor approach is compatible with all six attributes listed above for commercial POC applications of affinity biosensors. The remainder of this chapter focuses on our experiences and challenges in developing this sensor technology for POC applications. Specific details are also presented for several different immunoassays and molecular diagnostics assays formulated using our technology.

12.2 EXPERIMENTAL PROCEDURES

12.2.1 WAVEGUIDES

Planar waveguide sensors are injection molded from polystyrene by Opkor, Inc. (Rochester, NY). Being both disposable and relatively inexpensive, they are ideal for assay development purposes. The patented design[73–75] is shown in Figure 12.2. In addition to the waveguide, the sensor contains an integrated coupling lens for directing light into a waveguide and an exit lens for dumping light out of the waveguide. (In an alternate design, the rear of the waveguide is tapered to a knife edge that dumps light away from the sensing area.) Recognition molecules (either antibodies or oligonucleotides) were immobilized to the top of the planar waveguide and fluorescence emission was collected from the bottom of the waveguide (orthonormal to the plane of the waveguide) using a CCD camera as described in Section 12.2.3

12.2.2 FLOW CELLS

The biosensor assembly consisted of a planar waveguide sensor and a flow cell developed by Herron et al.[74,75] Flow cells comprised a top and a bottom plate fabricated from aluminum (see Figure 12.3). Top plates were milled with either two or three parallel flow channels, each with small inlet and outlet ports and sample volumes of 500 μl (two-channel version) or 200 μl (three-channel version). The interior of the bottom plate was milled to support the waveguide and provide a clear view of its bottom. The entire flow cell was anodized flat black. The top plate was sealed against the waveguide using a composite gasket with a low-refractive-index Teflon layer next to the waveguide and a silicon rubber layer next to the top plate. Mechanical pressure sealed the system by tightening four knurled bolts located at the four corners of the flow cell. A computer-controlled three-barrel syringe pump (Cavro Scientific Instruments, Inc., Sunnyvale, CA) injected specimens into the flow chambers.

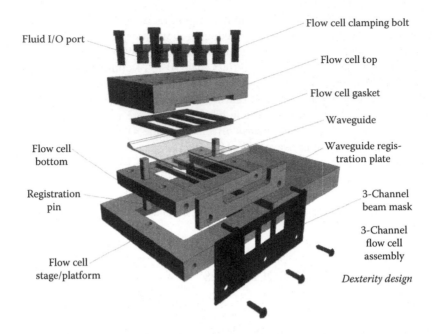

FIGURE 12.3 (See color insert following page 142.) Flow cell assembly. The flow cell top, gasket, and waveguide define three flow channels immediately above the waveguide. Samples injected into the flow channels (via fluid input/output [I/O] ports) sit statically in the flow channels during assays. The three-channel beam mask blocks light from entering the waveguide immediately under the longitudinal gasket ribs. Fluorescent emissions are imaged through the windows in the flow cell bottom and stage.

The availability of different top plates (two- or three-channel) afforded some flexibility in experimental design. For example, ratiometric measurements acquired with the two-channel version enabled sample measurements to be normalized to a reference, which was useful for endpoint measurements requiring long-term (1 h or more) instrument stability. The three-channel version facilitated triplicate kinetic measurements of a given sample, providing mean and standard deviation statistics from a 5-min assay of a single patient specimen on a single sensor assembly. Alternatively, assays could be calibrated in POC environments by injecting the patient's sample into one channel and positive and negative controls into the other two channels.

12.2.3 INSTRUMENTATION

Fluorescence measurements were taken with the Mark 1.5 evanescent wave imaging fluorometer constructed by Douglas Christensen at the University of Utah. The biosensor assembly was mechanically locked into a mounting plate on the Mark 1.5 fluorometer, providing tight registration of the waveguide to the exciting light. Red light from a diode laser (635 nm, approximately 12 mW) was formed into a sheet beam (20 mm × 1 mm) and coupled into the waveguide via its integrated coupling lens.

Light then propagated the length of the waveguide, reflected by TIR at the upper and lower surfaces of the waveguide, thereby creating a transverse standing wave within the waveguide (see Figure 12.4). This standing wave did not have a sharp boundary at the waveguide surface, tunneling instead into the surrounding medium. The intensity of this evanescent wave decayed exponentially over several hundred nanometers as it penetrated into the surrounding medium but retained enough intensity to excite Cy5-labeled analyte molecules bound to immobilized recognition molecules. Fluorescence emission emanated in all directions, but the portion that travels through the waveguide (and through a window in the bottom of the flow cell) was collected and imaged by a CCD camera (Santa Barbara Instrument Group, Santa Barbara, CA). This camera was equipped with a 55-mm f/2.8 macro lens (Nikon, Tokyo, Japan) to image the waveguide and a 670-nm band-pass interference filter (Orion, Santa Cruz, CA) to reject scattered light.

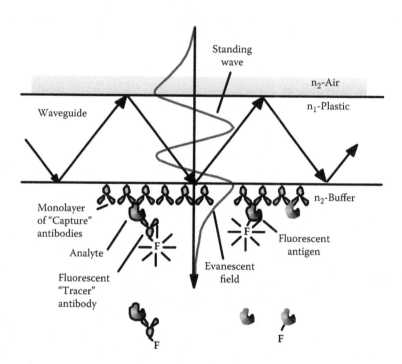

FIGURE 12.4 Sandwich and competitive immunoassays using a planar waveguide TIRF biosensor. Light propagates down the waveguide, reflecting from side to side and creating an evanescent field at each reflection point. A layer of capture antibodies is immobilized directly to the waveguide. The lower left side of the illustration depicts a typical sandwich assay wherein analyte binding to capture antibody is transduced by binding of a second, fluorescently labeled antibody to the analyte–capture antibody complex. The lower right side shows a typical competitive assay in which capture antibody binds a mixture of labeled and unlabeled analyte. In both assay formats, the evanescent field only excites captured fluorescent labels, leading to detection.

12.2.4 DATA ACQUISITION

The CCD camera was interfaced to a Power Macintosh 7600 computer (Apple Computer, Cupertino, CA) with a PowerForce 280 MHz G3 accelerator card (PowerLogix, Austin, TX). LabView version 4.0.1 (National Instruments Corp., Austin, TX) was used for all instrument data acquisition and control operations. Data collection involved four steps:

1. Collection of a "dark image" of the waveguide with shutter closed and light source off
2. Collection of a "light image" with shutter open and light source on
3. Subtraction of the dark image from the light image to produce the true signal
4. Summation ("binning") of individual pixels over the two or three channels of the waveguide to give an intensity value for each zone

This cycle was repeated at 6-s intervals, giving 51 data points in just over 5 min.

12.2.5 IMMOBILIZATION OF RECOGNITION MOLECULES ON WAVEGUIDES

12.2.5.1 Antibody Immobilization

Antibodies were immobilized to polystyrene waveguides by physical adsorption. This method was particularly convenient because antibodies adsorb strongly to the hydrophobic polystyrene. Also, a large body of knowledge exists about immobilizing antibodies by physical adsorption to polystyrene multititer plates, and much of that knowledge is transferable to immobilizing antibodies on polystyrene waveguides. Prior to immobilization, waveguides were washed in ethanol for 1 min with agitation to remove any residue, followed by rinsing in deionized water and drying. Immobilization reactions were performed in coating trays consisting of aluminum top and bottom plates that sandwich a waveguide with a silicon rubber gasket in between. The gasket was sealed against the waveguide by mechanical pressure produced by four knurled bolts located at the four corners of the tray. The gasket defines a 1.9 cm × 2.1 cm coating area on the upper waveguide face. Antibodies were immobilized by adding 1 ml of 50-nM capture antibody solution (phosphate buffered saline [PBS], pH 7.4) to the coating trays, which were then covered and incubated for 1 h at room temperature. The coating solution was discarded after incubation and the waveguide was rinsed twice with StabilCoat (Surmodics, Eden Prairie, MN), followed by postcoating with 1 ml of StabilCoat for an additional hour at room temperature. Postcoating minimizes NSB by masking any remaining adsorption sites on the waveguide. It also enhances shelf life during storage. After postcoating, the StabilCoat solution was discarded and waveguides were dried by vacuum desiccation for 1 to 2 h at room temperature and stored desiccated at 4°C until use.

12.2.5.2 Oligonucleotide Immobilization

Capture oligonucleotides were immobilized on waveguides, as described by Herron et al.[49] and Tolley et al.[51] Waveguides were first cleaned with a 5% chlorine bleach solution to remove layers of immobilized proteins and nucleic acids from previous experiments. Waveguides were then coated for 1 h with a 150-nM solution of neutravidin in PBS (pH 7.5, containing 0.02% sodium azide as a preservative) using the coating trays described above. Unadsorbed neutravidin was removed by twice washing each waveguide in 10-mM Tris buffer (pH 7.4) with 1-mM ethylenediaminetetraacetic acid (EDTA) (TE buffer). Then a 50-nM solution of 5' biotinylated oligonucleotide in TE buffer was allowed to react with the immobilized neutravidin for 1 h, followed by twice washing with TE buffer. Finally, waveguides were coated for 30 min with a 0.1% (w/v) solution of trehalose in TE buffer. Excess solution was decanted and the remaining trehalose was allowed to dry for several hours at 4°C. This final postcoating step protected the immobilized oligonucleotides and allowed coated waveguides to be stored at room temperature for at least a month.

12.2.6 ASSAY FORMAT

12.2.6.1 Immunoassays

A two-site sandwich immunoassay was employed in all clinical IVD assays. Antibodies fill two different roles in such assays: an immobilized antibody captures the analyte, whereas a soluble, fluorescently labeled antibody detects or "traces" analyte binding. To prevent competitive binding, capture and tracer antibodies must bind to different sites on the analyte molecule. For large analytes with repetitive epitopes (e.g., viruses and bacteria), a single antibody (specific for the repetitive epitope) can usually be employed in both capture and tracer roles. Smaller analytes (e.g., proteins and polysaccharides) expressing multiple, unique epitopes require two different antibodies, each specific for a unique epitope.

In a typical assay, equal volumes (0.3 ml) of sample and tracer antibody (10 nM, final concentration) were premixed and then injected into the flow cell. Binding was monitored for 5 min at room temperature using the Mark 1.5 fluorometer as described above (Section 12.2.3). Each assay was performed in triplicate using the three channels of the flow cell.

Assays were calibrated by constructing a standard curve of average reaction rate vs. analyte concentration in the undiluted sample. Analyte concentration varied over at least a 100-fold range (e.g., 20 pM to 2 nM), though the exact range depended on the clinical concentration range of the analyte being examined. Typically, two to three waveguides were measured at each concentration, giving sample sizes of four to six channels (n = number of waveguides × 3 channels/waveguide). Blank (i.e., no analyte present in the sample medium) measurements were typically repeated on 5 to 10 waveguides, giving samples sizes of 15 to 30. Standard curves were fit with a rate saturation model (see Equation 12.10).

12.2.6.2 Molecular Diagnostics Assays

Nucleic acid analytes were detected with a solid-phase hybridization assay. Single-stranded capture oligonucleotides were immobilized to planar waveguides as described in Section 12.2.5.2. The sample consisted of a soluble, single-stranded oligonucleotide analyte, labeled at its 5' end with Cy5. Clinically, such analytes may be prepared by polymerase chain reaction (PCR) initiated with Cy5-labeled primers; the resulting PCR product contains Cy5 dye at its 5' end. Following injection into the flow cell, labeled analyte diffuses through bulk solution, hybridizes with immobilized capture oligonucleotides, and fluoresces in the evanescent field.

A Peltier thermoelectric module attached to the top plate of the flow cell controlled hybridization reaction temperature precisely (tolerance of 0.05°C) over a range of 30°C to 60°C. The Peltier device could both heat and cool the reaction chamber using an electronic negative feedback controller. A thermistor embedded near the flow chamber gave continuous temperature readings. Samples (0.6 ml) were equilibrated for 5 min at reaction temperature and then injected into the flow cell to start the reaction. Hybridization was monitored with the Mark 1.5 fluorometer for at least 2 min, and up to 1 h in some cases, as described above (Section 12.2.3). Each assay was performed in triplicate using the three channels of the flow cell. Assays were calibrated using standard curves, as described above (Section 12.2.6.1).

12.3 EVOLUTION OF SENSOR DESIGN

In this section we discuss the evolution of our sensor design over the past decade.

12.3.1 SINGLE-BOUNCE TIRF

Early efforts involved the use of quartz microscope slides (or coverslips) that were coated with a layer of "capture" antibody and mounted in a flow cell.[60,76] Light was coupled into the slide using a dovetail quartz prism and reflected off the opposite face, as depicted in Figure 12.5. An evanescent field was set up at the reflection point and decayed over a distance of approximately 100 nm into the aqueous phase. This field was used to excite fluorescently labeled molecules that bind to the immobilized capture antibodies. Fluorescence emission was detected normal to the face of the sensor (through the flow cell) using the collection lens shown in Figure 12.5. A photomultiplier tube was typically used as the detector. Technically, this sort of sensor configuration is known as a "single-bounce total internal reflection fluorescence" or "single-bounce TIRF."

Although our early studies demonstrated that single-bounce TIRF could effectively be used to perform fluorescence immunoassays,[60,76] the configuration had a number of limitations that precluded its use in POC environments. The most serious of these was the small area of the slide (2 to 4 mm in diameter) illuminated by the single reflection. In fact, the illuminated area accounted for only about 1% of the slide's total surface area, and signal:noise ratio considerations prevented more than one assay from being performed at a time in the illuminated area. For this reason, we

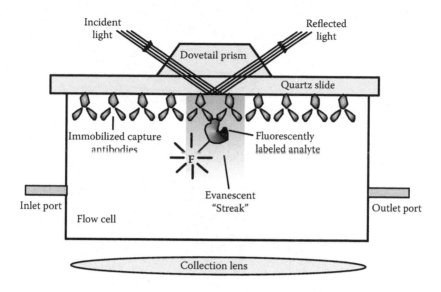

FIGURE 12.5 A single-bounce TIRF biosensor. A dovetail quartz prism couples light into a fused quartz microscope slide. Coupled light reflects off the opposite face of the slide, creating an evanescent field at the reflection point. The field decays exponentially over a few hundred nanometers into the aqueous phase and selectively excites fluorescently labeled molecules bound to immobilized capture antibodies. Fluorescence emission is detected normal to the face of the sensor (through the flow cell) using the collection lens.

concluded that single-bounce TIRF sensors were not suitable for the POC market, where assay calibration would undoubtedly be an issue. In our opinion, the best calibration strategy is to measure both unknown and several calibration standards side by side on the same sensor. Thus, a means of illuminating the whole microscope slide simultaneously was needed in order to put the sensor's entire surface area to work for us. It was this realization that led to the idea of using planar waveguides.

12.3.2 PLANAR WAVEGUIDES

Starting with our quartz microscope slide, the change from single-bounce TIRF to a planar waveguide was surprisingly easy to make—the only requirement was that the dovetail prism be replaced with a cylindrical lens to focus the light into the end of the slide. This configuration is shown schematically in Figure 12.6. Once light is coupled into the slide, it propagates down the waveguide by reflecting totally from side to side, setting up an evanescent field at each reflection point. We also opted to move the collection lens to the other side of the sensor so that the fluorescence emission did not have to travel through the assay solution and the flow cell. This modification gave us more latitude in flow cell design and also eliminated interfering optical effects from complex specimens such as whole blood. In addition, by using an imaging detector such as a CCD camera, we could perform multiple assays on a single sensor.

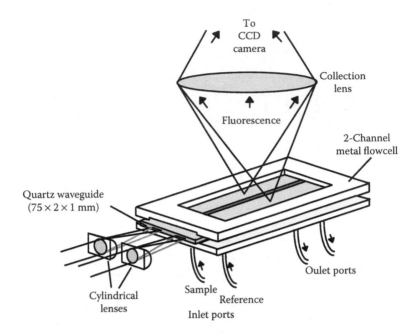

FIGURE 12.6 Early planar waveguide sensor design. The sensor assembly consists of a fused quartz microscope slide sandwiched between top and bottom plates of a two-channel metal flow cell. A rubber gasket inserted between the quartz slide and the bottom plate forms two flow channels. External cylindrical lenses couple light into the end of the slide, creating two optical channels. Fluorescently labeled molecules captured on the bottom of the slide fluoresce isotopically, but only the emissions passing through the waveguide and top plate window are collected and imaged by the CCD camera.

As shown in Figure 12.4, both sandwich and competitive immunoassays may be performed with the quartz waveguide sensor system. Our group has developed both types of assays for a number of different analytes, including bovine serum albumin (BSA), hCG, and creatine kinase MB (CK-MB).[41,64,77] In all cases, the sensitivity of a given sensor was in the high femtomolar to low picomolar range; however, this level of performance could not be realized when more than one sensor was used, because of high intersensor variability. This was a significant limitation, because a typical standard curve consists of five or six different analyte concentrations with at least three different sensors for each concentration, and intersensor variability, if too high, can obscure the low end of the curve. Our investigations showed that this problem was principally due to variabilities in sensor alignment—especially the relative positions of the waveguide and the cylindrical coupling lens. The solution to this problem was to integrate the coupling lens with the waveguide. First we considered various means of integrating the coupling lens with the quartz waveguide, but we eventually decided that a quartz sensor would be too expensive to be commercially viable in a disposable POC assay. This

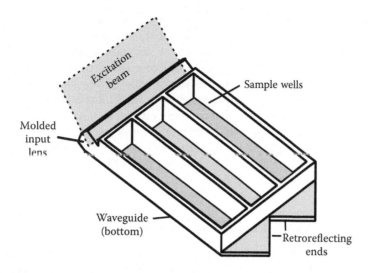

FIGURE 12.7 An archetypal injection-molded planar waveguide sensor. This disposable sensor unit incorporates both the coupling lens and waveguide for greater alignment tolerance. This particular design also includes assay wells for samples and retroreflecting ends for increasing light intensity. Exciting light generated by a 635-nm diode laser is collimated into a 20 mm × 2 mm sheet beam and steered into the coupling lens. Coupled light propagates down the waveguide, setting up the evanescent field. Recognition molecules are immobilized to the bottom of each well.

led to the concept of injection molding both a coupling lens and a waveguide into a disposable sensor unit.

12.3.3 DISPOSABLE INJECTION-MOLDED SENSORS

Figure 12.7 shows our archetypal injection-molded planar waveguide sensor. Molding of both the coupling lens and waveguide into the same unit allowed greater alignment tolerance.[65,78] This particular design also contained integral sample wells and retroreflecting ends for reflecting light back into the waveguide for a second pass (thus increasing the intensity of the evanescent field). Polystyrene was chosen as the waveguide material because its high refractive index (1.59) helps select for higher-order modes and it is compatible with the antibody immobilization chemistry employed in most present-day immunoassays.

Preliminary studies demonstrated that the integral coupling lens dramatically reduced intersensor variability and that antibodies could be immobilized successfully to the polystyrene surface, giving both high analyte binding activity and low levels of NSB.[65,77,78] Nevertheless, light scattering from the seams of the sample wells and the waveguide itself into the sample solution produced spurious signals by exciting unbound tracer antibody in the sample. In fact, sample solutions had to be decanted from sample wells to give reliable readings.

These observations suggested that the archetype waveguide design was too complex. Its sample well walls were essentially waveguide discontinuities that allowed light leakage and scattering. Moreover, its complex design caused uneven melt flow during injection molding that produced nonuniform waveguide thickness and submicron-scale surface roughness, both of which promoted light scatter. Simplifying the waveguide to the design shown in Figure 12.2 rectified these problems. In essence, sample wells were moved from the waveguide to the flow cell, allowing the former to be a featureless, planar structure. This design enabled a molding tool to be milled from a very hard stainless steel (STAVAX ESR [AISI 420 modified]; Uddeholm Tooling AB, Hagfors, Sweden) and polished to a surface finish of less than 2 nm (verified by profilometry).

12.3.4 Sensor Validation with Model Immunoassays

The next step was to ascertain whether our immunosensor behaved in a predictable fashion when binding analyte. Perhaps the best way to assess its response characteristics is to perform a binding isotherm over a wide range of analyte concentrations. In a traditional solid-phase immunoassay (e.g., RIA and ELISA), the binding of analyte to immobilized antibodies can be analyzed using the following binding equation:

$$f_b = \frac{C}{K_d + C} \qquad (12.1)$$

where f_b is the fraction of immobilized antibody active sites that contain bound analyte (referred to as "fractional saturation"), C is the bulk analyte concentration, and K_d is the apparent equilibrium constant for the binding of analyte to immobilized binding sites. Even though this equation contains only a single equilibrium constant, in most instances it fits data obtained from sandwich assays where at least two different equilibria are present (analyte can bind to both capture and tracer antibodies, each with a unique equilibrium constant). For this reason, K_d is referred to as an "apparent" equilibrium constant.

Bovine serum albumin labeled with fluorescein isothiocyanate (FITC) was used as a model analyte in a quality control assay used for evaluating the performance of our injection-molded planar waveguide sensor.[63,65,77-79] The assay is flexible in that it can be formulated in a number of different formats (direct, competitive, and sandwich) and detection can be accomplished using either radiolabels or fluorescent dyes. Furthermore, we have considerable experience with its performance under a wide variety of experimental conditions. A monoclonal antifluorescein antibody (BDC1) was immobilized to the waveguide by physical absorption followed by a casein postcoat; fluorescein$_5$-BSA was used as the analyte; and another monoclonal antibody (specific for BSA rather than fluorescein-BSA) was labeled with Cy5 and used as the tracer antibody. Experiments were performed using the injection-molded planar waveguides shown in Figure 12.2. Each sensor unit was mounted in a two-channel flow cell, where one channel was used for the actual assay (sample channel) and the other was used as a blank (reference channel) to control for factors such as background fluorescence, NSB, and variations in the intensity of the light source.

The actual experiment was a "step isotherm" in which increasing concentrations of fluorescein-BSA were injected in sequence to the sample channel (along with a 1-nM solution of tracer antibody), whereas only tracer antibody (1 nM) was injected into the reference channel. After each injection, images of the waveguide were recorded at 30-s intervals over a 5-min period using a CCD camera, but only those data collected at the 5 min time point were analyzed. This experiment was repeated on five different waveguides and the data were averaged.

Initially, data were reduced to a fluorescence ratio (X), which is defined as the ratio of the intensity of the sample channel (I_{sample}) to that of the reference channel $(I_{reference})$:

$$X = I_{sample}/I_{reference} \qquad (12.2)$$

Next, fluorescence ratio values obtained for each concentration of BSA were normalized to fractional saturation (f_b) values using the following equation:

$$f_b[BSA] = (X[BSA] - X_0)/(X_m - X_0) \qquad (12.3)$$

where X_0 is the fluorescent ratio in the absence of BSA and X_m is the fluorescence ratio in the presence of a saturating amount of BSA. Fractional saturation is plotted vs. BSA concentration (on a logarithmic scale) in Figure 12.8 and the data is fitted to Equation 12.1 using weighted nonlinear least squares. The model fit the data very well $(r^2 = 0.999, \chi^2 = 3.4)$ and gave a value of 1.1 nM for the apparent equilibrium constant. The precision of the five repeats was good (see error bars in Figure 12.8), and the sensitivity was about 50 pM.

12.4 CLINICAL IMMUNOASSAYS

The first step in evaluating our immunosensor system in clinical assays was to identify analytes for which there was an unmet need for POC testing. Qualitative tests for hCG have been sold in the consumer market for some years and women who test positive often have the results verified by their physician, so there is a need for quantitative tests in the physician's office. In a review of pregnancy tests, Chard[80] makes the point that the sensitivity of current hCG assays is about 25 mIU/ml or 5 ng/ml, which is certainly within the range of our immunosensor system.

Diagnosis of AMI is another area where there is a demonstrated need for fast POC testing.[11,12] Cardiac troponin subunits T and I (cTnT, cTnI) are the most cardiospecific biomarkers,[12,81–83] but their plasma levels do not rise until 4 to 6 h after infarction,[82–84] thus limiting their diagnostic value in the first few hours after onset of chest pain.[84] Also, some evidence suggests that cTnI is more cardiospecific than cTnT in patients with renal insufficiency or end-stage renal disease.[85] Both CK-MB and myoglobin are

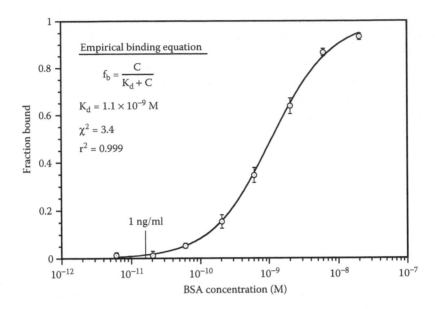

FIGURE 12.8 Performance evaluation of the injection-molded planar waveguide TIRF sensor. Sensor performance was appraised using an assay for fluorescein-labeled BSA (fluorescein-BSA). A monoclonal anti-fluorescein antibody was immobilized to the waveguide by physical adsorption. A monoclonal anti-BSA antibody was labeled with Cy5 and used as a tracer antibody (1 nM final concentration). Fluorescein-BSA was titrated over a concentration range of 6×10^{-12} to 2×10^{-8} M. Each titration was performed using a single waveguide sensor (of the type shown in Figure 12.2) mounted in a two-channel flow cell. A mixture of analyte and tracer antibody was injected into the sample channel, whereas tracer antibody alone was injected into the reference channel. Data were analyzed as described in the text. The data shown in the figure represent the average and standard deviation of five independent titrations ($n = 5$). The detection limit was about 50 pM.

useful early AMI markers, especially when repeated assessments are taken during the first few hours after emergency room admission.[84] However, CK-MB is only partially cardiospecific (elevated levels may also result from exercise-induced muscle trauma[86]), and myoglobin is not cardiospecific at all. For these reasons, the predominant POC testing strategy for AMI involves repeated assessment of cTnI, CK-MB, and myoglobin.[11,82]

According to Guest et al.,[87] the normal reference ranges for CK-MB and cTnI are 2.2 to 6.7 ng/ml and 1.5 to 3.1 ng/ml, respectively, although Antman et al.[88] states that cTnI levels of approximately 0.4 ng/ml are indicative of acute coronary syndromes. Myoglobin's normal range is 0 to 85 ng/ml.[89] Based on the above information, we concluded that sensitivities of approximately 1 ng/ml for CK-MB, 0.2 ng/ml for cTnI, and 10 ng/ml for myoglobin are required in order for an immunosensor system to be competitive with current assays. Unfortunately, the CK-MB and cTnI sensitivities were below the sensitivity observed in our BSA assay. Thus, some improvements were needed in order to perform cardiac enzyme assays.

Actually, this problem was not as daunting as it first seemed because assay sensitivity is directly related to the affinity of the capture and tracer antibodies employed in the assay and we were able to find high-affinity antibodies for both CK-MB and cTnI from commercial sources. Even using such antibodies, however, there was still a need for an additional sensitivity increment. Our approach to this problem is described below.

12.4.1 Rate-Based Detection

In the BSA assay, data were collected at 30-s intervals over a 5-min period, but only those data collected at the 5-min time point (about 10% of the collected data) were actually analyzed. Analyzing all the data should improve the statistical precision of our measurements as well as assay sensitivity. The challenge was to average all the data—which increased rapidly and nonlinearly with time—into a single parameter that could be plotted against analyte concentration in a standard curve. Our solution was to determine the analyte binding rate (R) at a particular point in time (t_i) from intensity vs. time data using the method of least squares. Being based on the entire dataset, this rate represents the average binding rate (R_{avg}) at evaluation time t_i:

$$R_{avg} = <R_{t_i}> = <(dI(t)/dt)_{t_i}> \qquad (12.4)$$

where $I(t)$ is a function (either analytical or empirical) that describes the sensor's fluorescence intensity response. Another advantage of rate measurements is their immunity to time-invariant signals, such as dark counts and the intrinsic waveguide fluorescence.

Binding reactions in our biosensor format are typically pseudo-first-order because solution analyte concentration is usually limiting. The following intensity response function (I_t) was derived based on the assumption of limiting analyte concentration:

$$I_t = I_0 + \Delta I_{max}(1 - e^{-kt}) \qquad (12.5)$$

where I_0 is initial fluorescence intensity, ΔI_{max} is the maximum change in fluorescence intensity at steady state, and k is the technical rate constant. Its derivative is

$$R_{t_i} = (dI(t)/dt)_{t_i} = (d(I_0 + \Delta I_{max}(1 - e^{-kt}))/dt)_{t_i} = \Delta I_{max} k \cdot e^{-kt_i} \qquad (12.6)$$

The strategy is to replace ΔI_{max} with R_{t_i} in Equation 12.5 because R_{t_i} is determinable with better precision than ΔI_{max} in rapid assays (5 min or less). This is because R_{t_i} can be evaluated at a time point within the data collection interval, whereas ΔI_{max} determination requires extrapolation to equilibrium conditions at long times. Solving Equation 12.6 for ΔI_{max} and substituting into Equation 12.5 gives:

$$I_t = I_0 + R_{t_i}(e^{kt_i}/k)(1 - e^{-kt}) \qquad (12.7)$$

When this equation is fit to fluorescence intensity vs. time data using nonlinear least squares, R_{t_i} is equivalent to average the binding rate R_{avg} defined in Equation 12.4.

In practice, R_{t_i} was evaluated at the midpoint of the data collection interval for immunoassays and at the end of the collection interval for nucleic acid hybridization assays. Although precision is slightly better at the midpoint, evaluation at the end of the collection interval improved discrimination of single base mismatches when screening for single nucleotide polymorphisms.[51] Intercept value (I_0) is the initial intensity response of the sensor (immediately after analyte injection). It is due to factors such as native waveguide fluorescence, leakage of scattered laser light through the interference filter, and excitation of unbound Cy5-labeled analyte in bulk solution by scattered laser light. Although not directly relevant to the binding reaction being monitored, it can be used to provide quality control information about the waveguides and/or light collection system. The technical rate constant k may be viewed as an empirical shape factor that describes the degree of curvature to the kinetics curve. As k approaches zero, the curve becomes linear.

Nonlinear equations such as Equation 12.7 are not suitable for real-time data fitting because numerical curve-fitting procedures require a few seconds of computer time to converge. Nevertheless, Equation 12.7 may be linearized by fixing the shape factor and defining a parametric time variable (Z), as shown in Equation 12.8 and Equation 12.9:

$$\text{For constant } k, \quad Z = e^{-kt} \quad \text{and} \quad Z_{t_i} = e^{-kt_i} \tag{12.8}$$

$$I_t = I_0 + \frac{R_{t_i}}{kZ_{t_i}}(1 - Z) = \left(I_0 + \frac{R_{t_i}}{kZ_{t_i}} \right) - \left(\frac{R_{t_i}}{kZ_{t_i}} \right) Z \tag{12.9}$$

In order for this linearization to be successful, shape factor k needed to be either fairly uniform between datasets or only weakly coupled to the average binding rate, R_{avg}. Although previous studies have shown that k is not particularly constant between data sets, these same studies showed that the hybridization rate is only weakly coupled to the shape factor.[51] For this reason, shape factor k was fixed at 0.3/min. With only two variables to calculate, it was possible to do real-time calculation of the average binding rate.

Assays were calibrated with a standard curve of the average binding rate (R_{avg}) vs. bulk analyte concentration (C). R_{avg} vs. C data were fit with the following rate saturation model:

$$R = R_0 + \frac{R_{max}C}{K_d + C} \tag{12.10}$$

where R_0 is the initial binding rate in the absence of analyte, R_{max} is the maximum binding rate (at saturation), and K_d is an apparent affinity constant defined as the concentration required for half saturation ($R_{max}/2$). The binding rate can saturate at high analyte concentrations due to depletion of either capture or tracer antibody.

The minimum detectable concentration (MDC; also referred to as "analytical sensitivity") of an assay was defined as the lowest analyte concentration distinguishable from zero with 95% confidence. The following equation was used to determine the MDC from a standard curve of binding rate vs. analyte concentration and assay background noise:

$$MDC = 2\sigma/m \tag{12.11}$$

where σ is the standard deviation of the zero concentration point and m is the initial slope of the standard curve. For analyte concentrations significantly less than K_d, the denominator of Equation 12.11 simplifies to K_d, giving a linear relationship between R and C:

$$R = R_0 + \left(\frac{R_{max}}{K_d}\right) C \tag{12.12}$$

with intercept of R_0 and slope R_{max}/K_d. (As an aside, IVD assays are usually optimized for linear standard curves within the clinical concentration range of the analyte—a property referred to as "linearity.") Substitution of the R_{max}/K_d ratio for the slope in Equation 12.5 gives

$$MDC = \frac{2\sigma}{\left(\dfrac{R_{max}}{K_d}\right)} = \frac{2\sigma K_d}{R_{max}} \tag{12.13}$$

12.4.2 CHORIONIC GONADOTROPIN

An assay for hCG was used as a test case to evaluate the rate-based analysis method. Two different monoclonal antibodies (anti-hCG A, anti-hCG D) that bind to different sites on hCG were obtained from Akzo-Nobel (Arnhem, The Netherlands). Anti-hCG A was immobilized to the waveguide by physical adsorption and postcoated with human serum albumin (HSA) and trehalose. Anti-hCG D was labeled with Cy5 and used as the tracer antibody (at 1 nM final concentration). A binding isotherm was performed over a concentration range of 0 to 300 ng/ml (samples were prepared by adding known amounts of purified hCG to pooled human plasma) using a different sensor unit for each concentration point (as opposed to a step isotherm, where the entire titration is performed on a single sensor). Five sensor units were used at each concentration point and assays were performed in triplicate on each sensor using a three-channel flow cell—giving a total of 15 replicate assays of each concentration. The average analyte binding rate was determined over a 5-min assay period for each concentration point and plotted (along with its standard error) vs. bulk hCG concentration (Figure 12.9).

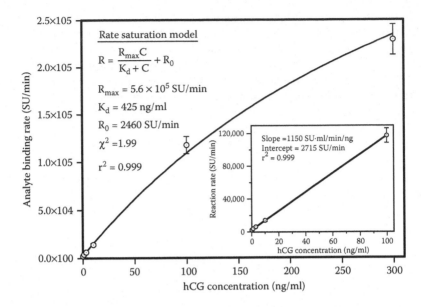

FIGURE 12.9 The hCG assay. A two-site sandwich immunoassay was formulated as described in the text. Analyte (hCG) was titrated over a range of 0 to 300 ng/ml. Triplicate 5-min measurements were obtained for each concentration using a single sensor assembly. The figure presents the average binding rate and standard deviation of five independent titrations ($n = 15$). The complete data set was analyzed using the rate saturation model shown in the figure. The inset shows good linearity at low concentrations (less than K_d). The analytical sensitivity of the assay was 0.85 ng/ml.

From the graph we see that the saturation model fits the entire dataset very well ($\chi^2 = 1.99$, $r^2 = 0.999$) and gives values of 5.6×10^5 SU/min and 425 ng/ml for R_{max} and K_d, respectively. In addition, data from the lower end of the concentration scale (0 to 100 ng/ml) are replotted in the inset. In this case, the hCG concentration is less than K_d and the plot becomes linear with a slope of 1150 SU·ml/min/ng and an intercept of 2715 SU/min. For this particular hCG assay, σ was 489 SU/min, which gives an analytical sensitivity of 0.85 ng/ml or 0.17 mIU/ml.

12.4.3 CK-MB ISOENZYME

Creatine kinase (86 kDa)[90] comprises two similar polypeptide chains referred to as B for brain and M for muscle that can dimerize into three isoenzymes (BB, MM, MB). Both the MM and MB isoenzymes are found in cardiac muscle, the latter constituting 10% to 30% of total creatine kinase in patients with chronic coronary disease.[91,92] Smaller amounts of MB are found in skeletal muscle—typically 1% or less of total creatine kinase, although MB composition may be 10% or more in marathon runners or in diseases such as muscular dystrophy or chronic renal disease.[91,92] Thus, both the MM and MB isoenzymes leak into the blood after myocardial infarction. However, MB is the more diagnostic of the two because of its higher prevalence in cardiac muscle. An effective assay requires strict specificity for MB, with little or no cross-reactivity with MM.

12.4.4 ASSAY DEVELOPMENT

Twenty-four antibodies specific for human creatine kinase were obtained from commercial or academic sources. Eight of these were nominally specific for CK-MB, 12 were nominally specific for CK-MM, and 4 were nominally specific for CK-BB. Our strategy was to employ a two-site sandwich assay giving stringent specificity for CK-MB (i.e., no cross-reactivity with either CK-MM or CK-BB). This could be accomplished with any of the following antibody specificity pairings: (1) anti-CK-MB with anti-CK-MB, (2) anti-CK-MB with anti-CK-BB, (3) anti-CK-MB with anti-CK-MM, and (4) anti-CK-MM with anti-CK-BB. The latter two pairings were less attractive due to possible competition with elevated plasma CK-MM levels in AMI patients.

Eighty-nine different antibody pairs were screened using the two-channel sensor assembly. Initial screens were for CK-MB binding activity. Each pair was evaluated with a step isotherm (1 to 100 ng/ml CK-MB). A mixture of CK-MB and Cy5-labeled tracer antibody (1 nM) was injected into the sample channel, whereas Cy5-labeled tracer antibody alone was injected into the reference channel. Ratiometric measurements were taken for at least 5 min at each concentration point. Surprisingly, only three antibody pairs exhibited strong-enough binding activity to meet our sensitivity target (1 ng/ml). One of these was subsequently eliminated due to cross-reactivity with CK-MM. The two remaining pairs performed comparably to the IMx CK-MB assay (Abbott Laboratories, Abbott Park, IL) for three different patient plasma samples containing normal, medium, and high CK-MB levels. One of these pairs (Conan-MB and Bill, Washington University)[93] has gained wide acceptance in commercial CK-MB assays and was selected for subsequent clinical studies. Being specific for CK-MB and CK-BB, respectively, Conan-MB and Bill fall within the second class of antibody specificity pairs discussed above, which may account for this pair's low CK-MM cross-reactivity. Finally, using Conan-MB and Bill as the respective capture and tracer antibodies gave better analytical sensitivity values than the opposite pairing.

12.4.5 SAMPLE MATRIX EFFECTS

Human plasma is the most common sample matrix for conventional IVD assays, although direct assay of whole blood specimens has certain advantages in POC or critical care settings, including elimination of sample preparation steps (e.g., centrifugation) and reduction of sample preparation time. Our biosensor's optical format should be relatively immune to sample matrix effects for the following reasons:

1. The penetration depth of the evanescent field (approximately 70 nm) is small compared to cells found in human blood or particulate matter sometimes found in plasma.
2. Collected fluorescence travels through the optically transparent waveguide rather than the sample, thus reducing absorption and light scatter that could occur within the latter.

3. Proteins and cells in the sample are transparent to the red light used for excitation (635 nm) and emission (670 ± 8 nm full width at half maximum [FWHM]).
4. The interference filter used to collect emission has a rejection of 0.01% at the excitation wavelength, thereby reducing light scatter by 10^4-fold.

Human plasma and whole blood matrix effects on our CK-MB assay are examined in the next two sections.

12.4.6 HUMAN PLASMA VS. BUFFER

Recombinant CK-MB (Genzyme, Cambridge, MA) was added to either PBS (pH 7.4) with 0.1% BSA or pooled human plasma to produce a series of samples with a concentration range of 0 to 300 ng/ml. Endogenous CK-MB had previously been depleted from the plasma pool by immunoadsorption with an anti-CK-MB-sepharose column. Binding isotherms were performed as described above for hCG, except that each concentration point was determined in quadruplicate (two sensor assemblies with two assays per sensor). CK-MB was titrated over a range of 0 to 300 ng/ml and the average binding rate was determined at each concentration point with a 5-min assay. The linear portions (0 to 100 ng/ml) of standard curves obtained for both buffer and plasma assays are plotted in Figure 12.10, along with a linear regression analysis for each.

As observed with hCG, the standard curve is nearly linear for CK-MB concentrations between 0 and 100 ng/ml. Also, the slope of the standard curve (286 SU·ml/min/ng) obtained for buffer assays is about 60% greater than the slope (178 SU·ml/min/ng) obtained for plasma assays, suggesting that analyte binding is somewhat faster in buffer than plasma. This difference is probably due to the greater viscosity of plasma, which will decrease the diffusion rate of the analyte/tracer antibody complex. Analytical sensitivity values of 0.40 ng/ml (4.65 pM) and 2.25 ng/ml (26.2 pM) were obtained for buffer and plasma, respectively. Although the buffer value was better than our 1 ng/ml target, the more relevant plasma value was not. Still, the plasma value is nearly identical to the sensitivities of commercially available assays.[87]

12.4.7 WHOLE BLOOD VS. HUMAN PLASMA

Being a soluble protein of cytoplasmic origin,[94] CK-MB partitions into the plasma fraction of whole blood. Thus, its endogenous concentration in the plasma fraction should be identical to that in clarified plasma derived from the same specimen. For this reason, direct assay of CK-MB plasma concentration in whole blood should be possible using an assay technology immune to blood cell interference. The 70-nm evanescent wave penetration depth of our planar waveguide biosensor is nearly 100-fold smaller than a red blood cell, which should impart some immunity to blood cell interference. This hypothesis was tested by measuring CK-MB standard curves in both whole blood and plasma. Recombinant human CK-MB was added to freshly

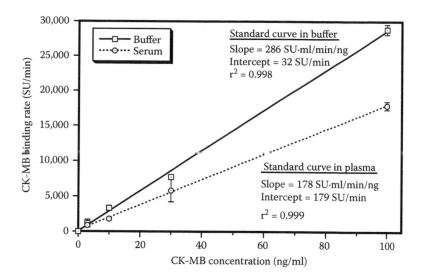

FIGURE 12.10 Comparison of CK-MB assays in buffer and human plasma. A two-site sandwich immunoassay was formulated as described in the text. Recombinant human CK-MB (Genzyme) was titrated over a range of 0 to 300 ng/ml, though only the linear portion (0 to 100 ng/ml) is shown. Each standard solution was assayed in duplicate for 5-min assays using a single three-channel sensor assembly (a buffer control was assayed in the third channel). The figure presents the average binding rate and standard deviation of two independent titrations ($n = 4$). Analytical sensitivity values of 0.4 and 2.25 ng/ml were determined for buffer and plasma standard curves, respectively.

drawn blood over a range of 0 to 200 ng/ml. Concentrations were determined by dividing the mass of CK-MB added to the whole blood sample by the volume of the plasma fraction (determined by hematocrit). Each sample was then split in two—half being measured directly in the sensor assembly and half centrifuged to produce plasma before being measured.

Results are shown in Figure 12.11. The similar slope values for whole blood (692 SU·ml/min/ng) and plasma (558 SU·ml/min/ng) indicated that the sensor response was comparable in these matrices. However, intercept values differed significantly between whole blood and plasma (−858 SU/min in whole blood, 1661 SU/min in plasma) suggesting that blood cells systematically affected measurements. Subsequent studies indicated that the negative slope value was due to settling of blood cells out of the evanescent field (which would reduce light scatter) during the 5-min assay period. Reorienting the sensor unit from a vertical to a horizontal position, where the waveguide was above the flow cell, minimized this phenomenon. Analytical sensitivity values of 4.51 ng/ml (52.4 pM) and 1.41 ng/ml (16.4 pM) were determined for the whole blood and plasma assays, respectively. Although the 4.51-ng/ml value falls within CK-MB's normal range, it is too high for a commercial whole blood assay. Still, the fact that such promising data were obtained using a home-built research instrument gave hope that 1-ng/ml sensitivities

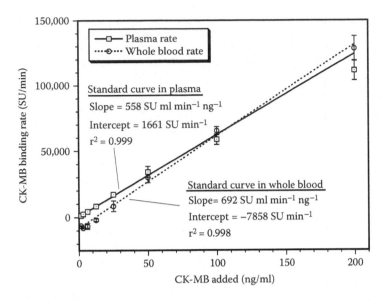

FIGURE 12.11 Comparison of CK-MB assays in human plasma and whole blood. Recombinant human CK-MB was added over a range of 0 to 200 ng/ml to fresh human whole blood (hematocrit 46, normal CK-MB 2.6 ng/ml). One milliliter of each standard solution was mixed with an equal volume of tracer antibody dissolved in the whole blood base (1-nM final tracer antibody concentration based on plasma volume). Samples were injected into two channels of a three-channel sensor assembly. A control solution (30 ng/ml CK-MB in human plasma) was injected into the third channel. Duplicate 5-min assays were performed at each concentration point using a single sensor assembly. The figure presents the average binding rate and standard deviation of two independent titrations ($n = 4$). The remainder of each whole blood CK-MB standard solution was clarified by centrifugation. One-milliliter volumes of each plasma solution were then mixed with an equal volume of tracer antibody dissolved in plasma base, giving a 1-nM final tracer concentration. These mixtures were assayed as described above. Analytical sensitivity values of 1.41 and 4.51 ng/ml were determined for the plasma and whole blood standard curves, respectively.

could be obtained with refined engineering designs of both the sensor assembly and instrument in a production prototype.

12.4.8 CLINICAL CORRELATION STUDY

The next step was to validate the CK-MB assay in a small clinical correlation study. Plasma was chosen as the sample matrix based on the results of the blood vs. plasma comparison above. Plasma samples were obtained from 62 anonymous chest pain patients who had previously been screened for CK-MB levels by the University of Utah Emergency Department using the IMx CK-MB assay (Abbott Laboratories). Each sample was measured in duplicate using the three-channel sensor assembly in a 5-min assay. A positive control (30-ng/ml recombinant human CK-MB in human plasma) was injected into the third channel. The results shown in Figure 12.12

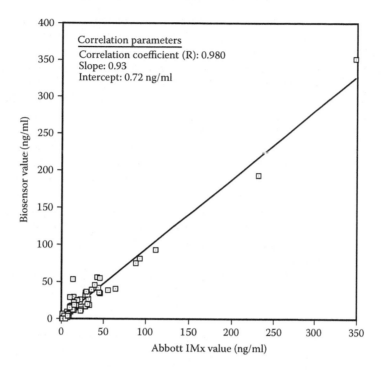

FIGURE 12.12 Clinical validation of the CK-MB assay. Our 5-min biosensor assay for CK-MB was validated in a small-scale clinical trial with plasma samples from 62 anonymous chest pain patients. Samples were obtained from the University of Utah Emergency Department, where they had previously been tested with the IMx CK-MB assay (Abbott Laboratories). The figure presents clinical correlation between the biosensor and reference assays. Good correlation ($R = 0.98$) was observed between the two assays. Slope and intercept values are discussed in the text.

indicated that the two assays were highly correlated ($R = 0.980$) and related by the following correlation function:

$$[\text{CK-MB}]_{\text{biosensor}} = 0.72 + 0.93[\text{CK-MB}]_{\text{IMx}} \qquad (12.14)$$

The slope value (0.93) indicates that the reference material (recombinant human CK-MB; Genzyme) used for calibrating our assays was comparable to Abbott Laboratories' reference material. The intercept is less than the analytical sensitivity of the assay (1.41 ng/ml) and is probably not statistically significant.

12.4.9 MYOGLOBIN

Human myoglobin is an oxygen storage protein (17,669 kDa)[95] expressed in both cardiac and skeletal muscle. Its structure is identical in both. It is also relatively abundant in both types of muscle, with an average content of 45.6 mg of myoglobin per gram of total protein (mg/g) in the ventricles, 20.5 mg/g in the atria, and 60.8 mg/g in skeletal muscle.[96] In contrast, cardiac troponin I content is 1.21 μg/g and 0.65 μg/g in the ventricles and atria, respectively, and less than 1 ng/g in skeletal muscle.[96] This 20,000-fold predominance of myoglobin over cTnI explains the former's utility as an early AMI marker; following infarction, myoglobin is the first cardiac marker to rise to detectable levels in the blood.

Myoglobin assay development and validation followed the same general strategy described above for CK-MB. Fortuitously, the first two antibodies tested proved to be a suitable pair (MBM0201 and MBM0202; Genzyme). Both were murine monoclonal anti-human myoglobin antibodies but bound to different myoglobin epitopes. MBM0201 was used as capture and MBM0202 was labeled with Cy5 and used as tracer. Recombinant human myoglobin (BiosPacific, Emeryville, CA) was used as a reference material in standard curves. The linearity of initial standard curves was poor above 100 ng/ml (reaching saturation by 500 ng/ml). Diluting the Cy5-labeled tracer antibody 1:10 with unlabeled MBM0202 solved the problem. Final MBM0202 concentrations were 1 and 9 nM for Cy5-labeled and -unlabeled MBM0202, respectively.

Standard curves obtained in whole blood and plasma samples are shown in Figure 12.13. The nearly parallel slopes indicate that the sensor response was

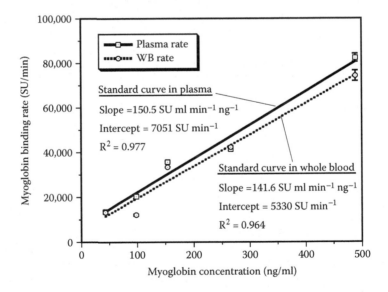

FIGURE 12.13 Comparison of myoglobin assays in human plasma and whole blood. Standard solutions of recombinant human myoglobin (BiosPacific) were prepared in whole blood and plasma as described in Figure 12.11. Myoglobin was titrated over a range of 0 to 1000 ng/ml, although linearity was poor above 500 ng/ml due to binding rate saturation (only the linear portions were plotted). Samples were injected into two channels of a three-channel sensor assembly. A control solution (blood or plasma blank) was injected into the third channel. Duplicate 5-min assays were performed at each concentration point using a single sensor assembly. Analytical sensitivity values of 3.30 and 5.64 ng/ml were determined for the plasma and whole blood standard curves, respectively.

comparable in the two sample matrices and the similar intercepts confirmed that orienting the flow cell so that blood cells settled out of the evanescent field virtually eliminated the systematic rate component observed with CK-MB. Analytical sensitivity values of 3.30 ng/ml (187 pM) and 5.64 ng/ml (319 pM) were determined from the whole blood and plasma curves, respectively.

Myoglobin assay performance was then evaluated in a small-scale clinical correlation study (Figure 12.14). Plasma samples were obtained from 27 anonymous chest pain patients who had previously been screened for myoglobin by ARUP Laboratories (Salt Lake City, UT) using the Stratus myoglobin assay (Dade Behring, Deerfield, IL). Samples with myoglobin levels greater than 500 ng/ml (in the Dade assay) were diluted before injection into the sensor unit. Each sample was measured in duplicate using the three-channel sensor assembly in a 5-min assay. A negative control (tracer antibody, but no antigen) was injected into the

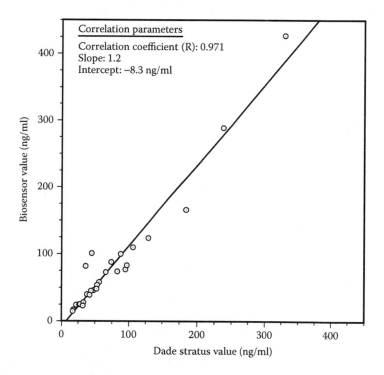

FIGURE 12.14 Clinical validation of the myoglobin assay. Our 5-min biosensor assay for myoglobin was validated in a small-scale clinical trial with plasma samples from 27 anonymous chest pain patients. Samples were obtained from ARUP Laboratories, where they had previously been tested with the Stratus myoglobin assay (Dade International). The figure presents the clinical correlation between the biosensor and reference assays. Good correlation ($R = 0.971$) was observed between the two assays. Slope and intercept values are discussed in the text.

third channel. The two assays were well correlated ($R = 0.971$) with the following correlation function:

$$[\text{Myoglobin}]_{\text{biosensor}} = 1.2[\text{Myoglobin}]_{\text{Stratus}} - 8.3 \qquad (12.15)$$

The slope value (1.2) suggests that our reference material (recombinant human myoglobin; BiosPacific) exhibits 20% greater mass units than Dade's reference material. The intercept's magnitude is somewhat greater than the assay's analytical sensitivity (5.64 ng/ml) and may be statistically significant, but it is probably not clinically relevant in an assay with a normal range of 0 to 85 ng/ml.

12.4.10 CARDIAC TROPONIN SUBUNIT I

The troponin complex contains three components (I, C, T) that help regulate muscle contraction.[97,98] In particular, troponin I inhibits actomyosin's adenosine triphosphate-ase (ATPase) activity, but complexation to troponin C can reverse this inhibition.[97,98] Calcium mediates the affinity of the troponin I–C complex by binding to troponin C's four calcium-binding sites.[97,98] Troponin T attaches the troponin I–C complex to actin–tropomyosin filaments.[97,98] Both troponin I and T have been used as AMI markers, although troponin I assays are the more common of the two, due to Roche's (Basel, Switzerland) exclusive patent position on troponin T. For that reason, we decided to focus assay development efforts on troponin I.

Humans express different isoforms of troponin I in cardiac and skeletal muscle. The cardiac form (29,000 Da)[99] contains a unique 32–amino acid segment at its N-terminus.[100] Thus, strict cardiospecificity may be achieved using an antibody that binds within this segment. Nevertheless, the epitope must be chosen with care because the N-terminal region is susceptible to proteolysis by enzymes released by necrotic cardiac tissue.[99] Fortunately, complexation to troponin C protects a region of troponin I (approximately residue 30 to 110) from proteolysis.[99] This 80–amino acid region provides sufficient steric separation for a two-site sandwich assay, as several different clinically relevant cTnI assays employ antibodies binding epitopes within this region.[99,101,102] In all cases, one epitope spanning residue 30 provides the cardiospecificity.

Lumenal Technologies, LP (Salt Lake City, UT) developed a cTnI assay for our planar waveguide biosensor with an analytical sensitivity of 0.2 ng/ml (6.9 pM) in human plasma. This two-site sandwich assay employed two monoclonal antibodies: Cy5-labeled HT3 (HTI Bio-Products, Ramona, CA; now part of Strategic BioSolutions) as tracer and 8E10 (Hytest, Turku, Finland) as capture. HT3 binds an epitope located within residues 24 to 40 of cTnI, whereas 8E10 binds an epitope spanning residues 87 to 91. Our laboratory evaluated assay performance in a clinical correlation study (Figure 12.15) using the Mark 1.5 evanescent wave imaging fluorometer. Plasma samples were obtained from 175 anonymous chest pain patients previously screened for cTnI by the University of Utah Emergency Department using the AxSYM cTnI assay (Abbott Laboratories). Samples were assayed for 5 min with an injection-molded, disposable three-channel sensor assembly developed by Lumenal. A different sample was injected into each channel, wherein eight independent

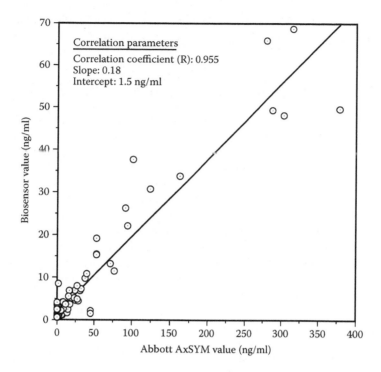

FIGURE 12.15 Clinical validation of the cTnI assay. Lumenal Technologies' 5-minute biosensor assay for cTnI was validated in a small-scale clinical trial with plasma samples from 175 anonymous chest pain patients. Samples were obtained from the University of Utah Emergency Department, where they had previously been tested with the AxSYM cTnI assay (Abbott Laboratories). The figure presents the clinical correlation between the biosensor and reference assays. Good correlation ($R = 0.955$) was observed between the two assays. The low slope value (0.18) suggests a significant difference in cTnI reference materials employed in the two assays (discussed in the text).

cTnI measurements were determined. The Lumenal and Abbott assays were well correlated ($R = 0.955$) with the following correlation function:

$$[cTnI]_{biosensor} = 0.18[cTnI]_{AxSYM} + 1.5 \qquad (12.16)$$

The low slope value (0.18) is due to a 5.5-fold difference in cTnI mass units between Abbott's and Lumenal's reference materials. Lumenal's cTnI assay was developed before the American Association of Clinical Chemistry (AACC) selected a candidate reference material (CRM) for human cardiac troponin I.[103] At that time, each manufacturer defined its own cTnI mass units, with variable results depending on whether their reference material contained pure cTnI, the I–C complex, or the I–C–T complex.[102] Lumenal's reference material was a native I–C–T complex purified from human cardiac tissue (HTI Bio-Products). The AACC later selected a comparable native I–C–T complex from Hytest as their CRM. Thus, Lumenal's mass units are consistent with the AACC's CRM. The intercept value (1.5 ng/ml) is statistically significant and may be due to different calibration standard formulations used in the two assays.

12.5 MOLECULAR DIAGNOSTICS ASSAYS

The literature contains several reports of molecular diagnostics applications of optical biosensor technology. The majority of these were either fiber-optic fluorescent sensors[48,104,105] or label-free SPR sensors,[106–110] but a few articles described other label-free evanescent wave formats, including interferometry,[111] diffractometry,[112] and evanescent-illuminated light scatter.[113,114] Two articles were particularly relevant because they reported clinical applications of evanescent wave biosensors. First, Nilsson et al.[107] detected the presence of the human tumor suppressor p53 gene in breast tumor biopsy material using SPR. These authors also showed that point mutations in clinical DNA specimens gave reduced levels of hybridization relative to the wild-type sequence, thereby demonstrating the utility of evanescent wave biosensor technology for detecting single nucleotide polymorphisms (SNPs). Second, Pilevar et al.[105] detected 25 pM levels of *Helicobacter pylori* RNA using a fiber-optic fluorescent sensor, demonstrating that high-sensitivity molecular diagnostics assays are achievable with evanescent wave fluorescent sensors.

Based on these two studies, we were optimistic that highly sensitive molecular diagnostics assays and SNP detection could both be achieved with planar waveguide biosensors. Nevertheless, our kinetics assay format represents a paradigm shift from conventional hybridization assays, where the extent of hybridization is measured at equilibrium and often as a function of temperature. Clearly, a feasibility study was needed to demonstrate that kinetic measurements could provide the same wealth of information as conventional assays.

12.5.1 FEASIBILITY STUDY

We selected the bacteriophage T3 RNA polymerase promoter site as a model sequence for the feasibility study. This site spans 20 bases (5′-AATTAACCCTCAC-TAAAGGG-3′) in the minus strand of the phage's DNA genome.[115] This sequence is commonly used as a transcription promoter in plasmids for expression vectors and genomic cloning.[116,117] Fluorescently labeled T3 primers are also used in automated nucleic acid sequencing applications.[118,119]

A professional DNA synthesis facility at the University of Utah synthesized a capture oligonucleotide (referred to as T3) with the above promoter sequence and biotinylated it at the 5′ end via a six-carbon spacer. It was immobilized to avidin- or neutravidin-coated waveguides using a self-assembly procedure described by Herron et al.[120] The same facility synthesized a complementary oligonucleotide analyte (referred to as cT3) and 5′-labeled it with Cy5. This oligo mimicked clinical molecular diagnostics analytes prepared by PCR using Cy5-labeled primers.

Figure 12.16 shows biosensor kinetic responses for T3/cT3 hybridization and three additional NSB controls. The same analyte solution (10 nM Cy5-labeled cT3)

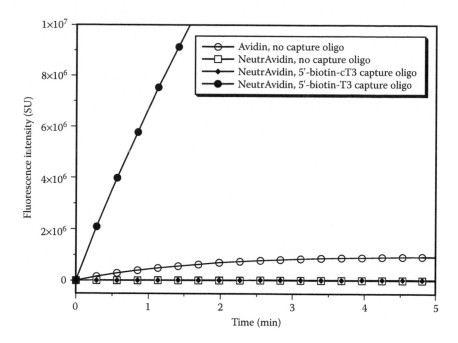

FIGURE 12.16 Immobilization chemistry evaluation for DNA hybridization assays. The bacteriophage T3 RNA polymerase promoter site was chosen as a model sequence for this evaluation. It spans 20 bases with sequence 5'-AATTAACCCTCACTAAAGGG-3'. A professional DNA synthesis facility at the University of Utah synthesized and 5' biotinylated capture oligonucleotide with this sequence (referred to as T3). Immobilization chemistry consisted of T3 self-assembly on avidin- or neutravidin-coated waveguides. The DNA facility also synthesized and 5' labeled (with Cy5) an analyte oligonucleotide with complementary sequence to T3 (referred to as cT3). The following four hybridization assays were performed with the same analyte solution (10 nM Cy5-cT3) but different immobilization chemistries: (open circles) control for NSB to immobilized avidin in the absence of capture oligonucleotide; (open squares) control for NSB to neutravidin in the absence of capture oligonucleotide; (closed diamonds) control for NSB to in the presence of a totally uncomplementary capture oligonucleotide (biotin-cT3); and (closed circles) hybridization to immobilized T3.

was used in all four cases. Each assay was performed for 5 min at room temperature. In the absence of capture oligonucleotide, significant NSB was observed with immobilized avidin (open circles) but not with immobilized neutravidin (open squares). We attribute this observation to avidin's positive charge at pH 7.4 (due to its high isoelectric point) that can lead to electrostatic interactions with negatively charged analyte. Having little net charge at neutral pH, neutravidin significantly reduced NSB and was used in subsequent studies. The third control (analyte binding to immobilized cT3) showed that binding was negligible between oligonucleotides with identical sequences (closed diamonds). Because such sequences are not expected to hybridize, this control is a good assessment of the

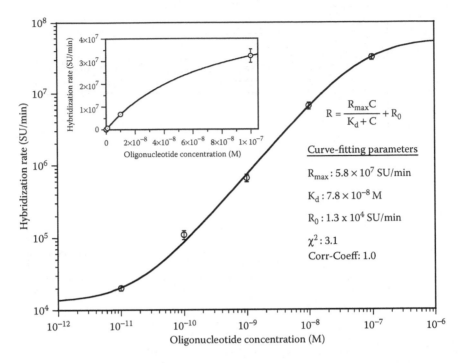

$$R = \frac{R_{max}C}{K_d + C} + R_0$$

Curve-fitting parameters

R_{max} : 5.8×10^7 SU/min

K_d : 7.8×10^{-8} M

R_0 : 1.3×10^4 SU/min

χ^2 : 3.1

Corr-Coeff: 1.0

FIGURE 12.17 Standard curve for the hybridization of Cy5-cT3 to immobilized T3. The sensitivity of our T3/cT3 hybridization assay was investigated by spiking Cy5-labeled cT3 into PBS. Analyte was titrated over a range of 0 M to 1×10^7 M. Each concentration was assayed for 5 min in triplicate using a single sensor assembly. The figure presents the average binding rate and standard deviation of two independent titrations ($n = 6$). The complete data set was analyzed using the rate saturation model shown in the figure. An analytical sensitivity value of 1.4 pM was determined for the assay.

assay's intrinsic NSB. Finally, strong hybridization (reaching 10^7 sensor units in less than 2 min) was observed between the analyte and its complementary capture oligonucleotide (closed circles).

Figure 12.17 shows a standard curve of T3/cT3 hybridization rate over an analyte concentration range of 0 to 100 nM. Each assay was performed for 5 min at room temperature. The wide dynamic range of the assay (four orders of magnitude) necessitated use of a double log plot (low-concentration data are plotted on linear axes in the inset). A saturation model with an apparent K_d of 78 nM fit the data. Assay analytical sensitivity was 1.4 pM.

Results presented in this section clearly demonstrate the utility of kinetic measurements for monitoring DNA hybridization. They also showed that rapid assays with low picomolar sensitivity are feasible using planar waveguide biosensor technology. Nevertheless, assay selectivity remained an open question. Specifically, how well do kinetic measurements in a biosensor format discriminate mismatched base pairs?

Such discrimination forms the basis of many molecular diagnostics assays, particularly ones detecting genetic polymorphism. This question is examined in the next two sections.

12.5.2 PROSTATE-SPECIFIC ANTIGEN GENE EXPRESSION ASSAY

We next developed a gene expression assay for prostate-specific antigen (PSA), suitable for measuring complementary DNA (cDNA) produced by reverse transcription polymerase chain reaction (RT-PCR). A high degree of homology exists between PSA and human glandular kallikrein (hGK2), which is also secreted by the prostate gland.[121–123] For this reason, it is important that the capture oligonucleotide recognizes cDNA derived only from the PSA message and not from the hGK2 message. We identified a region in exon 4 of the PSA gene (5′-CGAGCAG-GTGCTTTTGCCCC-3′) where the sequence differs from hGK2 in 7 of 20 positions (differences underlined). A complementary 5′ biotinylated capture oligonucleotide (5′-GGGGCAAAAGCACCTGCTCG-3′, referred to as complementary PSA [cPSA]) was synthesized by the aforementioned DNA synthesis facility. Two model cDNA analytes were also synthesized and labeled with Cy5. One of these (referred to as PSA) was identical to the exon 4 sequence given above, whereas the other (5′-CCAxCAAGTGTCTTTACCAC-3′, referred to as hGK2; "x" is a deletion relative to the PSA sequence) was derived from the comparable region of the hGK2 gene. Biotinylated cPSA was immobilized to a neutravidin-coated waveguide and a 1-nM solution of Cy5-labeled analyte (either PSA or hGK2) was injected into the flow cell. Hybridization kinetics curves are shown in Figure 12.18 for both of these reactions. The homoduplex (PSA/cPSA) gave a very high hybridization rate, whereas the heteroduplex's (hGK2/cPSA) rate did not differ statistically from background. Thus, the presence of six mismatched bases and one deletion totally abrogated duplex formation between hGK2 and cPSA, giving us hope that a heteroduplex containing a single mismatched base might also be distinguishable from the homoduplex. Detection of single mismatched bases is examined in the next section.

12.5.3 DETECTION OF SNPs IN LONG QT SYNDROME

The full elucidation of the human genome has spawned intense investigation of the genetic basis of disease. Genetic polymorphisms predispose certain individuals to specific types of disease, and SNPs account for more than 90% of known human genetic variation.[124] Thus, detection of a single mismatched base within a nucleic acid analyte is key to genetic screening. Our group recently demonstrated that planar waveguide biosensors are suitable for detecting single mismatched bases.[51,52] In particular, we developed genetic screening assays for determining whether an individual carries an SNP strongly correlated with long QT syndrome (LQTS).

Long QT syndrome is an electrical repolarization disorder of the heart characterized by a prolonged QT interval in the action potential. Affected individuals occasionally develop an acute polymorphic ventricular tachycardia called *torsade*

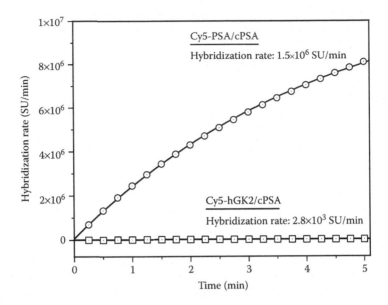

FIGURE 12.18 Hybridization assay for PSA cDNA. The prostate secretes both PSA and hGK2, which are highly homogenous. A gene expression assay was developed to distinguish the two at the molecular level (see text for details). The figure presents hybridization kinetics data for two assays. The first is hybridization of a Cy5-labeled model analyte oligonucleotide derived from the PSA gene sequence (Cy5-PSA) to a complementary capture oligonucleotide (cPSA). The second is hybridization of a Cy5-labeled model analyte derived from the hGK2 gene sequence (Cy5-hGK2). Analyte concentration was 1 nM in both experiments. Being perfectly complementary, Cy5-PSA and cPSA exhibited a high hybridization rate (1.5×10^6 SU/min), whereas Cy5-hGK2 hybridization with cPSA was much lower (2800 SU/min.). The latter rate was not significantly different from the NSB rate of this assay (513 ± 1416 SU/min).

de pointes that usually results in syncope, but 10% to 30% of cases degenerate into ventricular fibrillation and sudden death.[125] Either emotional stress or exercise may trigger these events. The disease causes 3000 to 4000 sudden deaths each year in the United States, mostly in children and young adults.[125] Presymptomatic diagnosis of LQTS is essential because sudden death may result from the first incident of syncope.[125] Thus, there is a compelling need for a rapid and inexpensive assay method for screening family members and other close relatives of an affected individual. Traditional diagnostic methods (e.g., electrocardiogram) are equivocal in about 40% of the cases because the affected individual exhibits either a normal or a borderline prolonged QT interval.[125] In many instances the affected individual's genotype is already known, making it desirable to know whether other family members carry this same mutation so that they can receive medical treatment (pre-symptomatic patients are often treated with beta blockers).[125]

Long QT syndrome inheritance is autosomal dominant and chromosomal linkage studies have shown that this form of the disease is linked to at least five different loci.[125] One of these (the *KVLQT1* gene located on chromosome 11) is responsible for about 50% of the LQTS cases that have been genotyped to date.[126] It encodes for a subunit of the cardiac I_{Ks} potassium channel.[125] We chose a common SNP (G760A) in the *KVLQT1* gene as a model system for developing rate-based assays for detecting this polymorphism. Two different approaches were examined, one based on differential melting temperature (T_m)[51] and the other based on enzymatic incorporation of fluorescently labeled dideoxynucleotides (single base extension).[52] Each of these is described below.

12.5.4 DIFFERENTIAL MELTING TEMPERATURE

We saw for PSA/hGK2 that six mismatches and one deletion in a 20-mer totally abrogated duplex formation at room temperature. Thus, duplex formation cannot tolerate more than a few mismatched bases in a given helical stretch. This begged the question of whether duplex formation for nearly complementary strands containing a single mismatched base could be totally abrogated at higher temperature. To investigate this question, we designed a capture oligonucleotide complementary to positions 750–770 of the wild-type *KVLQT1* gene (denoted as C_{wt}) and two Cy5-labeled model analyte oligonucleotides—one identical to positions 750–770 of the wild-type sequence (denoted as A_{wt}) and another identical to positions 750–770 of the G760A SNP sequence (denoted as A_{SNP}). All three oligonucleotides were synthesized by the aforementioned DNA synthesis facility. Biosensor kinetic measurements were used to determine the melting temperature (T_m) of both the perfectly matched C_{wt}/A_{wt} duplex and the C_{wt}/A_{SNP} duplex containing a single mismatched base. Both T_m values were in the range of 48°C to 55°C, with the T_m of the perfectly matched duplex being the higher of the two.[51] Because T_m also depends on monovalent and divalent cation concentration, we optimized Na^+, K^+, and Mg^{2+} concentrations to maximize the difference in melting temperature (ΔT_m) between the perfectly matched and single mismatched duplexes. Figure 12.19 shows the results of this optimization. The maximum ΔT_m (2.5°C) occurred at cation concentrations of 10 mM Na^+, 1 M K^+, and 1 mM Mg^{2+}. At these concentrations, the T_m values for the perfectly matched and single mismatched duplexes were 50.5°C and 48°C, respectively. Total abrogation of duplex formation for the single base mismatch could be achieved within this temperature range. Nevertheless, running hybridization reactions at 50°C, with only a 2.5°C tolerance in reaction temperature, presents challenges for practical deployment of such assays in POC environments. For that reason, we decided to investigate single base extension, an alternative method for identifying SNPs.

12.5.5 SINGLE BASE EXTENSION

Single base extension is routinely used in genotyping and SNP detection.[127,128] In this technique, a duplex forms between a single-stranded DNA analyte and a probe oligonucleotide. The 3′ end of the probe is immediately adjacent to the position of the suspected SNP in the analyte. DNA polymerase then incorporates a single base

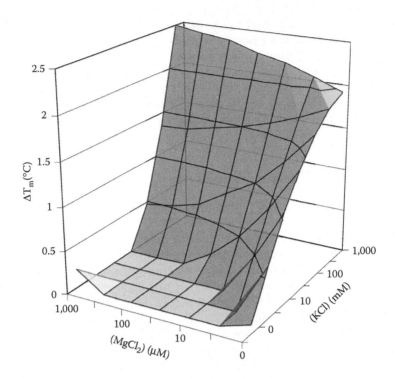

FIGURE 12.19 Topographic map depicting the difference in melting points (in °C) between a homoduplex and a heteroduplex with one base pair mismatch. The $MgCl_2$ and KCl concentrations were varied, whereas the NaCl concentration was held fixed at 10 mM.

(a labeled dideoxynucleotide) complementary to the suspected SNP. The polymerase ensures that this reaction proceeds with high fidelity. The label allows specific identification ("calling") of the incorporated base. ThermoSequenase polymerase I (Tpol-I; Amersham Biosciences, Piscataway, NJ) is the enzyme of choice because it incorporates fluorescently labeled dideoxynucleotides with high specificity and activity.[129]

Recently we reported the use of single base extension in conjunction with planar waveguide biosensors to identify the aforementioned G760A SNP.[52] In this assay we immobilized a 5′ biotinylated probe oligonucleotide to neutravidin-coated planar waveguide sensors. The probe's sequence was complementary to positions 761–781 of the *KVLQT1* gene. Four different model analyte oligonucleotides spanning positions 757–785 of the *KVLQT1* gene were also synthesized, each with a different nucleotide (A, T, G, or C) at position 760. Upon hybridization of the analyte to the immobilized probe, the base at position 760 serves as Tpol-I's template for adding a complementary Cy5-labeled dideoxynucleotide to the 3′ end of the probe oligonucleotide. Ligation of this dideoxynucleotide to the immobilized probe brings the Cy5 label within the evanescent field, where it fluoresces.

The single base extension reaction was initiated by injecting a mixture of analyte, Tpol-I, and a single Cy5-labeled dideoxynucleotide (either ddA, ddT,

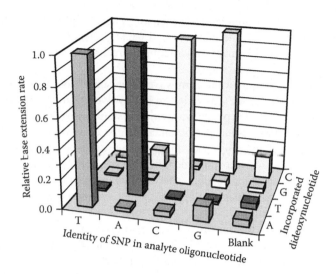

FIGURE 12.20 Validation of the single base extension rate assay for identifying point mutations. See the text for details.

ddG, or ddC) simultaneously into all three channels of a sensor assembly, allowing reaction kinetics to be monitored in triplicate. The reaction was carried out for 5 min at 40°C in 10 mM Tris buffer (pH 8.5) with 10 mM MgCl$_2$. The magnesium was required for Tpol-I activity.[52] Figure 12.20 shows a three-dimensional plot of single base extension reaction rates for the 16 possible permutations of the four analytes with four Cy5-labeled dideoxynucleotides as well as blank measurements for each dideoxynucleotide in the absence of analyte. The results indicated that only the dideoxynucleotide complementary to analyte position 760 was incorporated at an appreciable rate, although slight cross-reactivity was observed between the wild-type sequence (G at position 760) and ddA. They also confirmed our hypothesis that kinetic and equilibrium measurements provide comparable information, but the former are faster and better suited for biosensor applications. Moreover, this biosensor-based single base extension assay is applicable to base-calling reactions in general.

12.6 CONCLUSION

Nearly four decades have passed since Hirschfeld[21] first described TIRF, yet the first generation of evanescent wave fluorescence sensors is just emerging in the IVD market. There are probably several reasons for this. For one thing, solid phase immunoassays didn't become routine until the late 1970s, and the emphasis in research and development (R&D) over the next decade was on high-volume assays (e.g., enzyme-linked immunoassays run in microtiter plates) for use in clinical laboratories. In fact, a compelling need for biosensors in the IVD market arose only after POC testing became a reality in the 1990s. Another limiting factor has been the technology itself. For example, our biosensor system employs technologies such as red-emitting diode lasers, thermally cooled CCD cameras, optical-quality injection molding, and high-performance microcomputers that were either unavailable or prohibitively expensive 10 years ago.

Our group has worked on evanescent wave fluorescent biosensors for more than 15 years. During this time, our sensor design has evolved from single-bounce TIRF to the present injection-molded planar waveguide system, with concomitant improvements in system capabilities and performance. In the early days it took nearly an hour to assemble and align each sensor unit. Intersensor variability was so high that an entire titration was run on a single device. The present injection-molded planar waveguide sensor solved this problem, enabling development of clinical assays with sensitivity and precision values rivaling those of commercial immunoassays performed in clinical laboratories.

Several factors affect the sensitivity of a given optical biosensor system. First is the intrinsic light-gathering power of the sensor itself, which determines the ultimate sensitivity of the system, all other factors being optimal. Second is the registration or alignment of the sensor in the detection system. This can be a significant source of intersensor variability in cases in which a disposable sensor unit is used with a fixed detection system. Third is the statistical precision of the measurements—increasing the number of determinations is an inexpensive way to improve sensitivity, but at the expense of assay time. Finally, the antibodies themselves are an important determinant of sensitivity. For example, the sensitivity of our cTnI assay is 10-fold better than that of our CK-MB assay and the only significant differences between the two are the capture and tracer antibodies. Thus, it is worthwhile to screen a number of different antibody pairs to find the ones that give optimal sensitivity.

Planar waveguide fluorescent biosensors also possess ideal attributes (e.g., high sensitivity, rapid detection, and excellent selectivity) for molecular diagnostics assays. For instance, the 1.4-pM sensitivity observed for the T3/cT3 system compares very favorably to the 25-pM sensitivity reported by Pilevar et al.[105] for *H. pylori* RNA detection using fluorescence fiber-optic sensor technology. This higher sensitivity is due in part to our optical geometry, where fluorescence is excited by evanescent coupling and collected to the side of the waveguide through free space. Such free space collection is more efficient than the evanescent collection employed in fiber-optic sensors because only a small fraction (less than 10%) of emitted fluorescence couples evanescently back into the sensor.[64] Our biosensor's intrinsic ability to obtain kinetic measurements offers unique advantages over conventional technologies based on equilibrium measurements. Reactions such as hybridization or single base extension may be monitored in real time, enabling development of rapid molecular diagnostics assays for POC environments. Rapid kinetic measurements also lend themselves to high-throughput screening applications in gene expression, genomics, and SNP detection. Finally, the ability to detect single mismatched bases positions our biosensor technology for future generations of clinical POC assays for disease-associated SNPs, with far-reaching potential applications in both pharmacogenomics and personalized medicine.

ACKNOWLEDGMENTS

The cTnI assay was developed by Dr. Brent Burdick, Michael Mostert, and Cecily Vaughn, all formerly of Lumenal Technologies, LP (Salt Lake City, UT). We gratefully acknowledge support from the National Institutes of Health (grant HL 32132), the University of Utah Department of Pharmaceutics Summer Undergraduate Fellowship Program, and Lumenal Technologies, LP. Some of the technology reported herein is subject to patent applications assigned to the University of Utah Research Foundation and licensed to BioCentrex, LLC (Culver City, CA).

REFERENCES

1. Halasey, S. and Park, R., Making sense of the global IVD market. *IVD Technol,* September, 29–32, 2002.
2. Woo, J. and Henry, J.B., The advance of technology as a prelude to the laboratory of the twenty-first century. *Clin Lab Med,* 14, 459–471, 1994.
3. Wilkins, E. and Atanasov, P., Glucose monitoring: state of the art and future possibilities. *Med Eng Phys,* 18, 273–288, 1996.
4. Kost, G.J., Vu, H.T., Lee, J.H., Bourgeois, P., Kiechle, F.L., Martin, C., Miller, S.S., Okorodudu, A.O., Podczasy, J.J., Webster, R., and Whitlow, K.J., Multicenter study of oxygen-insensitive handheld glucose point-of-care testing in critical care/hospital/ ambulatory patients in the United States and Canada. *Crit Care Med,* 26, 581–590, 1998.
5. Bingham, D., Kendall, J., and Clancy, M., The portable laboratory: an evaluation of the accuracy and reproducibility of i-STAT. *Ann Clin Biochem,* 36(pt 1), 66–71, 1999.
6. Schaffar, B.P., Thick film biosensors for metabolites in undiluted whole blood and plasma samples. *Anal Bioanal Chem,* 372, 254–260, 2002.
7. Mendelson, Y., Pulse oximetry: theory and applications for noninvasive monitoring. *Clin Chem,* 38, 1601–1607, 1992.
8. Chapman, D.C., Stephens, M.A., Hamann, G.L., Bailey, L.E., and Dorko, C.S., Accuracy, clinical correlation, and patient acceptance of two handheld prothrombin time monitoring devices in the ambulatory setting. *Ann Pharmacother,* 33, 775–780, 1999.
9. Wong, C.M., van Dijk, P.J., and Laing, I.A., A comparison of transcutaneous bilirubinometers: SpectRx BiliCheck versus Minolta AirShields. *Arch Dis Child Fetal Neonatal Ed,* 87, F137–F140, 2002.
10. Apple, F.S., Christenson, R.H., Valdes, R., Jr., Andriak, A.J., Berg, A., Duh, S.H., Feng, Y.J., Jortani, S.A., Johnson, N.A., Koplen, B., Mascotti, K., and Wu, A.H., Simultaneous rapid measurement of whole blood myoglobin, creatine kinase MB, and cardiac troponin I by the triage cardiac panel for detection of myocardial infarction. *Clin Chem,* 45, 199–205, 1999.
11. Males, R.G., Stephenson, J., and Harris, P., Cardiac markers and point-of-care testing: a perfect fit. *Crit Care Nurs Q,* 24, 54–61, 2001.
12. Azzazy, H.M. and Christenson, R.H., Cardiac markers of acute coronary syndromes: is there a case for point-of-care testing? *Clin Biochem,* 35, 13–27, 2002.
13. McCullough, P.A., Omland, T., and Maisel, A.S., B-type natriuretic peptides: a diagnostic breakthrough for clinicians. *Rev Cardiovasc Med,* 4(2), 72–80, 2003.

14. Tarkkinen, P., Palenius, T., and Lovgren, T., Ultrarapid, ultrasensitive one-step kinetic immunoassay for C-reactive protein (CRP) in whole blood samples: measurement of the entire CRP concentration range with a single sample dilution. *Clin Chem,* 48, 269–277, 2002.

15. Abrams, J., C-reactive protein, inflammation, and coronary risk: an update. *Cardiol Clin,* 21, 327–331, 2003.

16. Libby, P. and Ridker, P.M., Inflammation and atherosclerosis: role of C-reactive protein in risk assessment. *Am J Med,* 116(suppl 6A), 9S–16S, 2004.

17. Collinson, P.O., Boa, F.G., and Gaze, D.C., Measurement of cardiac troponins. *Ann Clin Biochem,* 38(pt 5), 423–449, 2001.

18. Rogers, K.R., Principles of affinity-based biosensors. *Mol Biotechnol,* 14, 109–129, 2000.

19. Malmqvist, M. and Karlsson, R., Biomolecular interaction analysis: affinity biosensor technologies for functional analysis of proteins. *Curr Opin Chem Biol,* 1, 378–383, 1997.

20. Fivash, M., Towler, E.M., and Fisher, R.J., BIAcore for macromolecular interaction. *Curr Opin Biotechnol,* 9, 97–101, 1998.

21. Hirschfeld, T.B., Total reflection fluorescence (TRF). *Can Spectrosc,* 10, 128, 1965.

22. Axelrod, D., Burghardt, T.P., and Thompson, N.L., Total internal reflection fluorescence. *Annu Rev Biophys Bioeng,* 13, 247–268, 1984.

23. Thompson, N.L. and Lagerholm, B.C., Total internal reflection fluorescence: applications in cellular biophysics. *Curr Opin Biotechnol,* 8, 58–64, 1997.

24. Kronick, M.N. and Little, W.A., A new immunoassay based on fluorescence excitation by internal reflection spectroscopy. *J Immunol Methods,* 8, 235–240, 1975.

25. Ogert, R.A., Brown, J.E., Singh, B.R., Shriver-Lake, L.C., and Ligler, F.S., Detection of *Clostridium botulinum* toxin A using a fiber optic-based biosensor. *Anal Biochem,* 205, 306–312, 1992.

26. Cao, L.K., Anderson, G.P., Ligler, F.S., and Ezzell, J., Detection of *Yersinia pestis* fraction 1 antigen with a fiber optic biosensor. *J Clin Microbiol,* 33, 336–341, 1995.

27. Ligler, F.S., Taitt, C.R., Shriver-Lake, L.C., Sapsford, K.E., Shubin, Y., and Golden, J.P., Array biosensor for detection of toxins. *Anal Bioanal Chem,* 377, 469–477, 2003.

28. Sapsford, K.E., Shubin, Y.S., Delehanty, J.B., Golden, J.P., Taitt, C.R., Shriver-Lake, L.C., and Ligler, F.S., Fluorescence-based array biosensors for detection of biohazards. *J Appl Microbiol,* 96, 47–58, 2004.

29. Devine, P.J., Anis, N.A., Wright, J., Kim, S., Eldefrawi, A.T., and Eldefrawi, M.E., A fiber-optic cocaine biosensor. *Anal Biochem,* 227, 216–224, 1995.

30. Toppozada, A.R., Wright, J., Eldefrawi, A.T., Eldefrawi, M.E., Johnson, E.L., Emche, S.D., and Helling, C.S., Evaluation of a fiber optic immunosensor for quantitating cocaine in coca leaf extracts. *Biosens Bioelectron,* 12, 113–124, 1997.

31. Nath, N., Eldefrawi, M., Wright, J., Darwin, D., and Huestis, M., A rapid reusable fiber optic biosensor for detecting cocaine metabolites in urine. *J Anal Toxicol,* 23, 460–467, 1999.

32. Rogers, K.R., Cao, C.J., Valdes, J.J., Eldefrawi, A.T., and Eldefrawi, M.E., Acetylcholinesterase fiber-optic biosensor for detection of anticholinesterases. *Fundam Appl Toxicol,* 16, 810–820, 1991.

33. Fierke, C.A. and Thompson, R.B., Fluorescence-based biosensing of zinc using carbonic anhydrase. *Biometals,* 14, 205–222, 2001.

34. Shriver-Lake, L.C., Charles, P.T., and Kusterbeck, A.W., Non-aerosol detection of explosives with a continuous flow immunosensor. *Anal Bioanal Chem,* 377, 550–555, 2003.

35. DeMarco, D.R., Saaski, E.W., McCrae, D.A., and Lim, D.V., Rapid detection of *Escherichia coli* O157:H7 in ground beef using a fiber-optic biosensor. *J Food Prot,* 62, 711–716, 1999.

36. Maragos, C.M. and Thompson, V.S., Fiber-optic immunosensor for mycotoxins. *Nat Toxins,* 7, 371–376, 1999.

37. Demarco, D.R. and Lim, D.V., Detection of *Escherichia coli* O157:H7 in 10- and 25-gram ground beef samples with an evanescent-wave biosensor with silica and polystyrene waveguides. *J Food Prot,* 65, 596–602, 2002.

38. Lu, J. and Rosenzweig, Z., Nanoscale fluorescent sensors for intracellular analysis. *Fresenius J Anal Chem,* 366, 569–575, 2000.

39. Vo-Dinh, T., Nanobiosensors: probing the sanctuary of individual living cells. *J Cell Biochem Suppl,* 39, 154–161, 2002.

40. Hlady, V., Lin, J.N., and Andrade, J.D., Spatially resolved detection of antibody–antigen reaction on solid/liquid interface using total internal reflection excited antigen fluorescence and charge-coupled device detection. *Biosens Bioelectron,* 5, 291–301, 1990.

41. Herron, J.N., Christensen, D.A., Hlady, V., Janatova, V., Wang, H.-K., and Wei, A.-P., Fluorescent immunosensors using planar waveguides. In: *Advances in Fluorescence Sensing Technology,* Lakowicz, J.R. and Thompson, R.B., eds. *Proc SPIE,* 4252, 28–39, 1993.

42. Nath, N., Jain, S.R., and Anand, S., Evanescent wave fibre optic sensor for detection of *L. donovani* specific antibodies in sera of kala azar patients. *Biosens Bioelectron,* 12, 491–498, 1997.

43. Herron, J.N., Wang, H.-K., Terry, A.H., Durtschi, J.D., Tan, L., Astill, M.E., Smith, R.S., and Christensen, D.A., Rapid clinical diagnostics assays using injection-molded planar waveguides. In: *Systems and Technologies for Clinical Diagnostics and Drug Discovery,* Cohn, G.E., ed. *Proc SPIE,* 3259, 54–64, 1998.

44. Schult, K., Katerkamp, A., Trau, D., Grawe, F., Cammann, K., and Meusel, M., Disposable optical sensor chip for medical diagnostics: new ways in bioanalysis. *Anal Chem,* 71, 5430–5435, 1999.

45. Plowman, T.E., Durstchi, J.D., Wang, H.K., Christensen, D.A., Herron, J.N., and Reichert, W.M., Multiple-analyte fluoroimmunoassay using an integrated optical waveguide sensor. *Anal Chem,* 71, 4344–4352, 1999.

46. Garden, S.R., Doellgast, G.J., Killham, K.S., and Strachan, N.J., A fluorescent coagulation assay for thrombin using a fibre optic evanescent wave sensor. *Biosens Bioelectron,* 19, 737–740, 2004.

47. Piunno, P.A., Krull, U.J., Hudson, R.H., Damha, M.J., and Cohen, H., Fiber-optic DNA sensor for fluorometric nucleic acid determination. *Anal Chem,* 67, 2635–2643, 1995.

48. Abel, A.P., Weller, M.G., Duveneck, G.L., Ehrat, M., and Widmer, H.M., Fiber-optic evanescent wave biosensor for the detection of oligonucleotides. *Anal Chem,* 68, 2905–2912, 1996.

49. Herron, J.N., zumBrunnen, S., Wang, J.-X., Gao, X-L., Wang, H.-K., Terry, A.H., and Christensen, D.A., Planar waveguide biosensors for nucleic acid hybridization reactions. In: *In-Vitro Diagnostics Instrumentation,* Cohn, G.E., ed. *Proc SPIE,* 3913, 177–184, 2000.

50. Lehr, H.P., Reimann, M., Brandenburg, A., Sulz, G., and Klapproth, H., Real-time detection of nucleic acid interactions by total internal reflection fluorescence. *Anal Chem,* 75, 2414–2420, 2003.

51. Tolley, S.E., Wang, H.K., Smith, R.S., Christensen, D.A., and Herron, J.N., Single-chain polymorphism analysis in long QT syndrome using planar waveguide fluorescent biosensors. *Anal Biochem,* 315, 223–237, 2003.

52. Tolley, S.E., Long QT Syndrome G 760A Single Nucleotide Polymorphism Detection Using Planar Waveguide Technology, Ph.D. Dissertation, University of Utah, 2005.
53. Watterson, J.H., Raha, S., Kotoris, C.C., Wust, C.C., Gharabaghi, F., Jantzi, S.C., Haynes, N.K., Gendron, N.H., Krull, U.J., Mackenzie, A.E., and Piunno, P.A., Rapid detection of single nucleotide polymorphisms associated with spinal muscular atrophy by use of a reusable fibre-optic biosensor. *Nucleic Acids Res,* 32, e18, 2004.
54. Pandey, P.C. and Weetall, H.H., Evanescent fluorobiosensor for the detection of polyaromatic hydrocarbon based on DNA intercalation. *Appl Biochem Biotechnol,* 55, 87–94, 1995.
55. Hovius, R., Schmid, E.L., Tairi, A.P., Blasey, H., Bernard, A.R., Lundstrom, K., and Vogel, H., Fluorescence techniques for fundamental and applied studies of membrane protein receptors: the 5-HT3 serotonin receptor. *J Recept Signal Transduct Res,* 19, 533–545, 1999.
56. Wolfbeis, O.S., Fiber-optic chemical sensors and biosensors. *Anal Chem,* 74, 2663–2677, 2002.
57. Wadkins, R.M., Golden, J.P., and Ligler, F.S., Calibration of biosensor response using simultaneous evanescent wave excitation of cyanine-labeled capture antibodies and antigens. *Anal Biochem,* 232, 73–78, 1995.
58. Michael, K.L., Taylor, L.C., Schultz, S.L., and Walt, D.R., Randomly ordered addressable high-density optical sensor arrays. *Anal Chem,* 70, 1242–1248, 1998.
59. Anderson, G.P., King, K.D., Gaffney, K.L., and Johnson, L.H., Multi-analyte interrogation using the fiber optic biosensor. *Biosens Bioelectron,* 14, 771–777, 2000.
60. Ives, J.T., Reichert, W.M., Lin, J.-N., Hlady, V., Reinecke, D., Suci, P.A., Van Wagenen, R.A., Newby, K., Herron, J., Dryden, P., and Andrade, J.D., Total internal reflection fluorescence surface sensors. In: *Optical Fiber Sensors,* Chester, A.N., ed. Dordrecht: M. Nijhoff, 1987:391–397.
61. Weinberger, S.R., Morris, T.S., and Pawlak, M., Recent trends in protein biochip technology. *Pharmacogenomics,* 1, 395–416, 2000.
62. Pawlak, M., Schick, E., Bopp, M.A., Schneider, M.J., Oroszlan, P., and Ehrat, M. Zeptosens' protein microarrays: a novel high performance microarray platform for low abundance protein analysis. *Proteomics,* 2, 383–393, 2002.
63. Plowman, T.E., Reichert, W.M., Peters, C.R., Wang, H.-K., Christensen, D.A., and Herron, J.N., Femtomolar sensitivity using a channel-etched thin film waveguide fluoroimmunosensor. *Biosens Bioelectron,* 11, 149–160, 1996.
64. Christensen, D., Dyer, S., Fowers, D., and Herron, J., Analysis of excitation and collection geometries for planar waveguide immunosensors. In: *Novel FIA Chemiluminescence Fiber Optic Biosensor for Urinary and Blood Glucose,* Cattaneo, M.V. and Luong, J.H., eds. *Proc SPIE,* 1886, 2–8, 1993.
65. Christensen, D.A. and Herron, J.N., Optical immunoassay systems based upon evanescent wave interactions. In: *Laser Desorption Mass Spectrometry for Molecular Diagnosis,* Chen, C.H.W., Taranenko, N.I., Zhu, Y.F., Allman, S.L., Tang, K., Matteson, K.J., Chang, L.Y., Chung, C.N., Martin, S., and Haff, L., eds. *Proc SPIE,* 2680, 58–67, 1996.
66. Wadkins, R.M., Golden, J.P., Pritsiolas, L.M., and Ligler, F.S., Detection of multiple toxic agents using a planar array immunosensor. *Biosens Bioelectron,* 13, 407–415, 1998.
67. Rowe, C.A., Scruggs, S.B., Feldstein, M.J., Golden, J.P., and Ligler, F.S., An array immunosensor for simultaneous detection of clinical analytes. *Anal Chem,* 71, 433–439, 1999.

68. Rowe, C.A., Tender, L.M., Feldstein, M.J., Golden, J.P., Scruggs, S.B., MacCraith, B.D., Cras, J.J., and Ligler, F.S., Array biosensor for simultaneous identification of bacterial, viral, and protein analytes. *Anal Chem*, 71, 3846–3852, 1999.

69. Rowe-Taitt, C.A., Golden, J.P., Feldstein, M.J., Cras, J.J., Hoffman, K.E., and Ligler, F.S., Array biosensor for detection of biohazards. *Biosens Bioelectron*, 14, 785–794, 2000.

70. Rowe-Taitt, C.A., Hazzard, J.W., Hoffman, K.E., Cras, J.J., Golden, J.P., and Ligler, F.S., Simultaneous detection of six biohazardous agents using a planar waveguide array biosensor. *Biosens Bioelectron*, 15, 579–589, 2000.

71. Sapsford, K.E., Charles, P.T., Patterson, C.H., Jr., and Ligler, F.S., Demonstration of four immunoassay formats using the array biosensor. *Anal Chem*, 74, 1061–1068, 2002.

72. Taitt, C.R., Anderson, G.P., Lingerfelt, B.M., Feldstein, M.J., and Ligler, F.S., Nine-analyte detection using an array-based biosensor. *Anal Chem*, 74, 6114–6120, 2002.

73. Christensen, D.A., Herron, J.N., and Simon, E.M., Waveguide lens. U.S. patent no. 426,783, June 20, 2000.

74. Herron, J.N., Christensen, D.A., Pollak, V.A., McEachern, R.D., and Simon, E.M., Lens and associatable flow cell. U.S. patent no. 6,108,463, August 22, 2000.

75. Herron, J.N., Christensen, D.A., Pollak, V.A., McEachern, R.D., Simon, E.M., Lens and associatable flow cell. U.S. patent no. 6,356,676, March 12, 2002.

76. Andrade, J.D., Herron, J., Lin, J.-N., Yen, H., Kopecek, J., and Kopeckova, P., On-line sensors for coagulation proteins: concept and progress report. *Biomaterials*, 9, 76–79, 1988.

77. Herron, J.N., Christensen, D.A., Caldwell, K.D., Janatová, V., Huang, S.-C., and Wang, H.-K., Waveguide immunosensor with coating chemistry providing enhanced sensitivity. U.S. patent no. 5,512,492, April 30, 1996.

78. Herron, J.N., Christensen, D.A., Wang, H.-K., Caldwell, K.D., Janatova, V., and Huang, S.-C., Apparatus and methods for multi-analyte homogenous fluoro-immunoassays. U.S. patent no. 5,677,196, October 14, 1997.

79. Huang, S.-C., Caldwell, K.D., Lin, J.-N., Wang, H.-K., and Herron, J.N., Site-specific immobilization of monoclonal antibodies using spacer-mediated antibody attachment. *Langmuir*, 12, 4292–4298, 1996.

80. Chard, T., Pregnancy tests: a review. *Hum Reprod*, 7, 701–710, 1992.

81. Jaffe, A.S., New standard for the diagnosis of acute myocardial infarction. *Cardiol Rev*, 9, 318–322, 2001.

82. Plebani, M., Biochemical markers of cardiac damage: from efficiency to effectiveness. *Clin Chim Acta*, 311, 3–7, 2001.

83. Olukoga, A. and Donaldson, D., An overview of biochemical markers in acute coronary syndromes. *J R Soc Health*, 121, 103–106, 2001.

84. Karras, D.J. and Kane, D.L., Serum markers in the emergency department diagnosis of acute myocardial infarction. *Emerg Med Clin North Am*, 19, 321–337, 2001.

85. Watnick, S. and Perazella, M.A., Cardiac troponins: utility in renal insufficiency and end-stage renal disease. *Semin Dial*, 15, 66–70, 2002.

86. Shave, R., Dawson, E., Whyte, G., George, K., Ball, D., Collinson, P., and Gaze, D., The cardiospecificity of the third-generation cTnT assay after exercise-induced muscle damage. *Med Sci Sports Exerc*, 34, 651–654, 2002.

87. Guest, T.M., Ramanathan, A.V., Tuteur, P.G., Schechtman, K.B., Ladenson, J.H., and Jaffe, A.S., Myocardial injury in critically ill patients. A frequently unrecognized complication. *JAMA*, 273, 1945–1949, 1995.

88. Antman, E.M., Tanasijevic, M.J., Thompson, B., Schactman, M., McCabe, C.H., Cannon, C.P., Fischer, G.A., Fung, A.Y., Thompson, C., Wybenga, D., and Braunwald, E., Cardiac-specific troponin I levels to predict the risk of mortality in patients with acute coronary syndromes. *N Engl J Med,* 335, 1342–1349, 1996.
89. Myoglobin - serum. In: MedlinePlus Medical Encyclopedia; www.nlm.nih.gov/medlineplus/ency/article/003663.htm (accessed April 2, 2004).
90. Wood, T.D., Chen, L.H., White, C.B., Babbitt, P.C., Kenyon, G.L., and McLafferty, F.W., Sequence verification of human creatine kinase (43 kDa) isozymes by high-resolution tandem mass spectrometry. *Proc Natl Acad Sci USA,* 92, 11451–11455, 1995.
91. Apple, F.S., The specificity of biochemical markers of cardiac damage: a problem solved. *Clin Chem Lab Med,* 37, 1085–1089, 1999.
92. Apple, F.S., Tissue specificity of cardiac troponin I, cardiac troponin T and creatine kinase-MB. *Clin Chim Acta,* 284, 151–159, 1999.
93. Vaidya, H.C., Maynard, Y., Dietzler, D.N., and Ladenson, J.H., Direct measurement of creatine kinase-MB activity in serum after extraction with a monoclonal antibody specific to the MB isoenzyme. *Clin Chem,* 32, 657–663, 1986.
94. Schluter, T., Baum, H., Plewan, A., and Neumeier, D., Effects of implantable cardioverter defibrillator implantation and shock application on biochemical markers of myocardial damage. *Clin Chem,* 47, 459–463, 2001.
95. Witting, P.K., Douglas, D.J., and Mauk, A.G., Reaction of human myoglobin and H_2O_2. Involvement of a thiyl radical produced at cysteine 110. *J Biol Chem,* 275, 20391–20398, 2000.
96. Swaanenburg, J.C., Visser-VanBrummen, P.J., DeJongste, M.J., and Tiebosch, A.T., The content and distribution of troponin I, troponin T, myoglobin, and alpha-hydroxybutyric acid dehydrogenase in the human heart. *Am J Clin Pathol,* 115, 770–777, 2001.
97. Leavis, P.C. and Gergely, J., Thin filament proteins and thin filament–linked regulation of vertebrate muscle contraction. *CRC Crit Rev Biochem,* 16, 235–305, 1984.
98. Zot, A.S. and Potter, J.D., Structural aspects of troponin–tropomyosin regulation of skeletal muscle contraction. *Annu Rev Biophys Biophys Chem,* 16, 535–559, 1987.
99. Katrukha, A.G., Bereznikova, A.V., Filatov, V.L., Esakova, T.V., Kolosova, O.V., Pettersson, K., Lovgren, T., Bulargina, T.V., Trifonov, I.R., Gratsiansky, N.A., Pulkki, K., Voipio-Pulkki, L.M., and Gusev, N.B., Degradation of cardiac troponin I: implication for reliable immunodetection. *Clin Chem,* 44, 2433–2440, 1998.
100. Vallins, W.J., Brand, N.J., Dabhade, N., Butler-Browne, G., Yacoub, M.H., and Barton, P.J., Molecular cloning of human cardiac troponin I using polymerase chain reaction. *FEBS Lett,* 270, 57–61, 1990.
101. Ferrieres, G., Calzolari, C., Mani, J.C., Laune, D., Trinquier, S., Laprade, M., Larue, C., Pau, B., and Granier, C., Human cardiac troponin I: precise identification of antigenic epitopes and prediction of secondary structure. *Clin Chem,* 44, 487–493, 1998.
102. Katrukha, A.G., Bereznikova, A.V., Esakova, T.V., Pettersson, K., Lovgren, T., Severina, M.E., Pulkki, K., Vuopio-Pulkki, L.M., and Gusev, N.B., Troponin I is released in bloodstream of patients with acute myocardial infarction not in free form but as complex. *Clin Chem,* 43(8 pt 1), 1379–1385, 1997.
103. Troponin I standardization subcommittee chooses candidate reference material. In: *Clinical Laboratory Strategies,* vol. 7, Auxter, S., ed. Washington, DC: American Association of Clinical Chemistry, 2002:2.

104. Uddin, A.H., Piunno, P.A., Hudson, R.H., Damha, M.J., and Krull, U.J., A fiber optic biosensor for fluorimetric detection of triple-helical DNA. *Nucleic Acids Res,* 25, 4139–4146, 1997.
105. Pilevar, S., Davis, C.C., and Portugal, F., Tapered optical fiber sensor using near-infrared fluorophores to assay hybridization. *Anal Chem,* 70, 2031–2037, 1998.
106. Jensen, K.K., Orum, H., Nielsen, P.E., and Norden, B., Kinetics for hybridization of peptide nucleic acids (PNA) with DNA and RNA studied with the BIAcore technique. *Biochemistry,* 36, 5072–5077, 1997.
107. Nilsson, P., Persson, B., Larsson, A., Uhlen, M., and Nygren, P.A., Detection of mutations in PCR products from clinical samples by surface plasmon resonance. *J Mol Recognit,* 10, 7–17, 1997.
108. Persson, B., Stenhag, K., Nilsson, P., Larsson, A., Uhlen, M., and Nygren, P., Analysis of oligonucleotide probe affinities using surface plasmon resonance: a means for mutational scanning. *Anal Biochem,* 246, 34–44, 1997.
109. Bianchi, N., Rutigliano, C., Tomassetti, M., Feriotto, G., Zorzato, F., and Gambari, R., Biosensor technology and surface plasmon resonance for real-time detection of HIV-1 genomic sequences amplified by polymerase chain reaction. *Clin Diagn Virol,* 8, 199–208, 1997.
110. Niemeyer, C.M., Burger, W., and Hoedemakers, R.M., Hybridization characteristics of biomolecular adaptors, covalent DNA–streptavidin conjugates. *Bioconjug Chem,* 9, 168–175, 1998.
111. Schneider, B.H., Edwards, J.G., and Hartman, N.F., Hartman interferometer: versatile integrated optic sensor for label-free, real-time quantification of nucleic acids, proteins, and pathogens. *Clin Chem,* 43, 1757–1763, 1997.
112. Bier, F.F. and Scheller, F.W., Label-free observation of DNA-hybridisation and endonuclease activity on a wave guide surface using a grating coupler. *Biosens Bioelectron,* 11, 669–674, 1996.
113. Stimpson, D.I. and Gordon, J., The utility of optical waveguide DNA array hybridization and melting for rapid resolution of mismatches, and for detection of minor mutant components in the presence of a majority of wild type sequence: statistical model and supporting data. *Genet Anal,* 13, 73–80, 1996.
114. Stimpson, D.I., Hoijer, J.V., Hsieh, W.T., Jou, C., Gordon, J., Theriault, T., Gamble, R., and Baldeschwieler, J.D., Real-time detection of DNA hybridization and melting on oligonucleotide arrays by using optical wave guides. *Proc Natl Acad Sci USA,* 92, 6379–6383, 1995.
115. Basu, S., Sarkar, P., Adhya, S., and Maitra, U., Locations and nucleotide sequences of three major class III promoters for bacteriophage T3 RNA polymerase on T3 DNA. *J Biol Chem,* 259, 1993–1998, 1984.
116. Pruitt, S.C., Expression vectors permitting cDNA cloning and enrichment for specific sequences by hybridization/selection. *Gene,* 66, 121–134, 1988.
117. Evans, G.A., Lewis, K., and Rothenberg, B.E., High efficiency vectors for cosmid microcloning and genomic analysis. *Gene,* 79, 9–20, 1989.
118. Hung, S.C., Mathies, R.A., and Glazer, A.N., Optimization of spectroscopic and electrophoretic properties of energy transfer primers. *Anal Biochem,* 252, 78–88, 1997.
119. O'Shaughnessy, J.B., Chan, M., Clark, K., and Ivanetich, K.M., Primer design for automated DNA sequencing in a core facility. *Biotechniques,* 35, 112–116, 118–121, 2003.
120. Herron, J.N., Wang, H.-K., Janatová, V., Durtschi, J.D., Christensen, D.A., Caldwell, K.D., Chang, I.-N., and Huang, S.-C., Orientation and activity of immobilized antibodies. In: *Biopolymers at Interfaces,* vol. 110, Malmsten, M., ed. New York: Marcel Dekker, 2003:115–163.

121. Lundwall, A. and Lilja, H., Molecular cloning of human prostate specific antigen cDNA. *FEBS Lett,* 214, 317–322, 1987.

122. Lovgren, J., Piironen, T., Overmo, C., Dowell, B., Karp, M., Pettersson, K., Lilja, H., and Lundwall, A., Production of recombinant PSA and HK2 and analysis of their immunologic cross-reactivity. *Biochem Biophys Res Commun,* 213, 888–895, 1995.

123. Hoshi, S., Kobayashi, S., Takahashi, T., Suzuki, K.I., Kawamura, S., Satoh, M., Chiba, Y., and Orikasa, S., Enzyme-linked immunosorbent assay detection of prostate-specific antigen messenger ribonucleic acid in prostate cancer. *Urology,* 53, 228–235, 1999.

124. Twyman, R.M. and Primrose, S.B., Techniques patents for SNP genotyping. *Pharmacogenomics,* 4, 67–79, 2003.

125. Vincent, G.M., The molecular genetics of the long QT syndrome: genes causing fainting and sudden death. *Annu Rev Med,* 49, 263–274, 1998.

126. Maron, B.J., Moller, J.H., Seidman, C.E., Vincent, G.M., Dietz, H.C., Moss, A.J., Sondheimer, H.M., Pyeritz, R.E., McGee, G., and Epstein, A.E., Impact of laboratory molecular diagnosis on contemporary diagnostic criteria for genetically transmitted cardiovascular diseases: hypertrophic cardiomyopathy, long-QT syndrome, and Marfan syndrome: a statement for healthcare professionals from the Councils on Clinical Cardiology, Cardiovascular Disease in the Young, and Basic Science, American Heart Association. *Circulation,* 98, 1460–1471, 1998.

127. Vander Horn, P.B., Davis, M.C., Cunniff, J.J., Ruan, C., McArdle, B.F., Samols, S.B., Szasz, J., Hu, G., Hujer, K.M., Domke, S.T., Brummet, S.R., Moffett, R.B., and Fuller, C.W., Thermo Sequenase DNA polymerase and *T. acidophilum* pyrophosphatase: new thermostable enzymes for DNA sequencing. *Biotechniques,* 22, 758–762, 764–765, 1997.

128. Pastinen, T., Raitio, M., Lindroos, K., Tainola, P., Peltonen, L., and Syvanen, A.C., A system for specific, high-throughput genotyping by allele-specific primer extension on microarrays. *Genome Res,* 10, 1031–1042, 2000.

129. Syvanen, A.C., Aalto-Setala, K., Harju, L., Kontula, K., and Soderlund, H., A primer-guided nucleotide incorporation assay in the genotyping of apolipoprotein E. *Genomics,* 8, 684–692, 1990.

13 Fluorescence-Based Sensors for Bioprocess Monitoring

Leah Tolosa, Ph.D., Yordan Kostov, Ph.D., and Govind Rao, Ph.D.

CONTENTS

13.1 INTRODUCTION

Biotechnology and bioprocessing have been part of human activity since the first accidental discovery that the ingestion of fermented sugars leads to pleasurable effects and that moldy milk has a heady aroma and savory taste. For thousands of years, bioprocessing has been mainly an art. The quality of products is judged based on the instincts and experience of masters. Only in the last century, specifically with the first commercial production of antibiotics, was there a thrust toward more systematic and controlled bioprocessing.[1] Closed systems, or bioreactors, were designed to maintain the purity of cultures and allow some level of control over culture conditions. As genetic engineering progressed at a rapid pace, recombinant organisms producing therapeutically important proteins or by-products have become increasingly employed in the industrial production of so-called biologicals. With this development came considerable incentive to optimize and control process conditions to ensure high quality and yield. In addition, there is a need to abide by increasingly stringent government regulations in product purity and safety. Essential to these efforts is the development of methods to accurately measure important bioreactor parameters as close to real time as possible and with few or no deleterious effects on the growing cells. In this chapter we will discuss fluorescence-based sensing aimed specifically at applications in fermentation and cell culture. What makes the design of these sensors unique from other fluorescence sensors? What special requirements need to be achieved?

The design of sensors for bioprocess monitoring is not trivial. There are several considerations in adapting fluorescence sensors to cell culture. Foremost is robustness in harsh conditions. The sensor has to be sterilizable, which in most cases means being able to withstand the high pressures and high temperatures of an autoclave. Other sterilization methods, such as washing with 70% ethanol or gamma irradiation, may have detrimental effects on the dyes themselves or the materials used for dye immobilization. More than in other applications, sensor drift has to be minimized because bioprocesses may take days or even weeks. Sources of drift include dye decomposition, leaching of the immobilized dye, changes in the ionic strength of the medium, temperature fluctuations, etc. Fouling of the sensor may also occur once the multiplying cells adhere in large numbers to the sensor surface. In addition, the culture medium, a complex mixture of substances, scatters and absorbs light so strongly as to be effectively opaque and contains fluorescent components as well. Noise and background minimization are therefore major issues in developing fluorescence sensors for fermentation. An internal calibration scheme is always desirable because offline calibration midway through a process is never convenient and is often unacceptable.

Despite these difficulties, fluorescence sensors for bioprocess monitoring have become more available in recent years. In many instances, fluorescence-based sensing provides some advantages over the more traditional and often more costly electrochemical methods. Advances in fluorescent probe chemistry, optics, electronics, and materials have all contributed to practical and low-cost fluorescence sensors. Strategies for dealing with the high backgrounds of growth media have been developed with great success. These include gated detection,[2,3] lifetime-based sensing,[4] and the use of near-infrared (NIR) probes.[5] Contributions from other disciplines, such as molecular biology, and recombinant protein techniques have resulted in the development of protein-based biosensors,[6,7] some of which are nonenzymatic and therefore nonconsuming and reagentless. In addition, genetic engineering allows for the design of proteins with unique binding characteristics, high specificity, tunable sensitivity, and increased robustness.[8]

The advantages and limitations of fluorescence sensors in cell culture and fermentation processes are presented in this chapter. Considerations in the design of these sensors are also discussed. Specific examples of recent advances in fluorescence-based bioprocess sensors are presented. Finally, there is a section on future challenges in the field.

13.2 ADVANTAGES OF FLUORESCENCE-BASED SENSING

Fluorescence sensing is by its very nature versatile. The interaction of light with a fluorescent probe allows for the investigation of a number of optical effects that may be related to analyte concentration, substrate binding, cell metabolism, product formation, or any other phenomenon important to bioprocessing. Fluorescence intensity, fluorescence decay rate, spectral shape, and anisotropy of the probe emission may be addressed singly or in multiplex to obtain a more complete picture of bioreactor parameters. A single probe may be used to obtain a variety of information, or a single parameter may be measured redundantly using a variety of techniques. Another advantage of optical sensing is better noise immunity; that is, light is not subject to electromagnetic interferences except in cases of extremely strong fields.

Fluorescence measurements using dyes and indicators are generally performed when equilibrium with the analyte is achieved. Thus, the analyte need not undergo a chemical transformation. In contrast, electrochemical and enzymatic methods consume and transform the analyte itself. At very low concentrations, this may result in the depletion of the analyte in contact with the sensor, leading to erroneously low measurements.

In electrochemical sensing, detection of electrons requires direct contact of the sensor with the sample. In fluorescence sensing, detection of photons is possible through space, that is, through the walls of a transparent vessel or through light guides when the light needs to be delivered to a specific point. This allows for minimally or noninvasive monitoring of bioreactor conditions, thereby reducing or eliminating any disturbance to cell growth and propagation. For example, the response of an oxygen-sensitive patch to dissolved oxygen (DO) in media may be monitored noninvasively.[9,10] The DO sensing patch is attached to the interior of the bioreactor or shake flask (Figure 13.1), where it is in constant contact with the media, whereas the excitation

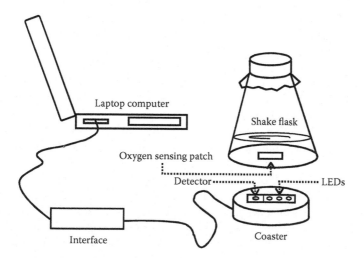

FIGURE 13.1 Setup of noninvasive oxygen-sensing system for shake flasks. Reprinted with permission from *Biotechnol Bioeng,* 80, 595–597, 2002. © 2002 Wiley Periodicals, Inc.

light source and detector are positioned outside the glass. The same configuration may be adapted for pH[11] and carbon dioxide[12] (CO_2) sensing. In addition, the intrinsic fluorescence of cells has been exploited with some success. Since the late 1970s, biomass has been correlated to the reduced form of nicotinamide adenine dinucleotide phosphate (NADPH) fluorescence.[13] A broader application is two-dimensional fluorescence spectroscopy, which can monitor NADPH, flavins, and amino acids, thereby providing information on cell mass as well as the metabolic state of the growing cells.[14]

13.3 POTENTIAL DRAWBACKS OF FLUORESCENCE SENSING

As in most analytical techniques, the fluorescence signal varies with scale. A decrease in sensor size results in a decrease in the number of photons reaching the photodetector per unit time. This in turn may decrease the signal:noise ratio. The decrease in signal amplitude is readily compensated for, typically by amplification and noise-reduction methods. Interestingly, these techniques were initially developed for radio transmission.

In addition, photodetectors are not capable of discriminating between light sources. Thus, they may detect ambient light as well as background fluorescence from the sample. This is particularly problematic in the complex media used to grow cells. A variety of proven methods for discrimination between useful and unwanted light exists. They include modulation of excitation light (which helps to distinguish between fluorescence and the ambient light), optical shielding (which decreases the amount of autofluorescence that reaches the photodetector), use of narrow band-pass emission filters (in cases when the emission maxima of the autofluorescence and the sensor are different), and gated detection (when the sensor's emission lifetime is considerably longer than the autofluorescence lifetime).

For most fluorescent probes, the transfer function (the relationship between the analyte concentration and the probes' output signal) is usually nonlinear, reflecting a narrower dynamic range in comparison with traditional electrochemical sensors. For example, the transfer function of all pH indicators has a sigmoidal shape (Figure 13.2), which limits the useful range to approximately ±1.5 pH units around the indicator's pK_a. However, the in-range sensitivity of the sensor is usually better than the sensitivity of the electrochemical sensors. In the case of oxygen sensors, the transfer function is hyperbolic; again, they exhibit better sensitivity at lower oxygen concentrations. These nonlinearities may be partially compensated for using digital signal treatment combined with proper model or lookup tables.

Until recently, the use of fluorescence sensing in bioprocess monitoring has been limited by costs. This is because traditional fluorescence sensing requires complex optics and electronics to generate excitation light and to detect the fluorescent signal. However, advances in optoelectronics and electronics (fed mostly by the development of the telecommunications industry) have helped in the design of smaller, less complicated, and better optoelectrical converters. Recent developments in integrated optics have also contributed to the lowering of the cost of fluorescence sensors, thus enabling further expansion of their use.

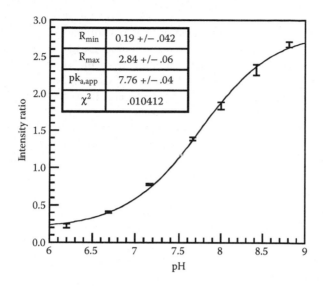

R_{min}	0.19 +/− .042
R_{max}	2.84 +/− .06
$pk_{a,app}$	7.76 +/− .04
χ^2	.010412

FIGURE 13.2 Calibration curve of HPTS pH sensor. Reprinted with permission from *Biotechnol Prog*, 18, 1047–1053, 2002. © 2002 American Chemical Society. Revised from Kermis et al.[11]

13.4 SENSOR DESIGN REQUIREMENTS

Optical sensor design is a complex task that requires addressing problems in sensor chemistry, dye immobilization, optics, optoelectronics, and computing. This is particularly important in designing sensors for bioprocess monitoring. Here we will briefly discuss the requirements for these aspects of sensor design as well as lay out some considerations about the even greater task of interfacing the sensor components with each other.

13.4.1 "INDICATOR" MOLECULE

In an ideal sensing situation, the analyte to be measured is intrinsically fluorescent. Indeed, as described previously, this has been used in the measurement of NADP(H), flavin, and protein fluorescence in cell culture.[14] Although the information from these measurements is useful, it presents a limited picture of bioreactor parameters. In addition, control and feedback are rather difficult. In most cases, the analytes that describe bioreactor conditions are nonfluorescent. Thus, recognition molecules (i.e., chemical sensors) are needed to interact with the analytes and report on their concentrations. The design and synthesis of such molecules for various analytes is the major undertaking in fluorescent probe chemistry. Fortunately, for bioprocess parameters there is already a variety of fluorescent pH/CO_2 indicators and oxygen-sensitive dyes.[5] Fluorescent sensors for glucose and glutamine that have been used for biotechnology measurements utilize polarity-sensitive dyes covalently attached to the corresponding binding proteins.[15,16]

The optical properties of the dye determine the type of optics used, the mode of operation of the optoelectronics, and the needed software. For example, a near-IR emitting dye with a long fluorescence lifetime greatly facilitates instrument design, as the detector can be easily shielded from the short-wavelength, short-lifetime autofluorescence of the culture media. In addition, low-cost, high-intensity laser diodes may be used as excitation light sources. In some cases, where there is a large choice of available dyes (e.g., pH indicators), screening allows finding the best dye for a particular application. Another issue is the functionalization properties of the dyes, as these strongly influence their subsequent immobilization.

Needless to say, it is important that the recognition molecule and/or dye and its functional groups are not affected by the sterilization procedures (autoclaving at 125°C, ethylene oxide, gamma sterilization, or 70% ethanol). Finally, leakage of the dye into the surrounding media (if unavoidable) should not result in interference with cell growth and survival (i.e., the dye should not be cytotoxic), and it should not result in the introduction of an unwanted impurity to the final product.

13.4.2 IMMOBILIZATION

The most important consideration in immobilizing an indicator molecule is to not affect its recognition properties. In many cases, an indicator that works perfectly in solution becomes insensitive or less sensitive after immobilization. For this reason, the immobilization technique should be carefully selected. For many low-molecular-weight indicators, covalent bonding to a surface or solid network is a suitable technique because leaching of the dye becomes less of a danger. However, the risk of altering the recognition properties of the indicator is greater. Attaching a long tether is often an alternative to minimize the effect. This is particularly true in protein sensor immobilization, where the protein needs to be in a solvated form and in the right configuration to remain active. A significantly milder means of immobilization is through electrostatic interactions (anion–cation-type bonding). A caveat is that the bond strength in this method is severely affected by the pH and by the ionic strength of the analyte and sample medium. Thus, it is not uncommon to observe a shift in the pK_a of an indicator dye when immobilized in this way.[17] Finally, protein-based sensors that depend on conformational changes for their signal-transducing mechanisms need to be physically entrapped in a matrix that does not hinder molecular motions.

From the bioprocess point of view, the immobilization matrix should provide sufficient separation between the sensing molecule and the compounds of the fermentation broth while still allowing the passage of the analyte. In this sense the immobilization matrix should provide some selectivity toward the measured compound. For example, a semipermeable membrane that allows the passage of small molecules, but not proteins and other macromolecules, is suitable for sensing ions and other small organic compounds. Furthermore, the effects of sterilization and exposure to the bioprocess environment are serious considerations in choosing the matrix. Because most immobilization matrices are polymers, it is important that sterilization does not result in the breakdown of the material into its monomers or into other products that could leak into the fermentation broth, thereby affecting cell growth and product purity. The matrix pores (if they are present) should be small enough to prevent the penetration of cells. In addition, the matrix surface should be neutral with respect to cells; that is, attachment of cells to the surface of the sensor should be minimal, as this could lead to biofouling and a decrease in the sensitivity of the sensor. However, more importantly, attachment of the cells on surfaces detrimentally affects cell growth in the bioreactor.

13.4.3 OPTICS

Optics selection and design depend on the optical properties of the sensors and the bioprocess to be monitored. When needed, transmission and collection optics, such as optical fibers and lenses, are utilized to increase the amount of light reaching the photodetector. Polarizers are used in the case of polarization sensors. However, wavelength selection devices—filters or monochromators—are the main optical components. They ensure the spectral separation between excitation and emission. They are selected to mainly conform to the spectral characteristics of the dye, but for bioprocess applications, additional care should be taken to use optics with minimum directional sensitivity. This is because the bioprocess medium is highly scattering, resulting in light coming from all directions. In contrast, most interference filters able to resolve the narrow wavelength difference between excitation and emission are designed to perform when the light reaches them at 90°. For smaller angles, they start transmitting shorter wavelengths (the so-called angular dependence). As a result, any scattered excitation light will always leak into the photodetector unless a second, nondirectional filter (colored glass) is used for its attenuation.

Another important consideration in optics design is the light path, particularly when no optical fibers are used. In this case, care should be taken to avoid direct illumination of the receiver (even when filtered) by the excitation light. The filters and monochromators are not perfect. They merely attenuate the light in the wavelength band(s) being blocked but do not completely exclude it. This again may result in unwanted leakage of the excitation light into the photodetector. Backscattered light from the fermentation medium poses an additional problem. For best results, the photodetector is positioned on the bottom of a deep narrow well with black walls.

13.4.4 OPTOELECTRONICS

Light-emitting diodes (LEDs) are preferable as excitation light sources for sensors because they are inexpensive, have relatively narrow emission bands, and are suitable for direct electronic modulation. In cases in which LEDs are not sufficiently bright, laser diodes or conventional high-power light sources are used. Photodetectors are usually silicon photodiodes. Alternatively, avalanche photodiodes or photomultipliers may be used for greater sensitivity. A careful match between the light source, the photodetector, and the optical properties to be detected is a must. For example, a fluorophore in a lifetime-based sensor with a nanosecond lifetime should not be paired with a detector that has a response time on the order of microseconds. Finally, bioprocess applications demand one additional requirement for optoelectronics: they must contained in a waterproof enclosure in case of leaks or spills.

13.4.5 Computing

The raw data from the sensor need to be converted into usable information. This is typically performed by the use of the transfer function of the sensor or by a series of calibration points presented in a lookup table. The values between the points are calculated using linear or some other type of approximation. As the rate of the optoelectronics conversion may be very high, a number of measurements are performed and averaged. In this way, signal fluctuations due to bubbles or particulates in the fermentation medium may be minimized. In addition, the software should account for working temperature, ionic strength, and other environmental variables that may alter the calibration of the sensor.

13.5 SENSORS FOR BIOREACTOR PARAMETERS

It is fascinating that relatively few conditions must be met to ensure the survival of cells in culture. Besides temperature, these include pH, DO, CO_2, and nutrients (such as glucose and glutamine). The development of sensors for monitoring these bioreactor parameters is described in this section. Other sensors that have been developed for more specific applications are mentioned only in passing. For example, the green fluorescent protein (GFP) fused to a recombinant protein may serve as a measure of the total correctly folded protein product.

13.5.1 pH

The acidity or pH of the cell culture media is typically measured by the use of indicators that exhibit distinct protonated and unprotonated species. There are two types of these fluorescent indicators. Dual excitation indicators are similar to the absorption-based indicators in that their basic and acidic forms possess different absorption maxima and extinction coefficients. The emission maxima, however, may be unaffected by pH, as the emission originates from the same transition state after internal conversion, but the quantum yield (i.e., emission intensity) is dependent on excitation wavelength. Thus, one may measure the emission intensities at two excitation wavelengths, and the ratio of the emission intensities becomes a function of pH. Examples of this type of fluorescent pH indicator include hydroxypyrene trisulfonic acid (HPTS), carboxydichlorofluorescein (CDCF), and carboxynaphthofluorescein (CNF).

A second type of pH sensor is the emission ratiometric indicator. In this case, not only the excitation maxima but also the emission spectra are different for the acid and base forms. One may then relate the ratio of intensities at two emission wavelengths to pH. Examples of these indicators are the seminaphthofluoresceins (SNAFLs) and seminaphthorhodafluors (SNARFs) (Molecular Probes, Inc., Eugene, OR). In both types of pH sensor, the relationship between the observed ratios and the proton concentration is given by the following equation[18]:

$$[H^+] = k_a \frac{(R_{max} - R)}{(R - R_{min})} \cdot \frac{\varepsilon_{A^-} \Phi_{A^1}}{\varepsilon_{HA} \Phi_{HA}} \qquad (13.1)$$

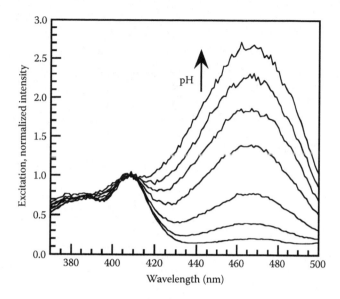

FIGURE 13.3 Excitation spectra of HPTS ($\lambda_{emission} = 515$ nm) immobilized in PEG-Dowex. The pH range is from 6.62 to 8.81. Reprinted with permission from *Biotechnol Prog*, 18, 1047–1053, 2002. © 2002 American Chemical Society. Revised from Kermis et al.[11]

where R_{min} and R_{max} are the ratios for the acid (HA) and conjugate base (A⁻), respectively; ε and Φ are the extinction coefficient and quantum yield of each species evaluated at λ_2; and k_a is the equilibrium dissociation constant. It follows from Equation 13.1 that the useful range comprises approximately ±1.5 pH units.

A good example of an optical pH sensor that has been successfully used in bioprocess monitoring is HPTS and its derivatives, as shown in Figure 13.3. This dye exhibits two excitation wavelengths, one ultraviolet (UV) ($\lambda_1 = 405$ nm) and one blue ($\lambda_2 = 457$ nm), that correspond to the acid and its conjugate base. The ratio of emission intensities at 530 nm, R, for these two wavelengths is related to the proton concentration, as shown in Figure 13.2. With a pK_a of 7.3, HPTS is suitable for ratiometric detection in the physiological range. It has also been used in monitoring *Escherichia coli* fermentation, which is carried out under neutral or slightly basic conditions (Figure 13.4). In addition, its low toxicity[19] and insensitivity to oxygen quenching[20] make HPTS an appropriate probe for physiological and bioprocess pH measurements. Finally, with the recent availability of low-cost UV LEDs, the dye may be measured with relatively inexpensive instrumentation that combines UV and blue LEDs and a photodiode module. A detailed description of the HPTS-based pH sensor can be found in Kermis et al.[11] This article provides an exhaustive account of the steps leading to a functional pH sensor and its characterization.

13.5.2 CARBON DIOXIDE

The design of optical CO_2 sensors essentially follows the Severinghaus principle,[21] in which a pH-sensitive electrode or indicator is in contact with a thin layer of

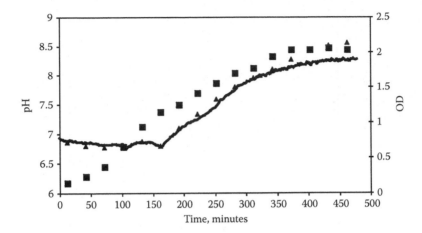

FIGURE 13.4 pH monitoring of *E. coli* fermentation. OD (■), pH measurements using conventional pH meter (▲), and output of the optical pH sensor (-----). Unpublished data.

bicarbonate solution encapsulated by a thin, gas-permeable membrane. As CO_2 diffuses into this membrane, it dissolves into the bicarbonate solution, resulting in a change in pH. In the original Severinghaus electrode, a pH electrode detects this pH change, but in an optical sensor, a fluorescent indicator dissolved in the same buffer is used to detect the change in pH. In practice, the CO_2 optical sensor with this design is not very suitable for bioprocess applications.[12] For this reason, the aqueous bicarbonate buffer is replaced with a phase transfer agent—a quaternary ammonium hydroxide (QMH)—and the gas-permeable membrane is replaced with a hydrophobic polymer. This phase transfer agent serves a number of purposes: it aids in dissolving the indicator dye in the hydrophobic polymer by forming an ion pair with the indicator dye anion[22,23]; it allows for the incorporation of moisture in the polymer, which is needed in the overall acid–base reaction; and it provides the basic environment essential for CO_2 sensing.

The sensing mechanism for such a film[24] is summarized by the following equation:

$$\{Q^+A^- \cdot xH_2O\} + CO_2 \Leftrightarrow \{Q^+A^- \cdot xH_2O \cdot CO_2\} \Leftrightarrow \{Q^+A^- \cdot (x-1)H_2O \cdot H_2CO_3\}$$
$$\Leftrightarrow \{Q^+HCO_3^- \cdot (x-1)H_2O \cdot HA\} \qquad (13.2)$$

where $\{Q^+A^- \cdot xH_2O\}$ represents the quaternary ammonium cation and the indicator ion pair associated with x water molecules and $\{Q^+HCO_3^- \cdot (x-1)H_2O \cdot HA\}$ represents the quaternary ammonium cation and the hydrogen carbonate anion ion pair associated with $x-1$ water molecules and the dye in its free acid form (HA). Again, an equation similar to Equation 13.1 may be used to relate the ratio R of the measured intensities (which depend on the concentration of A^- and HA) to describe its dependence on CO_2 concentration. The use of HPTS as the pH indicator for this sensor

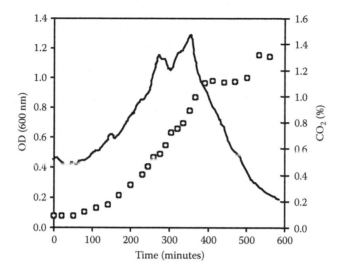

FIGURE 13.5 CO_2 monitoring of *E. coli* fermentation. OD (O) and output of the optical CO_2 sensor (-----). Reprinted with permission from *Biosens Bioelectron,* 18, 857–865, 2003. © 2002 Elsevier Science B.V.

is convenient because it allows use of the same optics, electronics, and software as in the pH sensor described in the previous section. Sample data obtained from a fermentation experiment using this sensor to monitor CO_2 are shown in Figure 13.5. Readers are advised to consult Ge et al.[12] for a more exhaustive account of the fabrication of this CO_2 sensor.

13.5.3 DISSOLVED OXYGEN

Optical sensing of DO is based on the principle that oxygen dynamically quenches the fluorescence of any fluorophore with a lifetime longer than 10 ns. Consequently, luminescent metal–ligand complexes with decay rates in the hundreds of nanoseconds are often the dyes of choice for DO sensing. The data shown in Figure 13.6 were derived from a sensor using [1,2-bis-(diphenylphosphino)ethane-Pt {S2C2(CH2CH2-*N*-2-pyridinium)}][BPh4] as the oxygen indicator immobilized in an autoclavable polymer film. This film (or so-called DO sensing patch) is affixed to a bioreactor or shake flask for noninvasive oxygen sensing.[9]

Fluorescence (or more accurately, luminescence) quenching by oxygen is described by the Stern-Volmer equation[5]:

$$\frac{F_0}{F} = \frac{\tau_0}{\tau} = 1 + K_{sv} \cdot [Q] \tag{13.3}$$

where F_0 and F are the emission intensities in the absence and presence of oxygen, respectively, and τ_0 and τ are the lifetimes with and without oxygen. K_{sv} is the

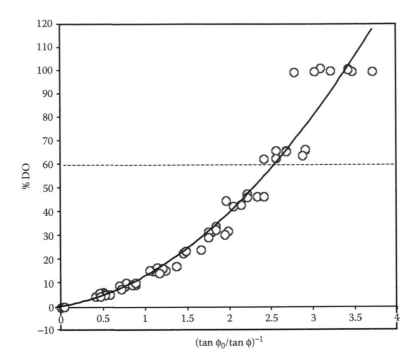

FIGURE 13.6 Calibration plot of the percent DO as a function of the phase angle, ϕ. The data were obtained by measuring the phase angle at various mixtures of air and nitrogen gas bubbled in water. The phase at 100% N_2 saturation is designated ϕ_0. The DO was measured with an Ingold oxygen electrode. Reprinted with permission from *Biotechnol Bioeng*, 80, 595–597, 2002. © 2002 Wiley Periodicals, Inc.

Stern-Volmer constant and [Q] is the oxygen concentration. It should be noted that $F_0/F = \tau_0/\tau$ applies only in the case of dynamic quenching.

The decay time, τ, of the dye is measured by frequency-domain fluorometry using electronics based on the lock-in principle.[25] Intensity-modulated light at a frequency ($\omega = 2\pi \times$ frequency [in Hz]) is generated from an excitation source (an LED in this example). The phase shift ϕ of the resulting emission is related to the decay time τ by the following equation:

$$\tan \phi = \omega \cdot \tau \qquad (13.4)$$

Thus, the DO concentration, [Q] = (DO), may be deduced from Equation 13.3 and Equation 13.4. The calibration plot of the phase shift, ϕ, as a function of DO (Figure 13.6), where 0% DO is N_2 gas saturated and 100% DO is air saturated, gives a nonlinear plot attributed to the inherent dual emitting properties of this dye.[26] The DO sensing system described here was used to measure

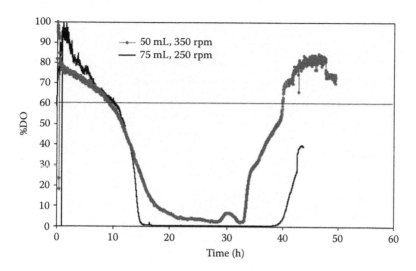

FIGURE 13.7 *Saccharomyces cerevisiae* fermentation in shake flasks. Above 60% DO, the sensor becomes less sensitive to increases in DO and is subject to larger errors. Reprinted with permission from *Biotechnol Bioeng*, 80, 595–597, 2002. © 2002 Wiley Periodicals, Inc.

oxygen levels during *Saccharomyces cerevisiae* fermentation. The results are shown in Figure 13.7.

13.5.4 GLUCOSE AND GLUTAMINE

Currently available sensors for glucose and glutamine for bioprocess applications utilize enzyme electrodes[27,28] or near-IR technology.[29] In recent years, proteins derived from the periplasm of *E. coli* bacteria that are known to function as binding proteins for various substances have drawn considerable attention in the design of optical sensors for their substrates.[6–8] These proteins are known to undergo a large conformational change when bound to the substrate. This change may be observed by introducing a point mutation to the protein chain for the attachment of a polarity-sensitive fluorophore. The spectral change of the dye is correlated to the concentration of the substrate/analyte, as shown in Figure 13.8 for the acrylodan-labeled S179C mutant of the *E. coli* glutamine binding protein.[30] A second label at the N-terminal end of this protein with an Ru metal–ligand complex (emission approximately 650 nm) acts as a reference. The ratios of the acrylodan emission (approximately 530 nm) and Ru complex emission are plotted against glutamine concentration (inset). In Figure 13.9, the glucose consumption of an *E. coli* culture was followed using the Q26C mutant of the glucose binding protein labeled with anilinonaphthalene sulfonate (ANS).[7] Note that, in contrast, the glucose oxidase electrode (YSI; Yellow Springs Instrument Co., Yellow Springs, OH) does not provide reliable data because the levels of glucose in the Luria-Bertani broth often used in *E. coli* culture are close to the lower detection limit of the enzyme electrode.

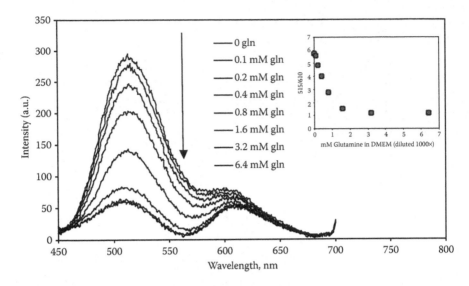

FIGURE 13.8 Emission spectra of 2.0 μM Ru–GlnBP–Acr in increasing concentrations of glutamine. Glutamine solutions were prepared in Dulbecco's modified eagle medium (DMEM) and diluted 1000× with phosphate buffered saline (PBS). Excitation wavelength 360 nm. Inset: Ratios of emission intensities at 515 and 610 nm plotted as a function of glutamine concentration. Reprinted with permission from *Anal Biochem*, 314, 199–205, 2003. © 2003 Elsevier Science B.V.

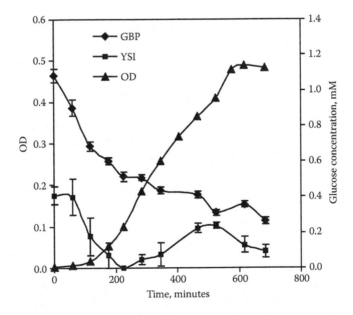

FIGURE 13.9 Glucose monitoring of *E. coli* fermentation in LB nutrient broth. OD (▲), YSI glucose oxidase electrode (■), and glucose binding protein sensor (♦). Reprinted with permission from *Biotechnol Bioeng*, 84, 723–731, 2003. © 2003 Wiley Periodicals, Inc.

13.6 FUTURE TRENDS

At this point, it is useful to recall that the main aim of developing sensors for bioreactor parameters is to be able to control the amount and quality of the bioprocess product. The ability to directly measure the concentration of a bioproduct in real time is therefore one of the main thrusts in the future development of sensors for bioprocessing. Baker et al.[31] provide a broad review of current technologies mainly for recombinant protein products (enzyme-linked immunosorbent assay [ELISA], electrochemiluminescence, surface plasmon resonance technology, nephelometry, etc.). These technologies are used offline and may require expensive equipment or reagents. The assays may entail a significant time lag between sampling and obtaining the result, precluding any feedback. The fusion of GFP or its variants to recombinant protein products has allowed for the quantitation of these products *in situ*.[32–34] However, not all proteins are amenable to fusion with GFP, or vice versa. The challenges that need to be addressed include low-cost equipment, real-time measurements, and sterilizable sensors for *in situ* measurements, as well as highly specific and highly sensitive sensors for the bioproduct.

The discovery of therapeutically important gene products is driving the need for more efficient ways to improve bioprocessing. Large-volume bioreactors often suffer from insufficient mixing and heterogeneity, necessitating multiple sensors. On the other hand, high-throughput method development is becoming inevitable, as more experiments are required to obtain more information. However, these can be done without sacrificing the quality of data only by miniaturization of the bioreactor and therefore of the sensors themselves. Nonetheless, multiple parameters must monitored in each of these miniaturized bioreactors, complicating the problem even more. Progress in sensor multiplexes for microbioreactors has already been reported.[35] Further improvement in solid-state technology should drive this development in the future.

REFERENCES

1. Humphrey, A., Shake flask to fermentor: what have we learned? *Biotechnol Prog,* 14, 3–7, 1998.
2. Rowe, H.M., Chan, S.P., Demas, J.N., and DeGraff, B.A., Elimination of fluorescence and scattering backgrounds in luminescence lifetime measurements using gated-phase fluorometry. *Anal Chem,* 74, 4821–4827, 2002.
3. Lakowicz, J.R., Gryczynski, I., Gryczynski, Z., and Johnson, M.L., Background suppression in frequency-domain fluorometry. *Anal Biochem,* 277, 74–85, 2000.
4. Bambot, S.B., Lakowicz, J.R., and Rao, G., Potential applications of lifetime-based, phase-modulation fluorimetry in bioprocess and clinical monitoring. *Trends Biotechnol,* 13, 1668–1674, 1995.
5. Lakowicz, J.R., *Principles of Fluorescence Spectroscopy,* 2nd ed. New York: Kluwer Academic, 1999.
6. De Lorimer, R.M., Smith, J.J., Dywer, M.A., Looger, L.L., Sali, K.M., Paavola, C.D., Rizk, S.S., Sadigov, S., Conrad, D.W., Loew, L., and Hellinga, H., Construction of a fluorescent biosensor family. *Protein Sci,* 11, 2655–2675, 2002.

7. Ge, X., Tolosa, L., Simpson, J., and Rao, G., Genetically engineered binding proteins as biosensors for fermentation and cell culture. *Biotechnol Bioeng*, 84, 723–731, 2003.
8. Looger, L.L., Dwyer, M.A., Smith, J.J., and Hellinga, H., Computational design of receptor and sensor proteins with novel functions. *Nature*, 423, 185–190, 2003.
9. Tolosa, L., Kostov, Y., Harms, P., and Rao, G., Noninvasive measurement of dissolved oxygen in shake flask. *Biotechnol Bioeng*, 80, 595–597, 2002.
10. Whitmann, C., Kim, H.M., John, G., and Heinzle, E., Characterization and application of an optical sensor for quantification of dissolved oxygen in shake flasks. *Biotechnol Lett*, 25, 377–380, 2003.
11. Kermis, H.R., Kostov, Y., Harms, P., and Rao, G., Dual excitation ratiometric fluorescent pH sensor for noninvasive bioprocess monitoring: development and application. *Biotechnol Prog*, 18, 1047–1053, 2002.
12. Ge, X., Kostov, Y., and Rao, G., High-stability non-invasive autoclavable naked optical CO_2 sensor. *Biosens Bioelectron*, 18, 857–865, 2003.
13. Zabrieski, D.W. and Humphrey, A.E., Estimation of fermentation biomass concentration by measuring culture fluorescence. *Appl Environ Microbiol*, 35, 337–343, 1978.
14. Marose, S., Lindemann, C., and Scheper, T., Two-dimensional fluorescence spectroscopy: a new tool for on-line bioprocess monitoring. *Biotechnol Prog*, 14, 63–74, 1998.
15. Tolosa, L., Gryczynski, I., Eichorn, L., Dattelbaum, J., Castellano, F.N., Rao, G., and Lakowicz, J.R., Glucose sensor for low cost lifetime-based sensing using a genetically engineered protein. *Anal Biochem*, 267, 114–120, 1999.
16. Dattelbaum, J.D. and Lakowicz, J.R., Optical determination of glutamine using a genetically engineered protein. *Anal Biochem*, 291, 89–95, 2001.
17. Price, J.M., Xu, W., Demas, J.N., and DeGraff, B.A., Polymer-supported pH sensors based on hydrophobically bound luminescent ruthenium (II) complexes. *Anal Chem*, 70, 265–270, 1998.
18. Tsien, R.Y., Fluorescent indicators of ion concentrations. *Methods Cell Biol*, 30, 127–156, 1989.
19. Lutty, G.A., The acute intravenous toxicity of stains, dyes, and other fluorescent substances. *Toxicol Pharmacol*, 44, 225–229, 1978.
20. Zhujun, Z. and Sitz, W.R., A fluorescence sensor for quantifying pH in the range from 6.5 to 8.5. *Anal Chim Acta*, 160, 47–55, 1984.
21. Severinghaus, J.W. and Bradley, A.F., Electrodes for blood pO_2 and pCO_2 determination. *J Physiol*, 13, 515–520, 1958.
22. Chang, Q., Randers-Eichhorn, L., Lakowicz, J.R., and Rao, G., Steam-sterilizable, fluorescence lifetime-based sensing film for dissolved CO_2. *Biotechnol Prog*, 14, 326–331, 1998.
23. Sipior, J., Randers-Eichhorn, L., Lakowicz, J.R., Carter, C.M., and Rao, G., Phase fluorometric optical carbon dioxide gas sensor for fermentation off-gas monitoring. *Biotechnol Prog*, 12, 266–271, 1996.
24. Mills, A. and Chang, Q., Modelled diffusion-controlled response and recovery behavior of a naked optical film sensor with a hydrophobic-type response to analyte concentration. *Analyst*, 117, 1461–1466, 1992.
25. Harms, P., Sipior, J., Ram, N., Carter, G.M., and Rao, G., Low cost phase-modulation measurements of nanosecond fluorescence lifetimes using a lock-in amplifier. *Rev Sci Instrum*, 70, 1535–1539, 1999.
26. Kostov, Y., Harms, P., Pilato, R.S., and Rao, G., Ratiometric oxygen sensing: detection of dual-emission ratio through a single emission filter. *Analyst*, 125, 1175–1178, 2000.

27. Moser, I., Jobst, G., Ashcauer, E., Svasek, P., Varahram, M., and Urban, G., Miniaturized thin film glutamate and glutamine biosensors. *Biosens Bioelectron,* 10, 527–532, 1995.
28. Renneberg, R., Trott-Kriegeskorte, G., Lietz, M., Jager, V., Pawlowa, M., Kaiser, G., Wollenberger, U., Schubert, F., Wagner, R., Schmid, R.D., and Scheller, F.W., Enzyme sensor-FIA-system for on-line monitoring of glucose, lactate and glutamine in animal cell cultures. *J Biotechnol,* 21, 173–186, 1991.
29. Riley, M.R., Arnold, M.A., Murhammer, D.W., Walls, E.L., and DelaCruz, N., Adaptive calibration scheme for quantification of nutrients and byproducts in insect cell bioreactors by near-infrared spectroscopy. *Biotechnol Prog,* 14, 527–533, 1998.
30. Tolosa, L., Ge, X., and Rao, G., Reagentless optical sensing of glutamine using a dual-emitting glutamine-binding protein. *Anal Biochem,* 314, 199–205, 2003.
31. Baker, K.N., Rendall, M.H., Patel, A., Boyd, P., Hoare, M., Freedman, R., and James, D.C., Rapid monitoring of recombinant protein products: a comparison of current technologies. *Trends Biotechnol,* 20, 149–156, 2002.
32. Li, J., Xu, H., Herber, W.K., Bentley, W.E., and Rao, G., Integrated bioprocessing in *Saccharomyces cerevisiae* using green fluorescent protein as a fusion partner. *Biotechnol Bioeng,* 79, 682–693, 2002.
33. DeLisa, M.P., Li, J., Rao, G., Weigand, W.A., and Bentley, W.E., Monitoring GFP-fusion protein expression during high cell density cultivation of *Escherichia coli* using an on-line optical sensor. *Biotechnol Bioeng,* 65, 54–64, 1999.
34. Kostov, Y., Albano, C.R., and Rao, G., All solid-state GFP sensor. *Biotechnol Bioeng,* 70, 473–477, 2000.
35. Kostov, Y., Harms, P., Randers-Eichhorn, L., and Rao, G., Low-cost microbioreactor for high-throughput bioprocessing. *Biotechnol Bioeng,* 72, 346–352, 2001.

14 Practical Aspects of Fluorescence Analysis of Free Zinc Ion in Biological Systems: pZn for the Biologist

Richard B. Thompson, Ph.D., Christopher J. Frederickson, Ph.D., Carol A. Fierke, Ph.D., Nissa M. Westerberg, Ph.D., Rebecca A. Bozym, B.S., Michele L. Cramer, M.S., and Michal Hershfinkel, Ph.D.

CONTENTS

14.1 INTRODUCTION

The biology of zinc has enjoyed a dramatic growth in attention over the last 20 years or so, as it has become apparent that zinc plays many roles in living systems in addition to serving as an enzymatic cofactor. At the same time, tools for studying zinc biology (mainly based on fluorescence) have been developed by us and others which have greatly aided our understanding of zinc in living systems of all kinds. However, zinc is maintained at very low free concentrations in biological systems, which makes the study of zinc biology challenging. In this chapter we confront some of these challenges and provide best practices for preparing zinc solutions and measuring zinc concentrations.

14.2 BASIC CHEMISTRY OF ZINC

As might be anticipated, aspects of the basic chemistry of zinc ions in aqueous solution impact experimental studies of its biology. Fortunately, it is unnecessary to compile a complete treatise on its chemistry to cover these issues; those interested may consult reviews on this topic.[1,2] There is a voluminous literature on the role of zinc in enzymatic catalysis[3–6] that constitutes the vast majority of our understanding of the biochemistry of zinc, but recently unique biological roles for zinc in the stabilization of protein structure[7] and neurochemical signaling[8–10] have been described. For studies involving optical imaging or analysis of zinc (especially "free" zinc), issues of solubility, zinc ion buffering, zinc speciation (what proportion of the total zinc ion is bound to which ligands), and purity of reagents are important.

Elemental zinc is found in Group IIB[1,2] of the periodic table, so zinc is not a transition element in that it does not form complexes with unfilled d shells. Zinc is diamagnetic, relatively electropositive, and almost exclusively found as the divalent cation Zn(II) (unlike copper or iron, which have multiple ionization states). Zn(II) is viewed as being on the borderline between "hard" and "soft," and although it exhibits coordination numbers ranging from 2 to 8 with a variety of ligands, tetrahedral coordination with sulfur and/or nitrogen ligands is most prevalent in biological systems.[11,12] Due to the filled d orbitals, a zinc ion does not exhibit electronic transitions in the near-ultraviolet (UV) or visible range in any of its ligand states, and consequently (again unlike iron and copper) optical absorption spectroscopy of metal-bound complexes has only been informative when using Co(II) as a surrogate.[13,14] Similarly, all of the naturally occurring isotopes of zinc, except ^{67}Zn at 4% natural abundance, have a nuclear spin angular momentum quantum number $I = 0$, making zinc effectively nuclear magnetic resonance (NMR)-silent; NMR studies of zinc biochemistry have mainly exploited ^{113}Cd as a surrogate,[15,16] although solid-state zinc NMR spectroscopy is being developed.[17] It is likely that the relative inaccessibility of zinc to these methods has hindered its study and perhaps allowed its importance to be underestimated; certainly the important class of zinc finger proteins would have been identified decades sooner if the bound zinc ion had a visible absorbance spectrum.

Most salts of Zn(II) are viewed as being soluble in water; important exceptions include zinc phosphate, carbonate, and pyrophosphate. However, an important caveat is that near neutrality, free zinc ions form essentially polymeric hydroxo complexes, which readily precipitate.[18] A practical consequence of this behavior is that a moderately dilute solution (say 1 mM) of zinc chloride ($ZnCl_2$) in water near pH 7 over time forms a visible precipitate and the soluble zinc concentration decreases significantly. Moreover, the kinetics of precipitate formation are slow, variable, and strongly pH- and zinc concentration–dependent, such that the precipitate may not be apparent immediately after mixing but substantial after some hours. Because of this phenomenon, it is futile to prepare zinc solutions near neutrality at just about any concentration range: at high zinc concentrations, precipitation occurs faster, but at lower zinc concentrations, other ligands normally present in biological specimens and solutions may bind the free zinc, changing its concentration. A corollary to this is that experiments involving the addition of zinc solutions of a given concentration to a specimen are likely to fail because the initial concentration of zinc will be suspect, and depend upon the concentration of ligands in the specimen itself (see Section 14.3), the free zinc concentration is likely to be unknown. The well-known solution to this problem is the use of a zinc "buffer," analogous to a pH buffer (see Section 14.4).

14.3 SPECIATION

In aqueous solution, no metal ion is truly "free" (except transiently); for instance, Zn(II) is found only in solution with coordinated water molecules (i.e., "aquo" zinc) as well as with anions and other ligands. Thus, "free" is an operational definition that refers to that proportion of the total zinc that is bound by weak, rapidly exchangeable ligands such as water or chloride. Although "pZn" may be defined as −log[free Zn], by analogy with pH, it effectively means the available zinc concentration that is not sequestered by tight binding to some complex. From the standpoint of biology, the free zinc is proposed to be "active," in that it is free to bind to other molecules in aqueous solution; however, other zinc complexes may also be "active" and able to exchange zinc with some other ligand, sometimes with more rapid kinetics. There is a time dimension to this, in that the proportion of zinc that is "available" is also limited by kinetics: in principle, over time a sufficiently tight binding ligand will ultimately bind all the available metal, but if the dissociation rate constants from the first complex are slow, this may take a very long time (see Section 14.6.3).

In complex media such as cytoplasm or blood, one may anticipate that the metal ion will become bound to a range of ligands; this is called the "speciation" of the metal ion. Marine chemists have studied this phenomenon more extensively than biochemists. It should be noted that the significant fraction of metal ions adsorbed to colloidal particles in sea water probably has no analog in terrestrial organisms.[18] Quantitatively, the fraction of total zinc bound by one ligand (X_i) among i ligands is proportional to its association constant (K_i) and the free ligand concentration [L]:

$$X_i = K_i[L]/1 + K_i[L] \qquad (14.1)$$

and the fraction of total zinc that remains free is

$$[Zn]free/[Zn]total = 1/(1 + K1[L1] + K2[L2] + K3[L3] + \ldots). \qquad (14.2)$$

Many molecules found in biological systems contain amine, thiol, and carboxylate functional groups that form quite stable complexes with Zn(II); among these

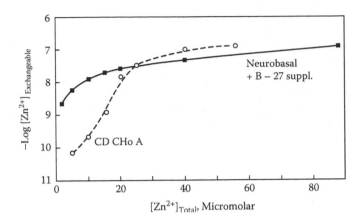

FIGURE 14.1 Log of [Zn]$_{exchangeable}$ as a function of total zinc calculated for Neurobasal plus supplement (squares) and CD CHO A (circles) tissue culture media.

are cysteine, histidine, and other amino acids; glutathione; proteins such as serum albumin; α_2-macroglobulin; metallothionein; and many others.[4,19] These ligands are in addition to hundreds of proteins, such as metalloenzymes and zinc fingers, whose zinc complexes are very stable indeed. The inescapable corollary to this is that free zinc added to a biological system is bound by a variety of ligands, and the remaining free (and biologically active) zinc is decreased. In fact, the fraction of total zinc that is free depends on the number, concentration, and affinity of the ligands (Equation 14.2). Figure 14.1 depicts calculated (using MINEQL+; Environmental Research Software, Hallowell, ME) concentrations of exchangeable zinc (operationally equivalent to pZn) as a function of total added zinc in a serum-free medium used to culture neurons (Neurobasal plus B-27 supplement)[20] as well as a proprietary protein-free medium (Invitrogen CD CHO A).

These data demonstrate that the concentration of free zinc may be very low under these conditions. For instance, addition of 25 μM total zinc to either medium results in a free zinc concentration of roughly 20 nM, or approximately 1000-fold less. At lower total zinc, the media exhibit much different pZns, with the CD CHO A being nearly 100-fold lower. This is attributable to the presence of a strong ligand in the CD CHO A formulation at just under 20 μM; when this ligand becomes saturated, their behavior is more similar. The Neurobasal medium (including supplement) contains substantial (known) amounts of serum albumin as well as cysteine, histidine, and glutamic acid and is unusual in having a predetermined level of total zinc. Most mammalian cell culture media contain serum, which has variable amounts of total zinc (8 μM to 16 μM), as well as unknown amounts of zinc ligands (especially albumin at 0.3 mM), so the zinc speciation cannot be readily calculated. By comparison, artificial cerebrospinal fluid without phosphate contains no strong ligands, and the same amount of total zinc results in almost 50% free zinc (results not shown). The inescapable conclusion to be drawn from these results is that, in general, free zinc does not equal total zinc, and adding even substantial quantities of total zinc to a cell culture medium will likely result in free zinc levels that are orders of magnitude different. In light of these results, a prudent observer would properly be skeptical of many results in the literature. As will be seen below, the correct approach to this problem (as for pH) is the use of zinc buffers.

14.4 METAL ION BUFFERS

Just as one would not perform experiments without a pH buffer, in general one should not attempt experiments at low free metal concentrations without the use of a metal ion buffer. The justifications for using metal ion buffers are similar to the reasons for pH buffers: there may be ligands present (either in solution or on the walls of the container) that will lower the free metal ion concentration (see above), and there may also be excess metal ions present either as impurities in the solution or which leach out from the material of the container. Any compound that coordinates metal ions — a chelator — may be used as a metal ion buffer, although the term is not exactly analogous to a pH buffer. Under certain conditions, the dilution of a metal ion buffer results in a change in the equilibrium free metal concentration, unlike pH buffers. In most cases, the free metal concentration is much lower than the total metal and total chelator concentrations. This technique allows experiments to be carried out at low free metal concentrations that would not be feasible without the reserve of chelator-bound metal ions present in the buffer. However, this convention also leads to properties that are different than the well-known pH buffers. Knowledge about the use and limitations of metal ion buffers is very important for measurements involving low metal concentrations.

In general, the common chelators used for metal ion buffers do not have pK_as that are useful for also buffering the pH of solutions near physiological conditions, so a pH buffer is included with the chelator. When selecting the pH buffer, both the pK_a of the buffer and its metal binding capability must be considered. Common biological buffers containing organic amines (such as Tris and bicine) readily coordinate metals near physiological pH.[21] Phosphate buffers should be avoided because of the extremely low solubility of metal–phosphate complexes — in particular, iron and zinc. Chlorides, hydroxides, and carbonates also form insoluble metal complexes.[22] Sulfonic acid derivative pH buffers (3-[N-morpholino]propanesulfonic acid [MOPS] and 2-(N-morpholino)ethanesulfonic acid [MES] are two examples) are good choices for maintaining the pH of metal ion buffers due to their lower metal ion binding affinities near physiological pH. Yu and others have proposed a series of noncomplexing pH buffers that span a wide pH range (3–11).[23,24] However, only complex formation with copper was investigated, so caution should be used in extrapolating these buffers to solutions with metals other than copper. In some cases, weak metal complexation behavior of a buffer may be explained by small amounts of impurities with strong metal affinities rather than weak complexation by the buffer.[25]

Knowledge of metal stability constants is essential in the selection of a chelator. As shown in Equation 14.3–Equation 14.5, stability constants (or stepwise formation constants, K) are chemical equilibrium constants for a single complexation step. Formation constants (β) refer to a product of stability constants and encompass all the steps necessary to form a particular complex. A conditional formation constant is simply the product of the formation constant with the fraction of the chelator species at a particular pH that is able to form the species of interest. This is especially useful when performing manual calculations.

$$M + L \xleftrightarrow{K_1} ML + L \xleftrightarrow{K_2} ML_2 + L \xleftrightarrow{K_3} ML_3 \qquad (14.3)$$

$$M + 2L \xleftrightarrow{\beta_2 = K_1 K_2} ML_2 \qquad (14.4)$$

$$M + 3L \xleftrightarrow{\beta_3 = K_1 K_2 K_3} ML_3 \qquad (14.5)$$

Commonly used chelators include ethylenediaminetetraacetic acid (EDTA), ethyleneglycol-bis(aminoethylether)-tetraacetic acid (EGTA), nitrilotriacetic acid (NTA), and citrate. The stability constants of a wide range of metal complexes are available in references such as *Critical Stability Constants*,[22] which is frequently updated and available online as "NIST Standard Reference Database 46" at www.nist.gov/srd/nist46.htm. These references compile values from literature reports and include references to the original publications. Stability constants depend upon conditions such as ionic strength and temperature. It is possible to correct stability constants for altered conditions to some degree, but such corrections are difficult at high ionic strength ($\mu > 0.5$ M), pH values that are less than 4 or greater than 10, or high temperatures (greater than 50°C).[26]

Care must also be taken to ensure that the metal–chelator complex does not form a ternary complex with the protein or peptide of interest. Ternary complex formation is indicated if altering the total concentration of the chelator or using a different chelator system results in different rate constants or apparent K_D values. Magyar and Godwin[27] recently showed that the major reason for the large differences (five orders of magnitude) in metal affinities reported for cobalt binding to a peptide was the formation of peptide–ligand–metal complexes. The formation of a ternary complex does not have to be totally avoided. In the case of carbonic anhydrase II (CA), dipicolinic acid (DPA) is able to form a transient ternary complex with the enzyme-bound zinc, greatly facilitating the removal of zinc to form apo enzyme.[28] This indicates, however, that the formation of a CA–DPA–zinc complex must be taken into account when using this buffer to measure metal affinity and kinetics.

For simple systems — one metal, one chelator, and one metal–chelator complex — calculations of the free metal concentration may be carried out manually. An overview of manual calculations is described by Baker.[29] The calculation process quickly becomes arduous if multiple metal–ligand complexes form, if the pH buffer also binds metals, if ionic strength or temperature corrections are desired, or if a metal ion buffer with multiple metals is desired. Aslamkhan et al.[26] recently offered a review of various freeware/shareware programs available for calculations and their limitations. We have used the commercially available Windows-based computer program MINEQL+. These programs include stability constants of many known species as well as the ability to add new species of interest.

Although these solutions are referred to as buffers, metal–chelator systems do not always act in the same way as pH buffers. In the region around the dissociation constant for a specific chelator–metal pair, the free concentration of that metal ion will be resistant to change with the addition or loss of small amounts of metal. However, the concentration region where the chelator functions to buffer the metal ion concentration in this classical fashion is limited ($pK_D \pm 1$ unit). Accessing a wider range of free metal ion concentrations using these conditions requires the use of either a combination of chelators with closely spaced K_Ds or different chelators for each concentration range. Both of these methods are used for obtaining a wide range of buffered pH solutions. However, the use of multiple chelators increases the risk of obtaining ternary complex formation that will complicate the data analysis.

A simpler alternative to working near the K_D is to use conditions in which the chelator concentration is significantly higher than both the K_D and the total metal concentration. Under these conditions, the free metal concentration (pM) is linearly dependent on the equivalents of added zinc (see Figure 14.2) over many orders of magnitude. Operating in this range allows for an extremely wide range of free metal concentrations that, although not as strongly buffered, provides a reservoir of metal ions. In the example in Figure 14.2, increasing the total zinc from 1 μM to 10 μM only raises the pZn ($-\log[\mathrm{Zn}]$) from 12.3 to 11.3, for example, from 0.5 pM to 5 pM, a significantly smaller change in concentration. Particularly at low free and total metal concentrations, binding of a metal ion to a biological ligand can significantly affect the free metal concentration. Although at higher metal concentrations, the biological ligand binds more moles of metal, the larger reservoir of metal ions present makes such changes less significant. It may also be useful — especially if anomalous results are observed — to determine the total metal ion concentration using an independent technique such as ICP-MS or ICP-AES. Once the total concentrations of the metals are known, the free metal concentrations may be recalculated and compared with the measured value. Although it is not necessary to do this for every metal ion buffer each time they are prepared, this step can identify major contaminants or aid in the understanding of anomalous results.

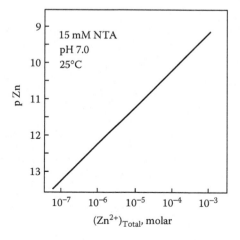

FIGURE 14.2 pZn as a function of total [Zn], for [Zn] << [NTA].

14.5 PURITY OF REAGENTS

A corollary to the determination or maintenance of free zinc ion at micromolar levels or less is the potential for interference by other metal ions. The chemistry of sodium, potassium, calcium, and magnesium cations in solution is sufficiently different that in biological systems they seldom interfere with the biological effects of zinc; that is, these ions will seldom bind to receptors with high-enough affinity to interfere with the biological effect of zinc. However, these ions can bind to chelators used to buffer or detect zinc (see above). By comparison, divalent copper, cadmium, cobalt, nickel, and especially mercury cations are likely to bind with high affinity (K_D tighter than micromolar) to many zinc sites and, thus, interfere. Indeed, with sulfur ligands, mercury will usually exhibit dramatically higher affinity. In the preparation of buffers and artificial cerebrospinal fluids, contaminating cations are thus a potential problem. "Analytical reagent"–grade salts, as specified by the American Chemical Society, permit some cation impurities at parts per million levels. Thus, a buffer with components at millimolar concentrations is likely to have impurities at nanomolar concentrations, which might obscure the effect of zinc at similar concentrations.

To avoid this problem, we used two expedients. To remove divalent cations from monovalent cation salt solutions, we used resins with immobilized iminodiacetate moieties (Chelex) which effectively filter divalent cations from aqueous solutions at pH 4 or higher. Although some of these resins can leach off iminodiacetate residues, we have not found this to be a problem. Of course, most salt solutions mimicking serum or cerebrospinal fluid will contain calcium and magnesium, which are also effectively retained on the Chelex column, and consequently the above procedure is infeasible. We have had good success preparing artificial cerebrospinal fluid by mixing all the components except the calcium and magnesium salts together, running the mixture over the iminodiacetate column, and then adding the calcium and magnesium salts. These salts must be essentially free of zinc and the other potential contaminants listed above. We have found "ultrapure" or "electronic" grades (e.g., $CaCl_2$ [EM Suprapur grade, catalog no. 2384-2] and magnesium sulfate [ALFA Aesar Puratronic grade, catalog no. 10801]) to be satisfactory. The water produced by laboratory water ultrapurification systems that remove cations by ion exchange and organics by adsorption and ultrafiltration (e.g., from models produced by Barnstead and Millipore) has usually proven adequate. Of course, making up the buffers using entirely high-purity reagents and water is possible but is seldom economical for most laboratories.

As a general rule of thumb, the procedures introduced by analytical chemists for ultratrace analysis are effective in avoiding contamination, but (except for laminar flow hoods) most biochemistry labs lack the equipment to perform them. Nevertheless, contamination can be minimized. Because of the possibility of contaminating samples by leaching metal ions out of glassware or metal implements, we eschew these things and use plasticware wherever possible. Most new plasticware we have tried has been satisfactory, for instance, plastic Pasteur pipets (Samco catalog no. 694 Q-PET, Samco Scientific, San Fernando, CA), 50-ml centrifuge tubes with caps and flat bottoms (Corning catalog no. 430921), 1.5-ml microfuge tubes (Costar catalog no. 3620), and VWR "TraceClean" HDPE 1-L bottles (catalog no. 15900-306). Other manufacturers doubtless make equivalents that will be acceptable. We have found it important to use "natural" or "metal-free" pipet tips (e.g., Bio-Rad BR-41or VWR catalog no. 53509-007) because the colored dyes used in pipet tips produce substantial metal contamination. Plasticware occasionally encounters zinc-containing mold-release compounds during processing, so all sources should be tested or briefly rinsed with dilute acid to avoid contamination. The use of metal ion buffers also minimizes the effect of contamination, because most of the metal ions of concern are also tightly bound by the buffers in use. Thus, using NTA to buffer zinc also nearly eliminates free Cu(II).

14.6 FLUORESCENCE ANALYSIS

A popular technique for analyzing metal ions and other constituents of biological systems is fluorescence, due to its sensitivity, simplicity, and versatility. In biomedical science, a key advantage is the ability to image fluorescence in the microscope and thus correlate analyte presence (or concentration) with morphological features of the cell or tissue. For instance, the ratiometric fluorescent indicators introduced by Grynkiewicz et al.[30] have revolutionized our understanding of the calcium ion's role in cell signaling by permitting accurate quantitation. We and others have developed a variety of fluorescence methods for zinc determination. Issues with all fluorescent indicator systems that act by reversibly binding zinc (whether small molecule- or protein-based (biosensors)) are their affinities, selectivities, and kinetics of binding. In addition, there are often issues having to do with the samples and their volume. Finally, the value of a particular approach for rapidly sequestering free zinc is considered.

14.6.1 AFFINITY

For the analysis of metal ions in biological systems, one popular approach has been the use of metallofluorescent indicators, which exhibit changes in fluorescence upon binding to the metal. For such indicators to work, they must have high-enough affinity as ligands for the metal ion to exhibit significant fractional saturation at the free metal ion concentrations present inside cells. To measure the nanomolar to micromolar levels of free calcium present inside cells, the indicators of Grynkiewicz et al. and others were based mainly on EGTA, which has high affinity ($K_d \sim 10$ pM) for calcium. For zinc in living systems, high affinity is generally necessary because free zinc levels appear to be low, in the picomolar to nanomolar range (with several notable exceptions).[8,31,32] The actual "resting" level of zinc in typical eukaryotic and prokaryotic cells is controversial.[33] For recent small-molecule zinc fluorescent indicators, published affinities have ranged from micromolar to nanomolar levels.[34-40] The fractional occupancy θ (for binding to a single site or class of sites) as a function of zinc concentration is given by the well-known equation

$$\theta = [Zn]K_\theta/(1 + K_A[Zn]) \qquad (14.6)$$

An important corollary to this equation is that θ ranges from 10% to 90% occupancy over slightly less than two log units of zinc concentration and that unless θ can be determined with unusual precision (e.g., better than a few percent[41]), this

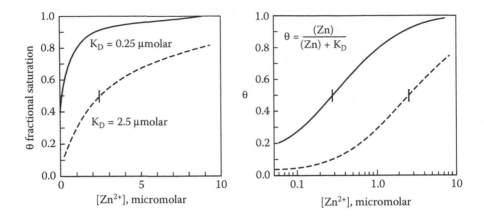

FIGURE 14.3 Fractional saturation θ as a function of free zinc for two different K_d values: left panel shows a linear plot, and the right shows a semilog (Bjerrum) plot.

effectively limits the dynamic range of this approach to an order of magnitude above and below the K_D (Figure 14.3).[42] This formalism requires that the binding be reversible and that all of the species remain in (aqueous) solution; this is not the case for the indicator TSQ,[43] because the TSQ–zinc complex is largely insoluble,[44] which confounds quantitation using this probe. The semilog plot of free zinc vs. fractional saturation θ in Figure 14.3 (right panel) introduced by Bjerrum[45] (and recognizable as the inverse of the common pH titration) is greatly preferable to linear plots because the data at low zinc concentrations are not compressed into an illegibly small space (Figure 14.3, left panel), which enables the whole binding curve to be observed. It is also superior to double reciprocal plots because the errors in the data points are not distorted at high and low values.[46]

14.6.2 Selectivity

For any fluorescence indicator used in natural samples, the issue of selectivity for the analyte is nearly as important as how great the affinity is: an indicator needs to have both adequate affinity to bind the concentrations that are present and sufficiently low affinity to other species so that they do not interfere. For free zinc in living systems (or natural waters), free calcium and magnesium ions may be present at million-fold higher concentrations, such that indicators displaying even modest affinity for these cations are easily saturated. This issue (the fractional saturation with a variety of metal ions of a single indicator or ligand) is essentially the inverse of that raised in speciation (the fractional saturation of different ligands with a single metal ion of interest). If the affinities for the ligand (K_i) and free concentrations of potential interferents [Me_i] are known, the degree of interference (fractional saturation of the ligand with any interferent Me_i) may be readily calculated using Equation 14.7:

$$X_i = K_i[Me_i]/1 + (\Sigma K_i[Me_i]) \tag{14.7}$$

Of the scores of small-molecule fluorescent zinc indicators described in the literature (for reviews of older literature, see White and Argauer[47] and Fernandez-Gutierrez and Munoz de la Pena[48]), the majority are unsatisfactory for determining zinc concentrations in biological systems because they have significant affinity for calcium, magnesium, or both. It was noted early on that the EGTA derivatives such as Quin-2 and Fura-2 introduced for calcium determination had 100-fold higher affinity for zinc than for calcium,[30] which led to some of these derivatives' being proposed for zinc determination.[40,49,50] However, having to correct for the potentially large and variable response from the calcium and magnesium present is unsatisfactory, and these probes have now been superseded by ones based on 2-methoxyphenyl iminodiacetate[37] (e.g., FuraZin-1) and *bis*-(2-methylpyridyl)amine (e.g., Newport Green DCF).[51] Although the affinities of these indicators are little better than nanomolar, they neither bind to nor exhibit fluorescence response for calcium or magnesium at physiological levels (Figure 14.4).[40] The distinction between binding and eliciting a fluorescence change is crucial, because several fluorescence indicators have been described as selective because they do not display fluorescence changes at some concentration of potential interferent. However, even if they bind the interferent without a change in fluorescence, the result is erroneous, as clearly shown in Equation 14.7. The correct approach is to examine the fluorescence response in sets of buffers formulated to provide the relevant range of zinc concentration in the absence and presence of the potential interferent; then compare the apparent K_Ds in Figure 14.4. In the case of Newport Green DCF, depicted in Figure 14.4, the apparent K_D for zinc differs only slightly in the presence and absence of calcium and magnesium at millimolar levels.

Because of the need for selectivity, most recent small molecule indicators have focused on the three chelating motifs listed above. The available indicators seem to have affinities in the nanomolar range. Unfortunately, greater sensitivity than nanomolar is needed for typical cells. Apart from perhaps Henary et al.'s recent

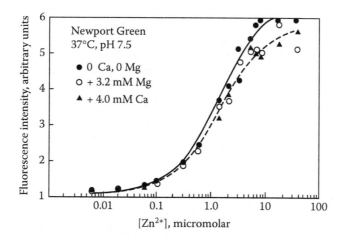

FIGURE 14.4 Interference testing. The intensity of the indicator Newport Green DCF is shown as a function of free zinc in the presence of 3.2 mM magnesium (open circles), 4.0 mM calcium (triangles), and no added metal (filled circles) (Redrawn from Reference 40.).

development,[52] the only indicator systems demonstrably selective and sensitive enough to quantitate zinc at subnanomolar levels in complex matrices are those based on carbonic anhydrase,[53] reviewed in Fierke and Thompson.[54] Briefly, these indicators bind zinc in the active site of the apoenzyme (as occurs *in vivo*) and transduce its presence as a change in fluorescence intensity, emission ratio,[53] anisotropy,[55] lifetime,[56] or excitation ratio.[57] A unique advantage of this approach is that by making subtle modifications in the protein structure, the affinity for zinc may be modulated over six orders of magnitude, the selectivity for zinc in comparison to other metals may be usefully modified, and the kinetics of zinc binding (and thus indicator response time) may be improved as much as 10,000-fold while losing less than a factor of 10 in affinity.[54] We shall discuss these systems below in the context of measuring free zinc in the brain.

14.6.3 KINETICS

The kinetics (rate) with which fluorescent indicator systems respond to an analyte are of central importance because this imposes a limit on how rapidly a cellular process can be observed with fidelity. A process completed in microseconds cannot be accurately measured by an indicator with a characteristic response time of seconds. Most fluorescence indicators for Zn(II) bind the metal reversibly, such that the equilibrium constant K_d may be expressed as the ratio of association (k_{on}) and dissociation (k_{off}) rate constants:

$$K_d = k_{off}/k_{on} \qquad (14.8)$$

We can calculate the approximate time for the system to respond (coming arbitrarily close to equilibrium) if we know the rate constants for association and dissociation and the concentrations of the ligand and analyte. As an example, we may examine apo-carbonic anhydrase binding to free Zn(II). Human apo-carbonic anhydrase II (apo-CA) binds zinc very tightly ($K_d = 4$ pM at pH 7), and the association rate constant (on rate) is rather slow (~3×10^5/M/s), about 10,000-fold slower than diffusion-controlled.[58] When a rapid response is useful, it is preferable to use variants of the protein developed in C.A. Fierke's laboratory, which exhibit more rapid kinetics. Among the fastest of these is E117Q CA,[59] which has a nearly diffusion-controlled on rate (3×10^8/M/s). This means it binds Zn(II) as fast as it encounters it and as fast as possible for a chemical agent of any kind (but a small molecule chelator would itself diffuse somewhat faster than the protein). The zinc affinity of E117Q CA is reduced compared with the wild type (to 4 nM), whereas E117A CA binds zinc substantially faster than wild-type CA (1×10^7/M/s) but loses only a factor of 10 in affinity.

To estimate how long it would take the apo-CA to scavenge free Zn(II), we use the following equation, assuming that the concentration of CA is five-fold or more greater than the free Zn(II) concentration and treating this as a pseudo-first-order rate equation (see chapter 6 of Stumm and Morgan[18] for a more detailed discussion):

$$k_{OBS} = k_{on}[\text{apo-CA}] + k_{off} \qquad (14.9)$$

where k_{OBS} is the observed rate constant and k_{on} and k_{off} are the association and dissociation rate constants. For E117Q CA, the off rate is slow enough (1.3/s) that we can neglect its contribution to the observed rate if the protein concentration [apo-CA] is much greater than K_D (4 pM); because micromolar protein concentrations are frequently used, this is surely the case. Therefore, the equation simplifies to

$$k_{OBS} \sim k_{on}[\text{apo-CA}] \qquad (14.10)$$

For 5 µM E117Q CA, k_{OBS} will thus be about 1500/s. The half time of the reaction (time it takes to reach 50% of completion) equals ln $2/k_{OBS}$, or about 0.46 ms. By comparison, the wild type exhibits a k_{OBS} of ~1.5 per second and a half time of about 0.5 s. Thus, the wild type is about 10^3-fold slower to respond under these conditions.

14.7 USE OF CaEDTA AS A SPONGE FOR FREE ZINC

Recently there has been substantial interest in the use of the calcium chelate of EDTA (CaEDTA) as a rapid and selective "sponge" for free zinc ion released by brain tissue in response to various stimuli.[60] The concept is that because EDTA's affinity for zinc is much higher than for calcium, the bound calcium will be rapidly replaced by zinc:

$$\text{Ca(II)-EDTA} + \text{Zn(II)} \rightleftarrows \text{Zn(II)-EDTA} + \text{Ca(II)} \qquad (14.11)$$

Marine chemists call this an exchange reaction, and it has elicited much study because of its importance in determining the speciation and bioavailability of metal ions in fresh water and sea water. The exchange of zinc and copper with CaEDTA has elicited study because EDTA is a pollutant present at submicromolar concentrations. Hering and Morel[61] have the clearest exposition of the issues in this type of reaction and describe the reaction occurring by one of two mechanisms: adjunctive or disjunctive. In the adjunctive mechanism, the zinc binds to the CaEDTA to form a ternary complex before the calcium falls off:

$$\text{Ca(II)-EDTA} + \text{Zn(II)} \rightleftarrows \text{Ca(II)-EDTA-Zn(II)} \rightleftarrows \text{Zn(II)-EDTA} + \text{Ca(II)} \qquad (14.12)$$

In the disjunctive mechanism, the Ca(II) falls off first and the Zn(II) binds to the free EDTA:

$$\text{Ca(II)-EDTA} \rightleftarrows \text{EDTA} + \text{Ca(II)} + \text{Zn(II)} \rightleftarrows \text{Zn(II)-EDTA} \qquad (14.13)$$

As expected, the kinetics are sensitive to excess free Ca(II) for the disjunctive mechanism because it reduces the free EDTA concentration, but this is not true not for the adjunctive. Hering and Morel found in studying the reaction of Cu(II) with CaEDTA that at low Ca(II) concentrations the copper exchange reaction is adjunctive, but at higher free Ca(II) (about 2 mM), the kinetics decreased and the reaction became disjunctive. The rate of the exchange reaction for copper with the adjunctive mechanism, expressed as a second-order rate constant, is

$$-d[\text{Cu}]/dt = k_{obs}[\text{CaEDTA}][\text{Cu}] \qquad (14.14)$$

and the half-life of free copper under these conditions is just

$$\ln 2/k_{obs} = \text{half-life} \qquad (14.15)$$

The rate constant and relative contributions of the two mechanisms are a function of pH and calcium concentration. For copper in sea water at pH about 8.4 and 10 mM calcium, k_{obs} is about 970/M/s. For 1.25 μM CaEDTA and nanomolar concentrations of copper (which we can therefore neglect in the rate expression), the half-life is about 570 s. Two groups have measured exchange reactions for Zn(II) and CaEDTA and they obtain second-order rate constants of 300/M/s and 1100/M/s, respectively.[62,63] Both groups carried out their studies in fresh water, so there might be some variations in their rates compared to the intracellular milieu, but Hering suggested that the rate of zinc exchange would generally be slower than for copper because bound water can be stripped from copper faster than zinc. Using the faster rate constant, with 10 mM CaEDTA and once again neglecting zinc (at micromolar concentrations), the half-time is about 60 ms, and at the slower rate it is 230 ms. These results suggest that CaEDTA at relatively high concentrations can rapidly scavenge free zinc. Although a concentration of calcium equal to the bound zinc is released, extracellularly this is negligible.

14.8   ISSUES IN QUANTITATING FREE ZINC IN SMALL VOLUMES AND COMPARTMENTS

When measuring free zinc levels within circumscribed volumes (such as the inside of a cell) or of small samples such as those obtained by dialysis probes within the brain (see below), other important but often overlooked issues come into play. Of course, reversible binding of zinc by a protein such as apo-carbonic anhydrase or a fluorescent indicator (generically a receptor, called P in this example) at a single class of sites may be described theoretically by the well-known equation

$$K_D = [Zn][P]/[PZn]. \qquad (14.16)$$

where [PZn] is the concentration of the complex and K_D is the dissociation constant. One may define a fractional occupancy of binding sites v, which can be expressed as a function only of K_D and the free zinc concentration [Zn]:

$$v = [Zn]/([Zn] + K_D). \qquad (14.17)$$

This is strictly true only when the species are all in solution and if binding to the receptor P does not appreciably change the concentration of free zinc. Thus, by measuring a change in intensity or the ratio of intensities at two different wavelengths to get v by comparison with some calibration curve, and knowing K_D, one may calculate [Zn]. Because no terms in P or PZn appear in the expression, it is often assumed that the receptor concentration is unimportant, which is perilous. The error frequently made is to presume that the concentration of free zinc does not change when P is added, particularly when the measurement is made in a limited volume or with an unknown concentration of P. Suppose we initially have in a cell a receptor P present at 1 μM concentration with a 1-pM affinity for zinc, but the free zinc is present only at 1 nM. Under these conditions (both [Zn] and [P] are well above K_D), virtually all the ion will be bound, but the fractional occupancy will still be only 0.1% because the receptor is present in 1000-fold excess. As Dinely et al. have pointed out,[64] such high receptor concentrations are easy to achieve intracellularly with trappable cell-permeable indicators (e.g., acetoxymethyl ester derivatives), and thus the values obtained with these methods should be scrutinized carefully. It is preferable to control, or at least know, the concentration of the receptor. Although this is sometimes feasible, it should also be noted that one can expect intracellular zinc to be buffered, because the vast majority is bound inside the cell. In this sense one may regard the fluorescent indicator or protein as yet another receptor competing for the exchangeable fraction of zinc and, unless present in overwhelming concentration compared to glutathione (millimolar) and metallothionein (micromolar), one whose fractional occupancy will reflect the free zinc ion concentration.

Similarly, measurements done on samples in limited volumes (such as those from dialysis probes; see below) may run the same risk. In these cases, the free zinc is again present at levels above K_D, but the excess of receptor in these cases ensures that nearly all will be bound. Unlike the interior of the cell, however, there are few sources of buffering zinc available because the dialysis membrane excludes polypeptides, and the amount of receptor can easily be known. Thus, it is quite feasible to measure even very small free zinc amounts in these samples by "capturing" effectively all the free zinc in the sample with a known concentration of receptor well above the K_D (this is the "stoichiometric regime"[65]) and measuring the saturation of the receptor by comparison to a calibration curve. Weber[65] provides a more rigorous mathematical treatment for interested readers.

A third regime is defined by sensors that introduce a very small amount of receptor but within a circumscribed volume. Examples are optical fiber sensors, which have very small amounts of receptor at the distal end of an optical fiber whose diameter is 100 μm or less,[66] and the nanometer-size PEBBLEs (probe encapsulated by biologically localized embedding) of Clark et al.,[67] which contain infinitesimal quantities of indicators. Under these circumstances, the very small amounts of receptors present may not alter the concentration of free zinc appreciably even in the absence of buffering, and Equation 14.17 is valid.

14.9 LOCALIZING AND MEASURING ZINC IN THE BRAIN

The importance of rapidly exchangeable zinc ions as physiological signals in the brain and as intracellular "death signals" in many cell types has forced attention onto methods to detect and quantify these free zinc signals. Currently, at least three "pools" or "compartments" of zinc have been identified in tissue: zinc in secretory granules $[Zn^{2+}]_{granule}$, zinc inside the cytosol of a cell $[Zn^{2+}]_{in}$, and zinc in the extracellular space

$[Zn^{2+}]_{out}$. Each of these compartments has a separate resting concentration, or pZn; each has a different signal value; and each can be visualized or quantified separately by the correct choice of methods.

14.9.1 Storage Granule Zinc: $[Zn^{2+}]_{GRANULE}$

In the brain, the concentration of free zinc in the secretory granules (neuronal vesicles) has been estimated to be in the 1-mM to 10-mM range.[68] This is similar to the estimates for other zinc storage granules, such as those found in the salivary glands or the Paneth cells of the intestine. Such estimates are highly indirect; to date there seems to be no direct measurement of this zinc pool. Whatever it is, the concentration of zinc in storage granules is so high that chromogenic[69,70] and fluorescent stains[43] vividly label the pool. Presumably, if the concentration of zinc in storage granules is really in the 1-mM to 10-mM range, then one could selectively stain (and quantify) the granule zinc by the use of a lipophilic probe with very low affinity for zinc, $K_D \sim 10^{-3}$ M. Probes with these properties are under development.[52]

Because storage granule zinc is always in a membrane-bound organelle (secretory granule), this zinc cannot be stained with any fluorescent probe that does not freely permeate membranes. Even after freezing and cutting on a microtome, the vesicles in the brain, for example, remain intact, and one cannot stain the storage granule zinc (vesicular zinc), even in frozen sections, unless one uses a lipophilic probe.[36,71] Therefore, fluorescent probes that are made (as many are) to be "trapped" inside cells by enzymatic cleavage of a lipophilic appendage will not stain zinc in vesicles or other storage granules because those probes cannot penetrate the second lipid membrane after the enzymatic cleavage. Therefore, to observe the granule or cytosolic zinc $[Zn^{2+}]_{granule}$, and $[Zn^{2+}]_{in}$, one can either stain the cytosolic zinc selectively (with a trappable probe) or stain both compartments with a lipophilic probe.

The first fluorescent probe used to visualize granule zinc was TSQ (N-6-methoxy-8-quinolyl)-p-toluenesulfonamide), introduced by Eschenko and Toropsev (cited in Frederickson et al.,[43]) who published one quite stunning picture. TSQ is especially useful as an histological stain on postmortem tissue because it is so lipophilic that it partitions preferentially from aqueous staining fluids into tissue membranes, giving a high focal concentration of TSQ:Zn in the membranes. TSQ also has an approximately 20-fold intensity increase on zinc binding, which makes it a very bright histological stain. However, TSQ cannot be used for fluorometry in aqueous solutions because it is barely water soluble at neutral pH, is highly pH sensitive, and changes partition coefficients on zinc binding[43] (see also Snitsarev et al.[72]). Newer lipophilic dyes that are more soluble than TSQ, excite and emit in the visible spectrum, and exhibit substantial intensity increases are now available,[36,71,73] and these are superior to TSQ for use in live tissue.

14.9.2 Extracellular Zinc Measurement by Microdialysis

Our measurements to date indicate that free zinc is present in the extracellular fluid at a baseline level of pZn ~ 8.3–9.3 but that transiently (and probably locally) it rises as high as pZn ~ 7 during some zinc signaling events (Zornow et al., submitted).

The baseline pZn for the resting brain may really only be measured *in situ*, within an intact and healthy brain, and this limits the possible methods.

We have used two methods to measure extracellular pZn. The first is to conduct microdialysis with an intracranial probe implanted in the brain of an anesthetized subject, collect the dialysate, and measure the free zinc concentration in the dialysate. The second is to employ a fluorescence-based optical fiber sensor[53,74] and insert it intracranially into anesthetized subjects. The results from the dialysis method and the preliminary results from the fiber-optic sensor (not shown) agree reasonably well, putting baseline pZn in the range of 8.3 to 9.3.

To perform the microdialysis experiments, one simply inserts a microdialysis probe stereotaxically into the brain of an anesthetized subject, collects dialysate sample for some period of time, then measures the free zinc in the dialysate (Zornow et al., submitted). For calibration of the recovery function of the probe, one uses "phantom" brains (i.e., beakers filled with zinc solutions) and microdialyses the beakers in the same way as the brain. This gives a percent recovery of free zinc by the system that is used to correct the calibration for losses. Most of the probes we have used to date have shown recovery in the 30% to 60% range.

The analytical complexities of this method are that the dialysate fluid must have less free zinc than the tissue being perfused, the materials used may neither extract (adsorb) nor add (leach) zinc into the dialysate, and the assay for free zinc must be sensitive and accurate enough to detect the zinc in the dialysate. Thus far, the available data indicate that the extracellular free zinc concentration in the brain of a healthy, anesthetized resting animal (rat, rabbit, dog) is near or below 10 nM. This means that the dialysate solution should contain no more than 0.1 nM of free zinc and (of course) that the measurement system must be capable of measuring accurately a "blank" of 0.1 nM free zinc. To attain the low free zinc level, all solutions are stripped of divalent metal ions by treatment with Chelex before adding (ultrapure) divalent cations (see above) to the final mix of artificial cerebrospinal fluid (ACSF) used to dialyze the brain. (Note that when using dialysis, one cannot simply subtract the blank from the final value as in other procedures. The blank must be below the to-be-measured concentration and *not* subtracted from the final observation.)

Once dialysate is collected, the act of measuring the free zinc fluorometrically involves the same problems encountered in intracellular measurement but with the comparatively greater ease of the cuvette vs., the cell. Thus, one must first decide whether to do measurements in the affinity mode or the percent occupancy mode, where the zinc is stoichiometrically bound. In fact, the choice is forced by the comparatively small number of zinc ions with which one has to work. Specifically, the microdialysis must be run at slow flow rates to get a decent percent recovery from the brain, perhaps 1 µl/min. That means more than 1.5 h is required to collect 100 µl of fluid. If the brain free zinc is 5 nM, and one gets 30% recovery, one has 1.5 nM zinc in the dialysate, or a total of 0.15 pmol of zinc per 100 min of experiment time. Measuring in an affinity mode requires that binding by the protein or indicator not appreciably change the free zinc concentration (see above), which in turn requires that a vanishingly small amount of protein or indicator be present throughout the sample, preferably no more than one-tenth the concentration of the

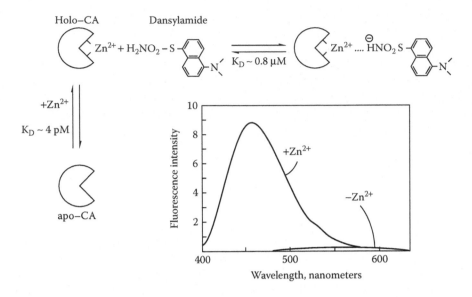

FIGURE 14.5 Schematic of fluorimetric zinc sensing by dansylamide and apo-CA. In the absence of zinc, dansylamide acid (DNSA) does not bind to apo-CA and exhibits weak orange fluorescence; when zinc binds, DNSA does, as well, and exhibits strong blue fluorescence.

to-be-measured zinc. If this concentration (perhaps picomolar) is much below K_D, or there is significant fluorescent background, these measurements become very difficult indeed.

The alternative (which we use) is to operate in the stoichiometric mode, in which the molarity of the protein is set to about 10-fold higher than both the expected molarity of the zinc (such as 10 nM or 100 nM) and the zinc K_D, such that virtually all the free zinc is "captured" by the protein. For this assay, the original CA/dansylamide system of Thompson and Jones[53] is used, wherein the binding of zinc to apo-CA permits binding of dansylamide, which is accompanied by a 100-nm blue shift and substantial emission increase (Figure 14.5). For dansylamide/CA-based determinations where there is a substantial shift and enhancement of the fluorescence upon binding of the zinc, even very low fractional occupancies may be discerned and relatively small amounts of zinc may be determined: one adds a fixed concentration of dansylamide (above K_D for dansylamide binding to holo-CA) (Figure 14.5). The intensity ratio of fluorescence is then measured, ranging from 7% of the peak value (the "off" intensity of the probe) to 100% of the saturated value (when [Zn] = [protein]). In the case of 1.5 nM of expected zinc signal, one might typically use something like 50 nM of probe and measure a fluorescence intensity of "off" + 3.0% ("on") as the net intensity. To translate that intensity to a true zinc concentration, one must use calibration standards that, like the sample, are composed of zinc salt in solution with no metal buffering. This is because buffered solutions (with the large pool reserve of ligand-coordinated zinc) will automatically saturate all probes that are present at concentrations that greatly exceed their K_D. Such a calibration curve is shown as Figure 14.6;

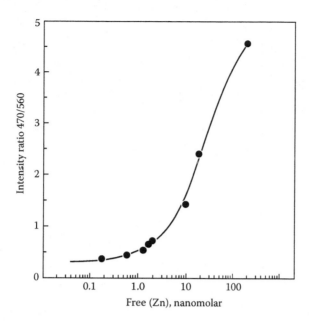

FIGURE 14.6 Calibration curve for zinc from dialysis probe obtained in "stoichiometric" mode using DNSA/apo-CA assay.

in this example, the protein is not saturated. In fact, the preparation and handling of nonbuffered calibration standards that are in the picomolar and nanomolar regime is exceedingly difficult because of contamination and loss (precipitation, adsorption) issues. This is especially daunting when one tries to add a free zinc salt (e.g., $ZnCl_2$ in solution) to a growth medium or an ACSF solution with tissue in it (see above). Under these conditions, the final free zinc in the calibration standard will be less (often several decades less) than the total zinc added (see Figure 14.1). This renders the calibration (and thus the calculated values) meaningless. However, an accurate calibration curve is essential if one is doing fluorimetric measurements using the percent occupancy mode of measurement.

When the brain is subjected to any of the "excitotoxic" insults (ischemia/reperfusion, trauma, seizure), a large increase in extracellular free zinc occurs, together with an increase in intracellular free zinc. The latter, intracellular increase, is cytolethal, but zinc chelation can rescue neurons in these excitotoxic crises.[9,75] The extracellular increase in zinc is probably both the cause and the result of increasing intracellular zinc, because zinc ions can flow in both directions across the membranes, especially in conditions (such as excitotoxicity) in which glutamate levels are high and cells are depolarized.[76] Such a data set is shown in Figure 14.7, which depicts time-dependent zinc determination by microdialysis for four rabbits serving as models for brain ischemia (stroke). Evidently, there is a substantial increase (to more than 100 nM) following the insult from a baseline of approximately 10 nM, followed by a slow, "noisy" decline in free zinc after reperfusion is established.

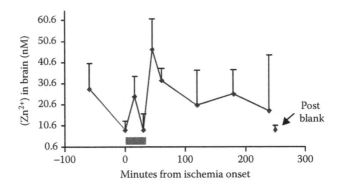

FIGURE 14.7 (See color insert following page 142.) Release of free zinc during ischemia and reperfusion. Fluorimetry was used to measure the free zinc concentration in brain dialysates collected from four rabbits. The zinc level decreases after recovery from probe insertion (to less than 10 nM), then rises with ischemia and especially upon reperfusion. The reperfusion release persists for hours.

The magnitude of the increase in extracellular zinc is measured by microdialysis and fluorometry (Figure 14.7) and by direct sensing with an optical fiber (data not shown) to be at least 100 nM and often as high as 600 nM. Local concentrations likely are much higher, because these measurements represent an average value sample across many millimeters of tissue. Thus, where release is expected to be regionally uneven, the averaging methods will miss the peaks of zinc release and show only the average. The same may be said for the temporal dimension in the microdialysis method. When taking samples every 20 to 100 min, one will obviously underestimate the magnitude of very short "spikes" in zinc release.

14.9.3 MEASUREMENT OF EXTRACELLULAR ZINC BY IMAGING

In principle, there is no reason why one cannot administer a membrane-impermeable fluorescent probe directly into tissue in a live animal, wait for it to diffuse and distribute throughout the extracellular fluid, and then illuminate the tissue with a suitable excitation wavelength and quantitate the fluorescent emissions. As far as we know, this has not been done for zinc, the obvious confounds being intrinsic optical changes and instabilities that would degrade the reliability of the fluorescence signal. Other difficulties with this approach include the absorption and scattering of light by skin and tissue, which are very wavelength dependent, as well as the pharmacological difficulty of getting the indicator to the right tissue and/or observing its fluorescence with spatial selectivity. Various expedients have been developed to permit such studies, including multiphoton excitation for transdermal measurement and infrared fluorometry to defeat scattering and absorbance.[77]

Measuring the concentration of free zinc in the extracellular fluid surrounding a piece of tissue *in vitro* is actually conceptually simpler than measuring intracellular zinc in the same *in vitro* situation. This is because the fluid into which the probe goes, the so-called extracellular fluid (which is actually the superfusate, perfusate, or subfusate), is a fluid over which one has direct control. Thus, if one puts 20 μM of a membrane-impermeant zinc probe into the bath, then that concentration is known. To measure the extracellular free zinc, one may then work in either the affinity mode or the stoichiometric mode, depending on how much free zinc is expected and what probes are available and convenient. For example, if one is expecting 20 μM of zinc to be released (e.g., in a localized region, not throughout the slice), one may load the bath with 50 μM of a very tight-binding indicator ($K_D \ll 0.1$ μM) and watch the changes in percent probe occupancy. This method has yielded an apparent release of 15 to 30 μM of zinc from a single brain slice in culture when the tight-binding probe apo CA ($K_D - 4$ pM) was used.[18] Similar results have been obtained using the synthetic probe ZnAF, also a tight binder ($K_D = 0.5$ nM) in the stoichiometric mode (Figure 14.8).

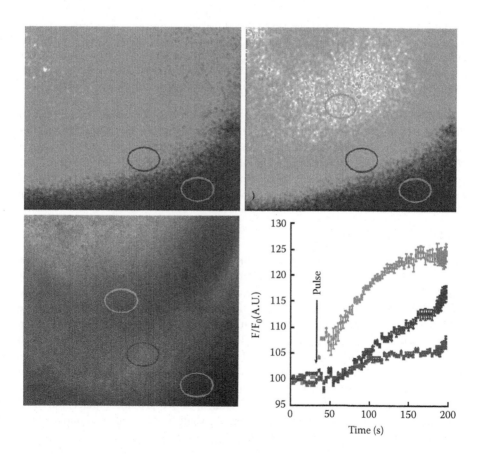

FIGURE 14.8 (See color insert following page 142.) Synaptic release of zinc. The lower two pseudocolor images show the CA3 region of the hippocampus before (left) and after (right) a one-second train of stimulus pulses (vertical line) delivered to the granule cell stratum, outside the field of view. The rise in free zinc concentration in the extracellular fluid (recorded by zinc:NG fluorescence) is shown for the stratum lucidum (red oval and graph), stratum pyramidale (blue) and stratum oriens (yellow). Static-pool slice, brightfield shown upper left.

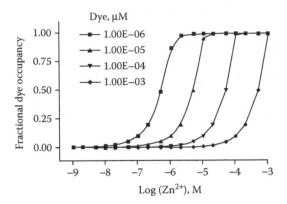

FIGURE 14.9 Dye fractional occupancy corrected for intracellular dye concentration. Assuming a K_d for Zn^{2+} of 20 nM, theoretical fractional occupancy curves were generated for various intracellular dye concentrations (1 μM, 10 μM, 100 μM, and 1000 μM). These curves model the saturation response of mag-fura-2 as a function of Zn^{2+}; increasing intracellular dye concentration results in a rightward shift. Reprinted with permission from Dinely et al.[64]

Unfortunately, many investigators have used a concentration of fluorescent probe that was substantially higher than the K_D for that probe (typically 10- or 100-fold)[60,79] and then unwittingly misused the Tsien equation ([Zn] = $K_D((F - F_{minimum})/(F_{maximum} - F))$)), thus generating potentially erroneous estimates of the free zinc concentration. As Dinely et al.[64] have shown, the more the total indicator concentration exceeds K_D, the worse will be the underestimate of true free zinc (Figure 14.9).

14.10 CONCLUSION

In this chapter we have summarized some of the issues and provided approaches for studying zinc ion in biological systems, particularly using fluorescence techniques. We will have succeeded (and been amply repaid) if there is a decline in the number of experimental miscues making their way into the literature. The points we have made are not limited to zinc ion (at least from the standpoint of theory) but are extendable to other metal ions as well (although their chemical behaviors will differ). The reader should not be daunted by the numerous caveats throughout this chapter: in our laboratories, and in many others around the world, these sorts of trace analyses are done routinely. Newton said, "If I have seen farther than others, it is because I stood on the shoulders of giants." If we provide a place for those elucidating the biology of zinc to stand on, we shall be very satisfied and perhaps even be taken for giants ourselves one day.

ACKNOWLEDGMENTS

The authors thank their sponsors at the National Institutes of Health, National Science Foundation, and the Office of Naval Research and their colleagues for numerous helpful discussions.

REFERENCES

1. Cotton, F.A. and Wilkinson, G., *Advanced Inorganic Chemistry*. New York: Wiley-Interscience, 1988.
2. Lippard, S.J. and Berg, J.M., *Principles of Bioinorganic Chemistry*. Mill Valley, CA: University Science Books, 1994.
3. Bertini, I., Sigel, A., and Sigel, H., eds., *Handbook on Metalloproteins*. New York: Marcel Dekker, 2001.
4. Vallee, B.L. and Falchuk, K.H., The biochemical basis of zinc physiology. *Physiolog Rev,* 73, 79–118, 1993.
5. Bertini, I., ed., *Zinc Enzymes*. Boston: Birkhäuser, 1986.
6. McCall, K.A., Huang, C.-C., and Fierke, C.A., Function and mechanism of zinc metalloenzymes. *J Nutr,* 130, 1437S–1446S, 2000.
7. Berg, J.M. and Shi, Y., The galvanization of biology: a growing appreciation for the roles of zinc. *Science,* 271, 1081–1085, 1996.
8. Frederickson, C.J., Suh, S.W., Silva, D., Frederickson, C.J., and Thompson, R.B., Importance of zinc in the central nervous system: the zinc-containing neuron. *J Nutr,* 130(5S suppl), 1471S–1483S, 2000.
9. Frederickson, C.J., Maret, W., and Cuajungco, M.P., Zinc and excitotoxic brain injury: a new model. *Neuroscientist,* 10, 18–25, 2004.
10. Ugarte, M. and Osborne, N.N., Zinc in the retina. *Prog Neurobiol,* 64, 219–249, 2001.
11. Rulisek, L. and Vondrasek, J., Coordination geometries of selected transition metal ions (Co^{2+}, Ni^{2+}, Cu^{2+}, Zn^{2+}, Cd^{2+}, and Hg^{2+}) in metalloproteins. *J Inorg Biochem,* 71, 115–127, 1998.
12. Glusker, J.P., Structural aspects of metal liganding to functional groups in proteins. *Adv Protein Chem,* 42, 1–76, 1991.
13. Lindskog, S. and Nyman, P.O., Metal-binding properties of human erythrocyte carbonic anhydrases. *Biochim Biophys Acta,* 85, 462–474, 1964.
14. Bertini, I., Luchinat, C., and Viezzoli, M.S., Metal substitution as a tool for the investigation of zinc proteins. In: *Zinc Enzymes,* Bertini, I., ed. Boston: Birkhauser, 1986:27–47.
15. Evelhoch, J.L., Bocian, D.F., and Sudmeier, J.L., Evidence for direct metal–nitrogen binding in aromatic sulfonamide complexes of cadmium(II)-substituted carbonic anhydrases by cadmium-113 nuclear magnetic resonance. *Biochemistry,* 20, 4951–4954, 1981.
16. Blackburn, G.M., Mann, B.E., Taylor, B.F., and Worrall, A.F., A nuclear magnetic resonance study of the binding of novel N-hydroxybenzenesulphonamide carbonic anhydrase to native and cadmium-111-substituted carbonic anhydrase. *Eur J Biochem,* 153, 553–558, 1985.
17. Lipton, A.S., Heck, R.W., and Ellis, P.D., Zinc solid state NMR spectroscopy of human carbonic anhydrase: implications for the enzymatic mechanism. *J Am Chem Soc,* 126, 4735–4739, 2004.
18. Stumm, W. and Morgan, J.J., *Aquatic Chemistry: Chemical Equilibria and Rates in Natural Waters*. New York: Wiley-Interscience, 1996.
19. Perkins, D.J., A study of the effect of amino acid structure on the stabilities of the complexes formed with metals of group II of the periodic classification. *Biochem J,* 55, 649–652, 1953.
20. Price, P.J. and Brewer, G.J., Serum-free media for neural cell cultures. In: *Protocols for Neural Cell Culture,* Federoff, S. and Richardson, A., eds. Totowa, NJ: Humana Press, 2001:255–264.

21. Wagner, F.W., Preparation of metal-free enzymes. *Methods Enzymol,* 158, 21–32, 1988.
22. Smith, R. and Martell, A.E., NIST selected critical stability constants of metal complexes, standard reference database 46. Gaithersburg, MD: National Institute of Standards and Technology, 1973.
23. Yu, Q.Y., Avoiding interferences from Good's buffers: a contiguous series of noncomplexing tertiary amine buffers covering the entire range of pH 3–11. *Anal Biochem,* 253, 50–56, 1997.
24. Kandegedara, A. and Rorabacher, D.B., Noncomplexing tertiary amines as "better" buffers covering the range of pH 3–11. Temperature dependence of their acid dissociation constants. *Anal Chem,* 71, 3140–3144, 1999.
25. Mash, H.E., Complexation of copper by zwitterionic aminosulfonic (good) buffers. *Anal Chem,* 75, 671–677, 2003.
26. Aslamkhan, A.G., Aslamkhan, A., and Ahearn, G.A., Preparation of metal ion buffers for biological experimentation: a methods approach with emphasis on iron and zinc. *J Exp Zool,* 292, 507–522, 2002.
27. Magyar, J.S. and Godwin, H.A., Spectropotentiometric analysis of metal binding to structural zinc-binding sites: accounting quantitatively for pH and metal ion buffering effects. *Anal Biochem,* 320, 39–54, 2003.
28. Hunt, J.B., Rhee, M.J., and Storm, C.B., A rapid and convenient preparation of apocarbonic anhydrase. *Anal Biochem,* 79, 614–617, 1977.
29. Baker, J.O., Metal-buffered systems. *Methods Enzymol,* 158, 33–55, 1988.
30. Grynkiewicz, G., Poenie, M., and Tsien, R.Y., A new generation of calcium indicators with greatly improved fluorescence properties. *J Biol Chem,* 260, 3440–3450, 1985.
31. Gee, K.R., Zhou, Z.-L., Qian, W.J., and Kennedy, R., Detection and imaging of zinc secretion from pancreatic beta-cells using a new fluorescent zinc indicator. *J Am Chem Soc,* 124, 776–778, 2002.
32. Bozym, R.A., Zeng, H.H., Cramer, M., Stoddard, A., Fierke, C.A., and Thompson, R.B., *In vivo* and intracellular sensing and imaging of free zinc ion. In: *Advanced Biomedical and Clinical Diagnostic Systems II,* Cohn, G.E., Grundfest, W.S., Benaron, D.A., and Vo-Dinh, T., eds. *Proc SPIE,* 5318, 34–38, 2004.
33. Finney, L.A. and O'Halloran, T.V., Transition metal speciation in the cell: insights from the chemistry of metal ion receptors. *Science,* 300, 931–936, 2003.
34. Budde, T., Minta, A., White, J.A., and Kay, A.R., Imaging free zinc in synaptic terminals in live hippocampal slices. *Neuroscience,* 79, 347–358, 1997.
35. Burdette, S.C., Walkup, G.K., Spingler, B., Tsien, R.Y., and Lippard, S.J., Fluorescent sensors for Zn^{2+} based on a fluorescein platform: synthesis, properties, and intracellular distribution. *J Am Chem Soc,* 123, 7831–7841, 2001.
36. Burdette, S.C., Frederickson, C.J., Bu, W., and Lippard, S.J., ZP4, an improved neuronal Zn^{2+} sensor of the ZinPyr family. *J Am Chem Soc,* 125, 1778–1787, 2003.
37. Gee, K.R., Zhou, Z.L., Ton-That, D., Sensi, S.L., and Weiss, J.H., Measuring zinc in living cells. A new generation of sensitive and selective fluorescent probes. *Cell Calcium,* 31, 245–251, 2002.
38. Hirano, T., Kikuchi, K., Urano, Y., Higuchi, T., and Nagano, T., Novel zinc fluorescent probes excitable with visible light for biological applications. *Angew Chem Int Ed Engl,* 39, 1052–1054, 2000.
39. Kimura, E. and Aoki, S., Chemistry of zinc(II) fluorophore sensors. *Biometals,* 14, 191–204, 2001.
40. Thompson, R.B., Peterson, D., Mahoney, W., Cramer, M., Maliwal, B.P., Suh, S.W., and Frederickson, C.J., Fluorescent zinc indicators for neurobiology. *J Neurosci Methods,* 118, 63–75, 2002.

41. Thompson, R.B. and Patchan, M.W., Fluorescence lifetime-based biosensing of zinc: origin of the broad dynamic range. *J Fluoresc,* 5, 123–130, 1995.
42. Weber, G., Energetics of ligand binding to proteins. *Adv Protein Chem,* 29, 1–83, 1975.
43. Frederickson, C.J., Kasarskis, E.J., Ringo, D., and Frederickson, R.E., A quinoline fluorescence method for visualizing and assaying histochemically reactive zinc (bouton zinc) in the brain. *J Neurosci Methods,* 20, 91–103, 1987.
44. Fahrni, C.J. and O'Halloran, T.V., Aqueous coordination chemistry of quinoline-based fluorescence probes for the biological chemistry of zinc. *J Am Chem Soc,* 121, 11448–11458, 1999.
45. Bjerrum, N., *Z Phys Chem,* 106, 219–242, 1923.
46. Klotz, I.M., *Ligand-Receptor Energetics. A Guide for the Perplexed.* New York: Wiley-Interscience, 1997.
47. White, C.E. and Argauer, R.J., *Fluorescence Analysis: A Practical Approach.* New York: Marcel Dekker, 1970.
48. Fernandez-Gutierrez, A. and Munoz de la Pena, A., Determinations of inorganic substances by luminescence methods. In: *Molecular Luminescence Spectroscopy, Part I: Methods and Applications,* vol. 77, Schulman, S.G., ed. New York: Wiley-Interscience, 1985:371–546.
49. Simons, T.J.B., Measurement of free zinc ion concentration with the fluorescent probe mag-fura-2 (furaptra). *J Biochem Biophys Methods,* 27, 25–37, 1993.
50. Canzoniero, L.M.T., Sensi, S.L., and Choi, D.W., Measurement of intracellular free zinc in living neurons. *Neurobiol Dis,* 4, 275–279, 1997.
51. Walkup, G.K., Burdette, S.C., Lippard, S.J., and Tsien, R.Y., A new cell-permeable fluorescent probe for Zn^{2+}. *J Am Chem Soc,* 122, 5644–5645, 2000.
52. Henary, M.M., Wu, Y., and Fahrni, C.J., Zinc(II)-selective ratiometric fluorescent sensors based on inhibition of excited state intramolecular proton transfer. *Chem Eur J,* 10, 3015–3025, 2004.
53. Thompson, R.B. and Jones, E.R., Enzyme-based fiber optic zinc biosensor. *Anal Chem,* 65, 730–734, 1993.
54. Fierke, C.A. and Thompson, R.B., Fluorescence-based biosensing of zinc using carbonic anhydrase. *Biometals,* 14, 205–222, 2001.
55. Elbaum, D., Nair, S.K., Patchan, M.W., Thompson, R.B., and Christianson, D.W., Structure-based design of a sulfonamide probe for fluorescence anisotropy detection of zinc with a carbonic anhydrase-based biosensor. *J Am Chem Soc,* 118, 8381–8387, 1996.
56. Thompson, R.B. and Patchan, M.W., Lifetime-based fluorescence energy transfer biosensing of zinc. *Anal Biochem,* 227, 123–128, 1995.
57. Thompson, R.B., Cramer, M.L., Bozym, R., and Fierke, C.A., Excitation ratiometric fluorescent biosensor for zinc ion at picomolar levels. *J Biomed Opt,* 7, 555–560, 2002.
58. Kiefer, L.L., Paterno, S.A., and Fierke, C.A., Hydrogen bond network in the metal binding site of carbonic anhydrase enhances zinc affinity and catalytic efficiency. *J Am Chem Soc,* 117, 6831–6837, 1995.
59. Huang, C.-C., Lesburg, C.A., Kiefer, L.L., Fierke, C.A., and Christianson, D.W., Reversal of the hydrogen bond to zinc ligand histidine-119 dramatically diminishes catalysis and enhances metal equilibration kinetics in carbonic anhydrase II. *Biochemistry,* 35, 3439–3446, 1996.
60. Li, Y., Hough, C., Suh, S.W., Sarvey, J.M., and Frederickson, C.J., Rapid translocation of Zn^{2+} from presynaptic terminals into postsynaptic hippocampal neurons after physiological stimulation. *J Neurophysiol,* 86, 2597–2604, 2001.

61. Hering, J.G. and Morel, F.M.M., Kinetics of trace metal complexation: role of alkaline earth metals. *Environ Sci Technol,* 22, 1469–1478, 1988.
62. Raspor, B., Nurnberg, H.W., Valenta, P., and Branica, M., Measurement of Ca–Zn EDTA exchange reaction. *J Electroanal Chem Interfac Electrochem,* 115, 293, 1980.
63. Xue, H., Sigg, L., and Kari, F.G., Speciation of EDTA in natural waters: exchange kinetics of Fe-EDTA in river water. *Environ Sci Technol,* 29, 59–68, 1995.
64. Dinely, K.E., Malaiyandi, L.M., and Reynolds, I.J., A reevaluation of neuronal zinc measurements: artifacts associated with high intracellular dye concentration. *Mol Pharmacol,* 62, 618–627, 2002.
65. Weber, G., *Protein Interactions.* New York: Chapman & Hall, 1992.
66. Zeng, H.H., Thompson, R.B., Maliwal, B.P., Fones, G.R., Moffett, J.W., and Fierke, C.A., Real-time determination of picomolar free Cu(II) in seawater using a fluorescence-based fiber optic biosensor. *Anal Chem,* 75, 6807–6812, 2003.
67. Clark, H.A., Hoyer, M., Philbert, M.A., and Kopelman, R., Optical nanosensors for chemical analysis inside single living cells. 1. Fabrication, characterization, and methods for intracellular delivery of PEBBLE sensors. *Anal Chem,* 71, 4831–4836, 1999.
68. Frederickson, C.J., The neurobiology of zinc and of zinc-containing neurons. *Int Rev Neurobiol,* 31, 145–238, 1989.
69. Frederickson, C.J., Perez-Clausell, J., and Danscher, G., Zinc-containing 7S-NGF complex. Evidence from zinc histochemistry for localization in salivary secretory granules. *J Histochem Cytochem,* 35, 579–583, 1987.
70. Frederickson, C.J., Howell, G.A., and Frederickson, M.H., Zinc-dithizonate staining in the cat hippocampus: relationship to the mossy-fiber neuropil and post-natal development. *Exp Neurol,* 73, 812–823, 1981.
71. Frederickson, C.J., Burdette, S.C., Frederickson, C.J., Sensi, S.L., Weiss, J.H., Yin, H.Z., Balaji, R.V., Truong-tran, A.Q., Bedell, E., and Prough, D.S., Method for identifying neuronal cells suffering zinc toxicity by use of a novel fluorescent sensor. *J Neurosci Methods,* 139, 79–89, 2004.
72. Snitsarev, V., Budde, T., Strickler, T.B., Cox, J.M., Krupa, D.J., Geng, L., and Kay, A.R., Fluorescent detection of Zn^{2+}-rich vesicles with Zinquin: mechanism of action in lipid environments. *Biophys J,* 80, 1538–1546, 2001.
73. Woodroofe, C.C. and Lippard, S.J., A novel two-fluorophore approach to ratiometric sensing of Zn(2+). *J Am Chem Soc,* 125, 11458–11459, 2003.
74. Thompson, R.B., Fluorescence-based fiber optic sensors. In: *Topics in Fluorescence Spectroscopy,* Vol. 2, *Principles,* Lakowicz, J.R., ed. New York: Plenum Press, 1991:345–365.
75. Suh, S.W., Chen, J.W., Motamedi, M., Bell, B., Listiak, K., Pons, N.F., Danscher, G., and Frederickson, C.J., Evidence that synaptically-released zinc contributes to neuronal injury after traumatic brain injury. *Brain Res,* 852, 268–273, 2000.
76. Weiss, J.H. and Sensi, S., Ca^{2+}–Zn^{2+} permeable AMPA or kainate receptors: possible key factors in selective neurodegeneration. *Trends Neurosci,* 23, 365–371, 2000.
77. Thompson, R.B., Red and near-infrared fluorometry. In: *Topics in Fluorescence Spectroscopy,* Vol. 4, *Probe Design and Chemical Sensing,* Lakowicz, J.R., ed. New York: Plenum Press, 1994:151–181.
78. Thompson, R.B., Whetsell, W.O., Jr., Maliwal, B.P., Fierke, C.A., and Frederickson, C.J., Fluorescence microscopy of stimulated Zn(II) release from organotypic cultures of mammalian hippocampus using a carbonic anhydrase-based biosensor system. *J Neurosci Methods,* 96, 35–45, 2000.
79. Kay, A.R., Evidence for chelatable zinc in the extracellular space of the hippocampus, but little evidence for synaptic release of Zn. *J Neurosci,* 23, 6847–6855, 2003.

Index

fluorescence-based detection methods for,
237–240
hybrid chips, 242
immunoassays derived from, 234–235
microscale challenges with, 237
nucleic acid analysis with, 235–236
plastic thin-film autofluorescence in, 243–249
polymers for, 241–243
silicon and glass in, 241
Microarrays, 263. *See also* Array biosensors
Microdevices, 234
Microdialysis, extracellular zinc measurement in
brain via, 366–370
Microfluidic technologies, 235, 274
Microwell biosensors, 68, 79–80, 86
MINEQL+ software, 356
Miniaturization
array biosensors, 266
fluidics system in Nanogen array biosensor,
272, 273
Minizymes, 30
miRNA treatment, 35
Modal dispersion, as limiting factor in metal ion
biosensing, 113
Molecular beacons, 12, 17
advantages of, 70–71
applications of, 69–71
array biosensors, 82–85
avidin-biotin binding mechanism, 73
for biosensor development, 75–86
carbodiimide chemistry for, 73
chemical binding methods for, 73–74
design of, 15–16
disulfide bonding for, 73
DNA probes derived from fluorescence
biosensing, 67–68
fluorophore selection for, 72
mechanism of elution and fluorescence
responsivities, 16
microwell biosensors, 79–80
nanoparticle biosensors, 80–82
optical fiber biosensors, 75–77
optical fiber bundle biosensors, 77–79
physical absorption methods for, 74–75
as point sensors, 77
selection of functional group, 72
similarity of TRAP structure to, 29
similarity to PtCP homogeneous hybridization
assays, 215
single molecule biosensors, 85–86
spacers for functional groups in, 72
stem sequences, 71
surface immobilization, 72
surface treatment for biosensor preparation,
73–75

synthesis and characterization of, 71–73
in vitro selection of, 15
Molecular diagnostics assays, 296, 316
differential melting temperature, 321
feasibility study, 316–319
prostate-specific antigen gene expression assay,
319
single base extension, 321–323
SNP detection in long QT syndrome, 319–321
Molecular Probes firm, v
Molybdate binding protein, 61
Monoclonal antibodies
main properties against PdCP, 209
obtaining against porphyrins, 208
optimal concentrations of PdCP and 5D, 210
Monosaccharide quantification, FRET-based
biosensors for, 99
mRNA inputs, sensing of, 37
Multiphoton excitation, 169, 175–176
and metallic nanoparticles, 165–168
potential uses of MEF in, 177
reducing background noise with, 238–239
Multiple target detection, 86
by array biosensors, 263
by MB DNA arrays, 82
Multiwavelength detection, 248
Mylar, 248
use in microscale devices, 243
Myoglobin assays, 311–314

N

NADPH fluorescence
biomass correlation to, 336
measurement by indicator molecules, 337
Nanogen array biosensor system, 271
CMOS chip in, 272
Nanoparticle biosensors, 68, 80–82, 86
Nanosensors, 3
Narrow band-pass emission filters, 336
Near-field excitation mode, 3
Near-infrared (NIR) probes, 334
in bioprocess monitoring, 338
in glucose and glutamine sensing, 345–346
Neurobasal medium, 354
Neurochemical signaling, role of zinc in, 352
Nickel binding protein (NBP), 52
Nickel ions, 115
Noble-metal substrates
for MEF, 126–128
and SiFs for MEF, 128–129
Nonspecific signals, importance of minimizing,
184
Nonspherical silver colloids, for MEF, 131–136
NOT functions, 33, 35